VOLUME EIGHTY

Current Topics in MEMBRANES

Sterol Regulation of Ion Channels

CURRENT TOPICS IN MEMBRANES, VOLUME 80

Series Editors

ROBERT BALABAN
National Heart, Lung and Blood Institute
National Institutes of Health
Bethesda, Maryland, USA

SIDNEY A. SIMON
Department of Neurobiology
Duke University Medical Centre
Durham, North Carolina, USA

VOLUME EIGHTY

CURRENT TOPICS IN MEMBRANES

Sterol Regulation of Ion Channels

Edited by

IRENA LEVITAN
University of Illinois at Chicago,
Chicago, IL, United States

ELSEVIER

ACADEMIC PRESS
An imprint of Elsevier

Academic Press is an imprint of Elsevier
50 Hampshire Street, 5th Floor, Cambridge, MA 02139, United States
525 B Street, Suite 1800, San Diego, CA 92101–4495, United States
The Boulevard, Langford Lane, Kidlington, Oxford OX5 1GB, United Kingdom
125 London Wall, London, EC2Y 5AS, United Kingdom

First edition 2017

ISBN: 978-0-12-809388-7
ISSN: 1063-5823

Publisher: Zoe Kruze
Acquisition Editor: Ashlie Jackman
Editorial Project Manager: Alina Cleju
Production Project Manager: Vignesh Tamil
Cover Designer: Greg Harris

Typeset by SPi Global, India

CONTENTS

Section 2
Structural Determinants of Cholesterol-Ion Channels Interactions

CONTRIBUTORS

Belinda S. Akpa
North Carolina State University, Raleigh, NC, United States

Manuela A.A. Ayee
University of Illinois at Chicago, Chicago, IL, United States

John E. Baenziger
University of Ottawa, Ottawa, ON, Canada

Carlos J. Baier
Laboratory of Molecular Neurobiology, Institute of Biomedical Research (BIOMED) UCA-CONICET, Faculty of Medical Sciences, Pontifical Catholic University of Argentina, Buenos Aires, Argentina

Andras Balajthy
Faculty of Medicine, University of Debrecen, Debrecen, Hungary

Nicolas Barbera
University of Illinois at Chicago, Chicago, IL, United States

Francisco J. Barrantes
Laboratory of Molecular Neurobiology, Institute of Biomedical Research (BIOMED) UCA-CONICET, Faculty of Medical Sciences, Pontifical Catholic University of Argentina, Buenos Aires, Argentina

Grace Brannigan
Center for Computational and Integrative Biology, Rutgers University–Camden, Camden, NJ, United States

Anna N. Bukiya
College of Medicine, The University of Tennessee Health Science Center, Memphis, TN, United States

Coralie Di Scala
EA-4674, Interactions Moléculaires et Systèmes Membranaires, Aix-Marseille Université, Marseille, France

Jaimee A. Domville
University of Ottawa, Ottawa, ON, Canada

Alex M. Dopico
College of Medicine, The University of Tennessee Health Science Center, Memphis, TN, United States

Luke S. Evans
Centre for Biological Sciences/Institute for Life Sciences, University of Southampton, Southampton, United Kingdom

Jacques Fantini
EA-4674, Interactions Moléculaires et Systèmes Membranaires, Aix-Marseille Université, Marseille, France

Rita Gutorov
Institute for Medical Research Israel-Canada (IMRIC), Edmond and Lily Safra Center for Brain Sciences (ELSC), Faculty of Medicine, The Hebrew University, Jerusalem, Israel

Peter Hajdu
Faculty of Dentistry, University of Debrecen, Debrecen, Hungary

Ben Katz
Institute for Medical Research Israel-Canada (IMRIC), Edmond and Lily Safra Center for Brain Sciences (ELSC), Faculty of Medicine, The Hebrew University, Jerusalem, Israel

Shaya Lev
Institute for Medical Research Israel-Canada (IMRIC), Edmond and Lily Safra Center for Brain Sciences (ELSC), Faculty of Medicine, The Hebrew University, Jerusalem, Israel

Irena Levitan
University of Illinois at Chicago, Chicago, IL, United States

Baruch Minke
Institute for Medical Research Israel-Canada (IMRIC), Edmond and Lily Safra Center for Brain Sciences (ELSC), Faculty of Medicine, The Hebrew University, Jerusalem, Israel

Sara L. Morales-Lázaro
Instituto de Fisiología Celular, Universidad Nacional Autónoma de México, Mexico City, Mexico

Ruth D. Murrell-Lagnado
Sussex Neuroscience, School of Life Sciences, University of Sussex, Brighton, United Kingdom

Gyorgy Panyi
Faculty of Medicine; MTA-DE-NAP B Ion Channel Structure-Function Research Group, RCMM, University of Debrecen, Debrecen, Hungary

Maximilian Peters
Institute for Medical Research Israel-Canada (IMRIC), Edmond and Lily Safra Center for Brain Sciences (ELSC), Faculty of Medicine, The Hebrew University, Jerusalem, Israel

Tamara Rosenbaum
Instituto de Fisiología Celular, Universidad Nacional Autónoma de México, Mexico City, Mexico

Avia Rosenhouse-Dantsker
University of Illinois at Chicago, Chicago, IL, United States

J.P. Daniel Therien
University of Ottawa, Ottawa, ON, Canada

Zoltan Varga
Faculty of Medicine; MTA-DE-NAP B Ion Channel Structure-Function Research Group, RCMM, University of Debrecen, Debrecen, Hungary

Philip T.F. Williamson
Centre for Biological Sciences/Institute for Life Sciences, University of Southampton, Southampton, United Kingdom

Rachel Zaguri
Institute for Medical Research Israel-Canada (IMRIC), Edmond and Lily Safra Center for Brain Sciences (ELSC), Faculty of Medicine, The Hebrew University, Jerusalem, Israel

PREFACE

Elevated levels of cholesterol are well known to be a key risk factor for the development of the cardiovascular disease. Multiple studies, therefore, explore the mechanisms by which cholesterol and other sterols regulate cellular functions. This volume focuses on sterol regulation of ion channels, a major class of membrane proteins responsible for controlled transport of ions across cellular membranes.

As all membrane proteins, ion channels exist and function within the milieu of a viscous lipid bilayer with cholesterol constituting a large portion of the membrane lipids. The first question that arises for each type of ion channel is whether cholesterol regulates the channel via changes in the physical properties of the membrane bilayer or by direct sterol–protein interactions. The answer to this question is not trivial and requires comparative analysis of multiple lipid species. Our current understanding is that both mechanisms may contribute to sterol-induced regulation of a specific channel. A compelling analysis of the dual cholesterol regulation of nicotinic acetylcholine receptor is presented by the group of Dr. Baenziger. Drs. Morales-Lazaro and Rosenbaum also describe in detail the two modes of cholesterol regulation for different types of transient receptor potential (TRP) ion channels. All other chapters of the volume focus on direct cholesterol–ion channel interactions that appear to be the dominant type of regulation for the majority of ion channels.

The next key question is the identification of the site of interaction or a "cholesterol-binding site." The group of Dr. Barrantes presents a comprehensive analysis of the most well-known cholesterol recognition motifs, CRAC and CARC, and their key role in cholesterol binding to nicotinic acetylcholine receptor. Interestingly, Dr. Rosenhouse-Dantsker in her chapter presents a new analysis based on the known crystal structures of multiple types of ion channels that highlight the possibility that cholesterol may bind to ion channels through other types of interactions, not defined by the known recognition motifs. Our chapter, Barbera et al., also presents additional consideration for the interpretation of the stereospecificity of cholesterol effects on ion channels, which is usually taken as evidence for the specificity of the binding site. In contrast, we present evidence that stereospecificity may arise not from the binding per se but from the ability of the channels to respond to cholesterol binding. This is a critical concept in understanding how cholesterol interacts with ion channels.

It is important to note that cholesterol is not the only sterol that regulates ion channels. Drs. Dopico and Bukiya present a unique comparative analysis of regulation of large-conductance K^+ channels by cholesterol and bile acids, showing that both types of lipids regulate the channels by direct interactions via distinct recognition sites. Interestingly, the group of Dr. Minke shows that cholesterol regulates TRP channels in *Drosophila* fruit flies even though *Drosophila* membranes do not have cholesterol. Dr. Minke proposes that in native *Drosophila* membranes, TRP channels are regulated by ergosterol, suggesting that regulation of ion channels by membrane sterols is a process that developed before the origin of the mammals.

In summary, it appears that we are currently on the verge of a good understanding of how cholesterol interacts with ion channels. Furthermore, one of the most important concepts that arise from this collection of assays is that elucidating the mechanisms by which sterols regulate membrane proteins requires a combination of computational and experimental approaches, as highlighted in many of the chapters in this volume. The next frontier, however, that is practically unexplored, is determining how the event of cholesterol binding is transduced to changes in the ion channel function. This question needs to be addressed in future studies.

Irena Levitan

ACKNOWLEDGMENTS

I would like to thank all the authors for contributing such interesting, thoughtful, and thought-provoking chapters. I also thank Dr. Sidney Simon for convincing me to edit this volume and Ms. Hannah Colford and Ms. Alina Cleju for their tremendous help in putting the volume together. Many thanks to all the members of my lab, Nik Barbera, Manuela Ayee, Sang Joon Ahn, Drew Fancher, Liz LeMaster, Yedida Bogachkov, Crystal Adamos, and Catherine Osborn who contributed with comments, discussions, and advice. And last but not least, I am very grateful to my husband, Gregory Kowalsky, for designing all my summary cartoons and for all his patience and understanding of all the working weekends.

SECTION 1

Membrane Structure and General Mechanisms of Sterol Regulation of Ion Channels

CHAPTER ONE

Relevance of CARC and CRAC Cholesterol-Recognition Motifs in the Nicotinic Acetylcholine Receptor and Other Membrane-Bound Receptors

Coralie Di Scala*, Carlos J. Baier†, Luke S. Evans‡, Philip T.F. Williamson‡, Jacques Fantini*, Francisco J. Barrantes†,[1]

*EA-4674, Interactions Moléculaires et Systèmes Membranaires, Aix-Marseille Université, Marseille, France
†Laboratory of Molecular Neurobiology, Institute of Biomedical Research (BIOMED) UCA-CONICET, Faculty of Medical Sciences, Pontifical Catholic University of Argentina, Buenos Aires, Argentina
‡Centre for Biological Sciences/Institute for Life Sciences, University of Southampton, Southampton, United Kingdom
[1]Corresponding author: e-mail address: rtfjb1@gmail.com

Contents

Abstract

Cholesterol is a ubiquitous neutral lipid, which finely tunes the activity of a wide range of membrane proteins, including neurotransmitter and hormone receptors and ion channels. Given the scarcity of available X-ray crystallographic structures and the even fewer in which cholesterol sites have been directly visualized, application of in silico computational methods remains a valid alternative for the detection and thermodynamic characterization of cholesterol-specific sites in functionally important membrane proteins. The membrane-embedded segments of the paradigm neurotransmitter receptor for acetylcholine display a series of cholesterol consensus domains (which we have coined "CARC"). The CARC motif exhibits a preference for the outer membrane leaflet and its mirror motif, CRAC, for the inner one. Some membrane proteins possess the double

Current Topics in Membranes, Volume 80
ISSN 1063-5823
http://dx.doi.org/10.1016/bs.ctm.2017.05.001

CARC–CRAC sequences within the same transmembrane domain. In addition to in silico molecular modeling, the affinity, concentration dependence, and specificity of the cholesterol-recognition motif–protein interaction have recently found experimental validation in other biophysical approaches like monolayer techniques and nuclear magnetic resonance spectroscopy. From the combined studies, it becomes apparent that the CARC motif is now more firmly established as a high-affinity cholesterol-binding domain for membrane-bound receptors and remarkably conserved along phylogenetic evolution.

1. INTRODUCTION

Cholesterol modulates the activity of a wide range of membrane receptors and ion channels in multiple ways, i.e., via general effects on the bulk bilayer lipid, altering membrane fluidity (el Battari, Ah-Kye, Muller, Sari, & Marvaldi, 1985; Lazar & Medzihradsky, 1992; Maguire & Druse, 1989) or curvature (Lee, 2004; Yesylevskyy, Demchenko, Kraszewski, & Ramseyer, 2013) or through direct binding to these proteins (Baier, Fantini, & Barrantes, 2011; Barrantes, 2004; Dopico & Bukiya, 2014; Fantini & Barrantes, 2013; Levitan, Singh, & Rosenhouse-Dantsker, 2014; Picazo-Juarez et al., 2011; Popot, Demel, Sobel, van Deenen, & Changeux, 1977; Posada et al., 2014; Singh, Shentu, Enkvetchakul, & Levitan, 2011; Rosenhouse-Dantsker, Noskov, Durdagi, Logothetis, & Levitan, 2013; Singh et al., 2012). The availability of the crystal structures of the β2-adrenergic receptor represented an important milestone in the identification of direct interactions between a paradigmatic transmembrane (TM) protein and member of the most abundant and functionally important superfamily of receptors in eukaryotic cells, i.e., the G-protein-coupled receptors (GPCRs) on the one hand and cholesterol (Cherezov et al., 2007; Rosenbaum et al., 2007) on the other. The X-ray data of the β2-adrenergic receptor rapidly catalyzed a cascade reaction in the crystallography of membrane-embedded proteins, which led to the identification of cholesterol-binding sites in various other GPCRs like the β1-adrenergic receptor (Warne et al., 2008) and several other members of the GPCRs (reviewed, e.g., in Vaidehi, Bhattacharya, & Larsen, 2014). Cholesterol–β-adrenergic receptor interactions have been reported to increase the compactness of the receptor structure and to enhance the conformational stability toward active or inactive receptor states (Gimpl, 2016). The available crystal structures of these macromolecules represent, however, only a minor fraction of the genome coding for functionally relevant membrane protein targets. Viable alternative methodologies for identifying cholesterol-recognition motifs in hormone or neurotransmitter receptors and ion channels are therefore currently being sought in membrane biology. Various groups, including ours, have resorted to computational methods to explore sequences in protein data banks and detect the presence of putative cholesterol-binding linear domains in the TM regions of proteins, with special emphasis on neurotransmitter receptors such as GPCRs (Baier et al., 2011; Fantini, Di Scala, Evans, Williamson, & Barrantes, 2016; Jafurulla, Tiwari, & Chattopadhyay, 2011), ion channels (Fantini, Di Scala, Baier, & Barrantes, 2016), and transporters (Clay, Lu, & Sharom, 2015; Gal et al., 2015; Sharpe et al., 2015). This approach resulted in the definition of consensus motifs with predictive value, which can be further applied for identifying cholesterol-binding linear domains (Baier et al., 2011; Epand, 2008; Epand, Thomas, Brasseur, & Epand, 2010; Epand et al., 2006; Jamin et al., 2005). The first such consensus motif to be identified was defined by the sequence array (L/V)-X_{1-5}-(Y)-X_{1-5}-(K/R) and coined "cholesterol-recognition amino acid consensus" (CRAC) (Jamin et al., 2005; Li & Papadopoulos, 1998). This motif was readily found in several proteins known to bind cholesterol, including both viral and host membrane proteins (Epand, 2006, 2008; Epand et al., 2010, 2006). We subsequently introduced the linear sequence (K/R)-X_{1-5}-(Y/F)-X_{1-5}-(L/V), which is essentially the reverse or mirror version of the CRAC algorithm, and hence referred to it as the "CARC" consensus motif (Baier et al., 2011).

The CARC motif was originally explored in greater detail using the nicotinic acetylcholine receptor (nAChR), the paradigm of the pentameric ligand-gated ion channels (Barrantes, 2015). The free energy of interaction between cholesterol molecules and the nAChR is about $-510/-530$ kJ mol^{-1}, i.e., more than -100 kJ mol^{-1} per subunit. The particularly favorable fit between the "CARC-like" γTM4 segment from human nAChR (428RVCFLAML435) and cholesterol is noteworthy, with an energy of interaction of about -60 kJ mol^{-1}, i.e., ~60% of the total energy of interaction of the entire γ subunit, which exhibits the highest affinity for cholesterol among all nAChR subunits (cf. Table 1).

Thermodynamic analyses of the energy of interaction of cholesterol with an assortment of membrane proteins (Fig. 4) revealed that the CARC motif generally exhibits more affinity for cholesterol than the CRAC motif (Fantini & Barrantes, 2013). We provided physicochemical arguments to account for the difference in the predictive value of the two linear algorithms, i.e., the snorkeling effect of Lys/Arg residues (the apolar part of the amino acid chain buried in the membrane and cationic group outside)

Table 1 Energetics of Interaction of Cholesterol and the CARC and CARC-Like Motifs in the Transmembrane Domains of Human Muscle-Type AChR

AChR TM Domain	Energy of Interaction ($kJ\ mol^{-1}$)
αTM1	−35.129
αTM3	−31.729
αTM4	−27.903
Total α subunit	−94.761
βTM1	−52.332
βTM3	−20.808
βTM4	−26.241
Total β subunit	−99.453
γTM1	−30.542
γTM3	−37.066
γTM4	−59.961
Total γ subunit	−127.569
δTM1	−46.184
δTM3	−33.083
δTM4	−29.197
Total δ subunit	−108.464
εTM1	−44.438
εTM3	−24.421
εTM4	−44.050
Total ε subunit	−112.909
Embryonic AChR[a] ($\alpha_2\beta\gamma\delta$)	−525.008
Adult AChR[a] ($\alpha_2\beta\varepsilon\delta$)	−510.348

[a]The stoichiometric contribution of two α subunits is taken into account in the estimation of the total energy of interaction of the AChR pentamer with cholesterol molecules.
From Baier, C. J., Fantini, J., & Barrantes, F. J. (2011). Disclosure of cholesterol recognition motifs in transmembrane domains of the human nicotinic acetylcholine receptor. Scientific Reports, 1, 69.

(Strandberg & Killian, 2003) and cholesterol structure (Fantini & Barrantes, 2009; as reviewed in Fantini & Barrantes, 2013; Fantini, Di Scala, Baier, et al., 2016).

2. EXPERIMENTAL VALIDATION OF THE CARC MOTIF

A series of sequence analyses and molecular dynamics studies describing the occurrence of the CRAC mirror sequence ("CARC") in several membrane (Baier et al., 2011; Fantini & Barrantes, 2013) and soluble (Morrill & Kostellow, 2016; Morrill, Kostellow, & Gupta, 2014, 2016) proteins have provided substantial evidence for the possible functional relevance of these linear amino acid sequences in cholesterol–protein interactions. However, direct experimental demonstration of physical interactions between the two partners was lacking for proteins other than GPCRs. In a recent work the working hypothesis was subjected to experimental test by studying the interaction of a prototype CARC domain with cholesterol employing lipid monolayer strategies and nuclear magnetic resonance (NMR) spectroscopy.

Selection of a representative CARC domain for these experimental validations was based on previous photolabeling studies of the *Torpedo* nAChR with the cholesterol analogue probe [^{3}H]azicholesterol, which led to the identification of a (predominant) cholesterol-binding domain in the fourth transmembrane domain (TM4) of the human nAChR γ subunit (Hamouda, Chiara, Sauls, Cohen, & Blanton, 2006). Subsequent in silico computational approaches (Baier et al., 2011) led us to identify a typical CARC motif: 455-**R**VC**F**LAM**L**-462 (the characteristic Arg, Phe, and Leu amino acid residues outlined bold and underlined) in the human γTM4 that incorporated most of the label in the photoaffinity studies. Furthermore, the molecular modeling simulations disclosed that this TM segment displayed the highest energy of interaction (in the order of -60 kJ mol^{-1}) with cholesterol when compared to all other subunits of the nAChR in *Homo sapiens* and other species (Baier et al., 2011). The homologous γTM4 segment in the *Torpedo* nAChR possesses a CARC motif similar to the human form: 449-**K**AC**F**WIA**L**-456. In *Torpedo* TM4 the highest [^{3}H]azicholesterol labeling is observed in Asp448, the second residue after Lys449, which is the first amino acid of the N-term CARC motif.

To further refine the correlation between the [^{3}H]azicholesterol photolabeling data with the in silico calculations, additional molecular dynamics studies were carried out on the cholesterol derivative-*Torpedo* γTM4 segment (445–460), that is, comprising the CARC domain plus a few upstream and downstream amino acid residues. We found that azicholesterol interacted with the amino acid residues defined by the CARC algorithm, namely Lys449, Phe452, and Leu456, and that the azi-group was at a distance of only 2 Å from the side chain of Asp448. The CARC motif present within the γTM4 exhibited a tight cholesterol binding, although it seemed unlikely that the strength of the interaction sufficed to account for the reduction in mobility observed in the molecular simulations. This led us to propose that the reduction in mobility arises from the cholesterol-induced oligomerization of γTM4, which is consistent with previous fluorescence studies from our laboratory showing the cholesterol-dependent oligomerization of γTM4 in POPC bilayers (de Almeida et al., 2004). The molecular dynamics studies further revealed that the cholesterol contact with the CARC motif within the γTM4 was established with the β-face, leaving the α-face exposed, suggesting that the cholesterol-mediated oligomerization of the peptide is dictated by cholesterol–cholesterol interactions rather than by protein–protein interactions (Fantini & Barrantes, 2009). When we compared the interaction between cholesterol and either wild-type or mutant TM4 peptides, the WT CARC motif was found to exhibit a high affinity for cholesterol. All three amino acids defining the CARC domain were found to interact with cholesterol, especially the central Phe452 residue. Replacement of this aromatic residue with alanine (F-452/A mutant) resulted in a significant loss of affinity. This important result was fully confirmed by physicochemical lipid monolayer studies (Di Scala, Chahinian, Yahi, Garmy, & Fantini, 2014) of CARC–cholesterol interactions (Fig. 1). In these experiments, a lipid monolayer is prepared at the air–water interface at a controlled surface pressure (π_0), after which the peptide is injected in the aqueous phase. The interaction of the peptide with the lipid is assessed by surface pressure measurements (basically a surface pressure increase $\Delta\pi$ measured in mN m^{-1}) (Di Scala et al., 2014). The conservative F-452/W mutation had no effect, indicating that it is the aromatic nature of Phe452, and not its specific structure, that is required for optimal binding. This is in line with previous studies, suggesting that CARC motifs could contain any of the three aromatic residues, i.e., Phe, Trp, or Tyr (Baier et al., 2011; Fantini & Barrantes, 2013). The other major outcomes of our physicochemical studies are the demonstration of lipid specificity (CARC recognized cholesterol but not phosphatidylcholine) and the concentration dependency of the binding (saturation was reached for peptide concentrations <10 μM) (Fig. 1).

The second line of experimental validation stemmed from NMR studies (MAS triple resonance magic-angle spinning deuterium NMR using deuterated Ala471) (Fantini, Di Scala, Evans, et al., 2016). Inclusion of cholesterol to phospholipid bilayers containing a synthetic $^{13}C/^{15}N$-labeled peptide corresponding to Asp464 to Val492 in the intact *Torpedo* γTM4 peptide caused a reduction in the rotational motion of the peptide within the bilayer, a result consistent with the cholesterol-mediated peptide oligomerization, as discussed in the preceding paragraph. The functional significance of this in the intact receptor remains to be elucidated, but the location of the

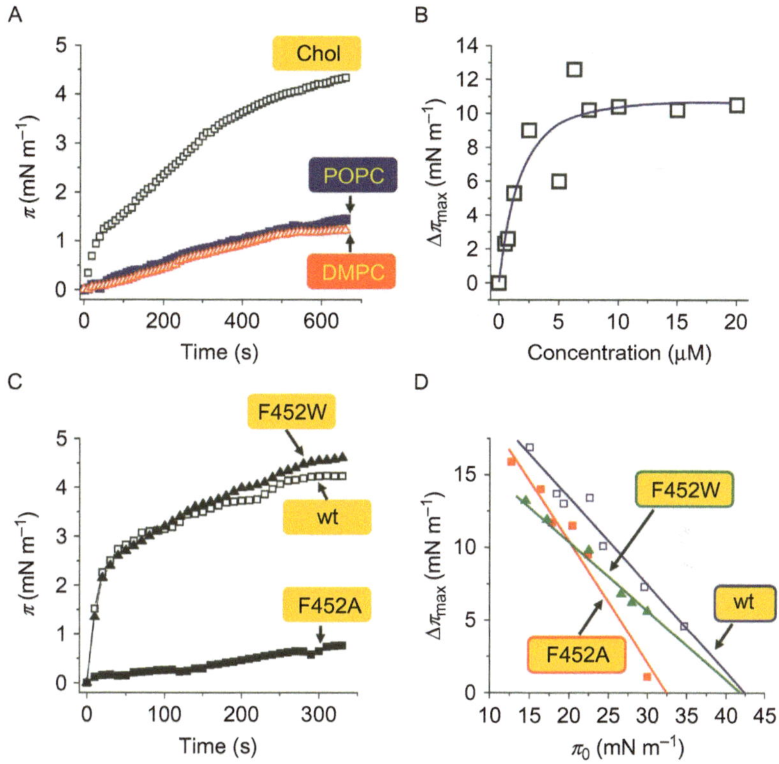

Fig. 1 Lipid monolayer studies of CARC–cholesterol interactions. (A) Kinetics of interaction of a synthetic γTM4 peptide corresponding to fragment 445–460 of the *Torpedo* nAChR γ subunit with a monolayer of cholesterol (Chol) or phosphatidylcholine (palmitoyl-oleoyl phosphatidylcholine, POPC or dimyristoyl phosphatidylcholine, DMPC). The peptide is injected in the aqueous phase underneath the lipid monolayer, and the interaction is measured by the surface pressure increase (π) induced by the peptide. (B) Effect of peptide concentration on the interaction between nAChR γTM4 and cholesterol (surface pressure increase $\Delta\pi_{max}$ induced by the peptide after 30 min of incubation). (C) Kinetics of interaction of wild-type (wt) and mutant peptides (F-452/A; F-452/W) derived from γTM4 with cholesterol monolayers ($\pi_0 = 30 \pm 3.5$ mN m^{-1}). (D) Maximal surface pressure increase ($\Delta\pi_{max}$) induced by wild type and mutant peptides on cholesterol monolayers at various π_0 values. *From Fantini, J., Di Scala, C., Evans, L. S., Williamson, P. T. F., & Barrantes, F. J. (2016). A mirror code for protein-cholesterol interactions in the two leaflets of biological membranes.* Scientific Reports, *6, 21907.*

CARC domain on the lipid-facing surface of the helix may play a role in cholesterol-mediated clustering of the receptor. In summary, the experimental approaches combining lipid monolayer data, NMR spectroscopy, and in silico molecular modeling simulations provided much stronger evidence for the direct physical interaction of cholesterol with a CARC cholesterol-recognition motif. The interaction was found to be of high affinity, lipid specific, and saturable (Fantini, Di Scala, Evans, et al., 2016).

3. COEXISTENCE OF CARC AND CRAC SEQUENCES WITHIN THE SAME TM DOMAIN

We have recently conducted a search for CARC/CRAC cholesterol-recognition motifs over a large series of membrane proteins (listed in Fig. 4) in combination with molecular dynamics simulations of the whole TM regions of various membrane-embedded proteins in order to determine whether two cholesterol molecules could actually be docked onto these domains (Fantini, Di Scala, Evans, et al., 2016). This was indeed the case. As shown in Fig. 2, the CRAC and CARC motifs are vectorial ("apolar" Leu/Val→ "basic" Lys/Arg for CRAC and "basic" Lys/Arg→ "apolar" Leu/Val for CARC, from the N-terminus to the C-terminus sequence).

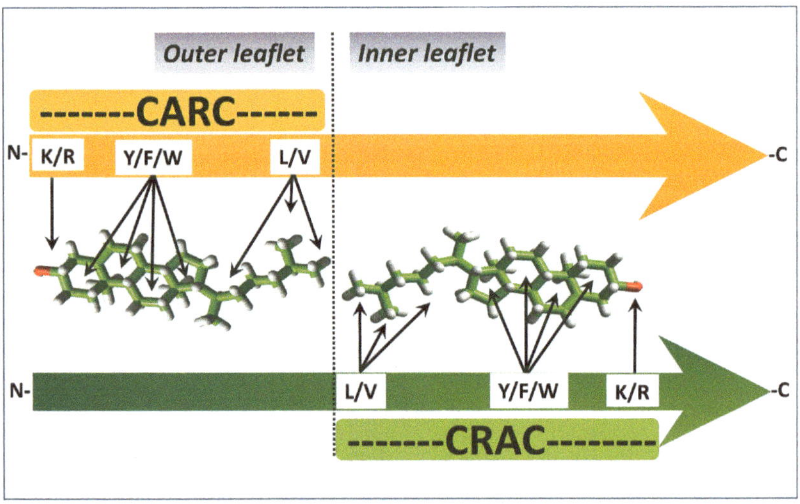

Fig. 2 The vectorial arrangement of CARC and CRAC motifs. For proteins whose N-terminus is extracellular, the CARC domain is located in the outer leaflet and the CRAC domain is in the inner one. This topology applies for type-1 membrane proteins as well as for TM domains 1, 3, 5, and 7 of G-protein-coupled receptors (GPCRs) with seven transmembrane domains. For type-2 membrane proteins (extracellular C-terminus) and domains 2, 4, and 6 of GPCRs, the algorithms still apply, but CARC is located in the inner leaflet and CRAC in the outer one. *From Fantini, J., Di Scala, C., Evans, L. S., Williamson, P. T. F., & Barrantes, F. J. (2016). A mirror code for protein-cholesterol interactions in the two leaflets of biological membranes.* Scientific Reports, *6, 21907.*

The CARC sequence starts with a basic residue (Arg or Lys), and this feature makes the CARC motif ideally suited for interaction with cholesterol in the outer leaflet of biological membranes. Indeed, the N-terminal domain of type I membrane proteins is extracellular, such that the carbon chain enters the membrane bilayer in the N- to C-terminus direction (Fig. 2). This is also the case for TM domains 1, 3, 5, and 7 in the GPCRs (see below).

The three main amino acid residues defining the CARC and CRAC motifs were always involved in the interaction. In addition, in both cases the central aromatic residue could be either Phe or Tyr (and even Trp in the case of CARC). This finding is consistent with the nature of the interaction between cholesterol and aromatic rings, i.e., the CH–π stacking interaction (Nishio, Umezawa, Fantini, Weiss, & Chakrabarti, 2014). The branched aliphatic residues (Leu/Val) are well suited to accommodate the protruding methyl groups of cholesterol. Furthermore, the terminal basic residue of the motif often forms a hydrogen bond with the oxygen atom of the –OH group of cholesterol (Fantini & Barrantes, 2013). Thus, the selection of CARC and CRAC as cholesterol-binding motifs is justified by robust physicochemical rules. These rules can be put into practice via a combination of London, CH–π stacking, and hydrogen bonding that cooperate to control protein–cholesterol interactions in the membrane environment (Fantini & Barrantes, 2013).

Since the consensus CRAC sequence starts with an aliphatic residue (Leu or Val), its N-terminal is expected to interact with the apolar groups of cholesterol (sterane, methyl, and isooctyl) in the inner leaflet (Fig. 2). In other words, the sequential chaining of CARC and CRAC motifs in the amino acid sequence of a TM domain, starting from the N-terminus, is consistent with the binding of a cholesterol molecule in each leaflet.

The coexisting presence of CARC and CRAC motifs, one in each leaflet of the membrane, has an important consequence: the host TM protein segment can accommodate two opposite (tail-to-tail) cholesterol molecules. This ensures that the polar amino acid residues of the motif (Lys/Arg) face the intra- and extracellular milieu, whereas the apolar ends of the motifs (Leu/Val) are deeply buried in the most hydrophobic region of the lipid bilayer (Fig. 2).

This information has provided a new twist in the interpretation of the mode of action of the mitochondrial translocator protein, TSPO, and its interaction with cholesterol. In a recent work, Jaremko, Jaremko, Giller, Becker, and Zweckstetter (2014) had speculated that the binding of cholesterol at the outside of the TSPO structure and the ability of cholesterol to dimerize were the two factors determining the oligomeric state of the transporter. Our studies showed that TSPO possesses not only a CRAC domain but also a CARC motif in the same TM region (Fig. 3). Furthermore, the CARC motif has an exceptionally high energy of interaction with cholesterol, in the order of -62 kJ mol^{-1} (Fantini, Di Scala, Evans, et al., 2016). The additional cholesterol site on the same membrane-embedded surface provides further energetic grounds for the cholesterol-mediated oligomerization hypothesis.

If one searches protein sequence databases for the occurrence of both CARC and CARC motifs in the same TM domain, one finds that for type I membrane proteins and for domains 1, 3, 5, and 7 of GPCRs, the CARC motif is always located in the outer leaflet, whereas the CRAC sequence is found in the inner leaflet (Fig. 4). The examples include signaling membrane receptors like the somatostatin, GABA, serotonin, adenosine, VIP, and cannabinoid receptors, as well as the voltage-dependent TRVP1 channel. Overall, the mean energy of interaction was in the order of -58 kJ mol^{-1}

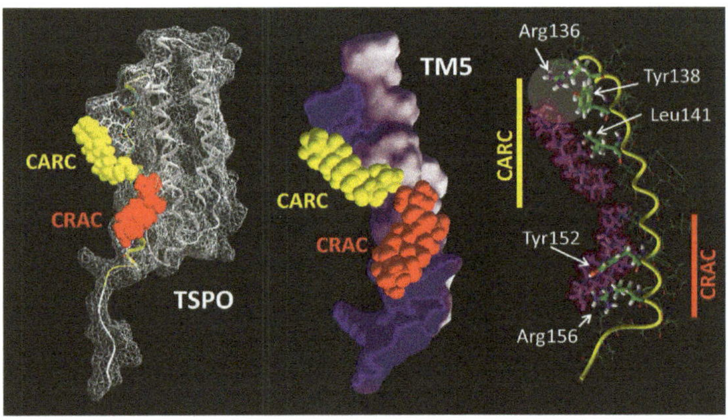

Fig. 3 Coexistence of both CARC and CRAC motifs in the transmembrane domain of the mitochondrial translocator protein TSPO. On the *left*, the docking of cholesterol on the CARC motif (cholesterol in *yellow*) and the CRAC motif (cholesterol in *red*). The 3D structure of cholesterol was retrieved from PDB entry 1MT5. The model in the *middle* represents the fifth transmembrane domain (TM5) of TSPO with cholesterol bound to both CARC and CRAC motifs. The model on the *right* shows the molecular interactions between cholesterol (in *purple*) and each cholesterol-binding motif of TM5. The hydrogen bond network stabilizing the cholesterol–CARC interaction is indicated by a *disk*. *From Fantini, J., Di Scala, C., Evans, L. S., Williamson, P. T. F., & Barrantes, F. J. (2016). A mirror code for protein-cholesterol interactions in the two leaflets of biological membranes.* Scientific Reports, *6, 21907.*

```
N-ter/Extracellular→CARC---CRAC→C-ter/Intracellular
----------------------------------------------------

Type 3 somatostatin receptor  Human #P32745
      TM5 203- RAGFIIYTAALGFFGPLLVICLCYLLIVVK-232
      Energy:        -54.0              -43.6
Mitochondrial translocator protein Mouse #P50637
      TM5 135- RLLYPYLAWLAFATVLNYYVWR-156
      Energy:   -62.4          -44.4
Neuropeptide FF receptor 2 Human #Q9Y5X5
      TM5 318-RKIYTTVLFANIYLAPLSLIVIMYGR-343
      Energy:   -32.7              -52.5
Metabotropic glutamate receptor 5 Human #P41594
      TM7 798-KIITMCFSVSLSATVALGCMFVPK-821
      Energy:      -60.6       -39.0
GABA type B receptor subunit 2 Human #O75899
      TM1 475-RKISLPLYSILSALTILGMIMASAFLFFNIK-555
      Energy:     -64.1                 -31.3
Cannabinoid receptor 1 Human #P21554
      TM3 186-RNVFLFKLGGVTASFTASVGSLFLTAIDR-214
      Energy:     -56.7                -60.6
      TM7 375-KTVFAFCSMLCLLNSTVNPIIYALR-400
      Energy:     -53.5            -55.3
5-hydroxytryptamine receptor 7 Human #P34969
      TM7 367-RTFLWLGYANSLINPFIYAFFNRDLR-392
      Energy:   -78.1                -53.2
Adenosine receptor A1 #P30542
      TM7 265-KPSILTYIAIFLTHGNSAMNPIVYAFRIQK-294
      Energy:     -54.9                  -45.9
VIP receptor 1 Human #P32241
      TM1 143-KTGYTIGYGLSLATLLVATAILSLFRK-169
      Energy:     -54.2                -46.4
Prolactin-releasing peptide receptor Human #P49683
      TM5 221-RQLYAWGLLLVTYLLPLLVILLSYVR-247
      Energy:     -55.3              -27.3
Oxytocin receptor Human #P30559
      TM5 198-KAYITWITLAVYIVPVIVLAACYGLISFK-226
      Energy:     -50.1              -46.4
TRPV1 Human #Q8NER1
      TM2 468-KTGDYFRVTGEILSVLGGVYFFFRGIQYFLQRR-500
      Energy:     -83.2                  -63.2
Corticotropin-releasing factor receptor 1 Human #P34998
      TM1 111-KSKVHYHVAVIINYLGHCISLVALLVAFVLFLRLR-145
      Energy:     -52.6                       -60.8
```

Mean energy:	CARC – 58.03 ± 12.10 kJ. mol^{-1}
	CRAC – 47.85 ± 10.68 kJ. mol^{-1}

Fig. 4 Energetics of cholesterol binding to TM domains displaying *both* CARC and CRAC motifs. The CARC motif is framed in *yellow*, and the CRAC motif in *green*. The calculated energy of interaction (in kJ mol^{-1}) is indicated under each motif. The UniProt entry is indicated for each protein after the # symbol. *From Fantini, J., Di Scala, C., Evans, L. S., Williamson, P. T. F., & Barrantes, F. J. (2016). A mirror code for protein-cholesterol interactions in the two leaflets of biological membranes.* Scientific Reports, *6, 21907.*

for CARC and $-48\ \text{kJ mol}^{-1}$ for CRAC, indicating that the CARC domain generally exhibits more affinity for cholesterol than a CRAC domain (Baier et al., 2011; Fantini & Barrantes, 2013; Fantini, Di Scala, Evans, et al., 2016). Exceptions to this rule can be found, e.g., for neuropeptide FF and corticotrophin-releasing factor receptors (Fig. 4).

In assessing the possible biological implications of the coexisting CARC–CRAC motifs within the same membrane-embedded peptide domain, the heterologous protein expression in the yeast system poses a singular case, since the functional expression of, e.g., human receptors is not always sustained (Opekarova & Tanner, 2003). In this context, it is interesting to note that yeasts have an essential requirement for ergosterol for cell growth, and cholesterol is not a valid substitute for the former sterol. However, ligand binding to the human μ-opioid receptor was found to increase in transfected *Saccharomyces* cells when ergosterol was replaced by cholesterol (Lagane et al., 2000). We have speculated (Fantini, Di Scala, Evans, et al., 2016) that human CARC/CRAC domains might exhibit species specificity for cholesterol and cannot mediate functional ergosterol binding in yeast, perhaps due to subtle conformational differences between the two sterol molecules, cholesterol being more flexible due to the presence of several extra-double bonds in ergosterol (Czub & Baginski, 2006; Baginski et al., 1989).

4. RELIABILITY OF THE CARC AND CRAC ALGORITHMS

Despite the robust biochemical rules that explain why CARC and CRAC domains exhibit specific cholesterol-binding properties (Fantini & Barrantes, 2013), it remains the case that the algorithms used for the detection of these domains are very general (basically a vectorial triad of key amino acid residues). As a matter of fact, multiple copies of CARC and/or CRAC may be detected in the same protein (Palmer, 2004). However, this drawback should not be regarded as insurmountable. In the specific case of membrane proteins, which have been extensively studied (Fantini & Yahi, 2015), additional criteria might be considered to determine whether the presence of a consensus CARC/CRAC motif is likely to constitute a functional cholesterol-binding domain. From a molecular point of view, the reliability of the CARC/CRAC algorithm is excellent. This statement is based on the experimental demonstration that (i) all synthetic peptides derived from CARC/CRAC motifs tested so far display specific cholesterol-binding properties, and (ii) mutations affecting the motif, especially the central aromatic residue, always decrease the binding of cholesterol (Fantini, Di Scala, Evans, et al., 2016). These data of course only prove that these consensus motifs are able to bind cholesterol, not that the protein displaying the motifs actually interacts with this lipid. In addition, the motif has to be located in a TM domain (Fantini, Di Scala, Evans, et al., 2016). The bioavailability of cholesterol in the membrane area surrounding the TM domain displaying a CARC/CRAC motif will determine whether cholesterol interacts with this protein, and when it does so. Combining all these criteria led us to propose a step-by-step method for identifying linear cholesterol-binding motifs in membrane proteins (Fantini, Di Scala, Evans, et al., 2016).

Finally, one should note that besides the CARC and CRAC motifs, cholesterol can bind to three-dimensional pockets that combine several TM domains and thus might remain undetected by the CARC/CRAC algorithms. Specifically, it was shown that Kir2 channels have a functionally important cholesterol-binding pocket with residues that do not contain any of the previously identified cholesterol-recognition motifs (Levitan et al., 2014; Rosenhouse-Dantsker et al., 2013). Nevertheless, the key amino acid residues that define the CARC/CRAC motif (e.g., an aromatic one) are generally present in those three-dimensional motifs, as is the case for instance with the cholesterol consensus motif (CCM) (Hanson et al., 2008). In any case, it appears that the interaction of a TM domain with cholesterol is controlled by general biochemical rules that determine a series of fully predictable van der Waals interactions and hydrogen bonding (Fantini & Yahi, 2015).

5. CHOLESTEROL AND ITS KEY ROLE IN LIQUID-ORDERED LIPID DOMAINS

Cholesterol occurs in the inner or outer leaflet of the membrane bilayer, or in both leaflets. In the inner leaflet, it may interact with phosphatidylserine, whereas in the outer leaflet it associates predominantly with sphingomyelins. In general, the latter possess a more rigid apolar surface than glycerophospholipids, and this facilitates the preferential interaction with cholesterol. The cholesterol–sphingomyelin association is further enriched in glycerophospholipids with saturated fatty acyl chains (relative to the average saturation in the rest of the bilayer). The lipid raft hypothesis proposes that these ternary complexes formed by specific self-associated lipid species constitute microdomains or platforms that can intervene in protein partition, signaling, and other functional events in cell physiology (Anderson & Jacobson, 2002; Lingwood & Simons, 2010; Simons & Ikonen, 1997;

Simons & van Meer, 1988). The physicochemical basis for the formation of these domains probably stems mostly from the peculiar and still not fully understood thermodynamic properties of biological membranes: favorable and unfavorable lipid–lipid interactions result in transient lateral heterogeneities that join or segregate their constituent molecules, respectively. Above/below certain critical concentrations and/or temperatures, these lateral heterogeneities generate transiently separated lipid phases, the two most prominent of which are the liquid-disordered (Ld) and liquid-ordered (Lo) phases (Marsh, 1991; Yeagle, 1989). The temperature and compositional range over which these lateral separations into liquid phases occur is rather large (Veatch & Keller, 2002, 2003a, 2003b; Veatch, Polozov, Gawrisch, & Keller, 2004). The domains enriched in cholesterol–sphingomyelin–saturated glycerophospholipids constitute the Lo phase, a more condensed, rigid, and thicker fraction of the membrane. Outside these Lo domains, cholesterol associates with other glycerophospholipids (mainly phosphatidylcholines) in a rather loose manner, and at relatively lower concentrations (Fig. 5).

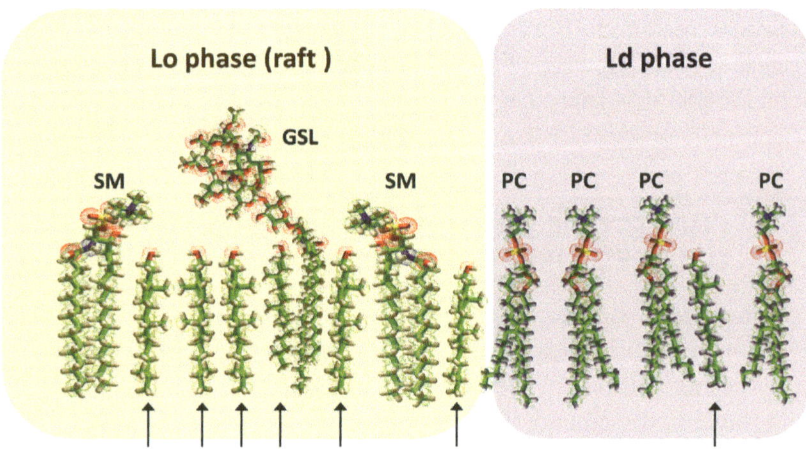

Fig. 5 Cholesterol accessibility in Lo and Ld phases. In the Lo phase (cholesterol–sphingomyelin enriched, "raft" domains), cholesterol (*arrows*) is masked by sphingolipids such as sphingomyelin (SM) or glycosphingolipids (GSL). In the Ld phase, cholesterol molecules (*arrows*) are more sparsely distributed among glycerophospholipids such as phosphatidylcholine (PC). In this case the polar –OH group of cholesterol is accessible to extracellular ligands. *From Fantini, J., Di Scala, C., Baier, C. J., & Barrantes, F. J. (2016). Molecular mechanisms of protein-cholesterol interactions in plasma membranes: Functional distinction between topological (tilted) and consensus (CARC/CRAC) domains.* Chemistry and Physics of Lipids, 199, *52–60.*

Lipid domains apparently cover a wide range of sizes, from assemblies with <5 nm radius ("ultrananodomains"; Pathak & London, 2015), comprising a couple of hundred lipid molecules per bilayer, to micron-sized platforms with thousands of molecules readily observable by conventional wide-field light microscopy (Griffie, Burn, & Owen, 2015; Maxfield, 2002; Rao & Mayor, 2014; van Zanten & Mayor, 2015). Functionally, lipid domains play important roles in the cell by way of the lateral separation of chemical species in the plane of the membrane. The chemical analysis of the postsynaptic apparatus in the peripheral nervous system shows that cholesterol is a very abundant component (see review in Barrantes, 1989) of this specialized membrane. This sterol is an essential partner of the nAChR, affecting its distribution and several of its functional properties in the peripheral synapse, the neuromuscular junction (Barrantes, 2010, 2012). The lateral heterogeneity of lipids in the postsynaptic membranes of the *Torpedo* electrocyte was an early biophysical finding: protein-associated lipids were shown to be immobilized with respect to bulk membrane lipid (Marsh & Barrantes, 1978), and subsequent work has shown that cholesterol-like molecules form part of this protein-immobilized pool (Barrantes, 2007). The functional implications of this finding became apparent when it was demonstrated that cholesterol is an essential component for maintaining nAChR agonist-dependent state transitions in the postsynaptic membrane (Criado, Eibl, & Barrantes, 1982). It has been proposed that there are two cholesterol populations in nAChR-rich membranes from *Torpedo*: an easily extractable fraction that influences the bulk fluidity of the membrane and a tightly bound receptor-associated fraction (Leibel, Firestone, Legler, Braswell, & Miller, 1987).

In muscle cells, cholesterol was found to influence the formation of micron-sized nAChR clusters induced by agrin (Campagna & Fallon, 2006). Signaling via the agrin/MuSK complex and interaction between the receptor and rapsyn appears to involve lipid platforms (Zhu, Xiong, & Mei, 2006). Using Laurdan two-photon fluorescence microscopy (Stetzkowski-Marden, Gaus, Recouvreur, Cartaud, & Cartaud, 2006), it was concluded that nAChR clusters reside in Lo membrane domains. Another study (Willmann et al., 2006) proposed that these cholesterol-rich lipid microdomains and Src-family kinases both contribute to stabilizing nAChRs and the postsynaptic apparatus. We often resort to an experimental clonal cell line, CHO-K1/A5, which is devoid of nAChR-clustering proteins such as rapsyn and tyrosine kinases, and therefore, homophilic protein–protein interactions, heterophilic protein–lipid interactions, and links with the actin cytoskeleton are more likely candidates for maintaining the nAChR nanocluster assemblies. Membrane-embedded proteins with preferential affinities for Lo or Ld domains could influence both the lifetime and size of the domains in which they are located by selecting their local lipid environment. A fraction of nAChRs has indeed been found in Lo domains in mammalian cells (Bruses, Chauvet, & Rutishauser, 2001; Campagna & Fallon, 2006; Marchand, Devillers-Thiery, Pons, Changeux, & Cartaud, 2002; Stetzkowski-Marden et al., 2006; Willmann et al., 2006; Zhu et al., 2006). On the other hand, when reconstituted in a sphingomyelin–cholesterol–POPC (1:1:1) model system, purified nAChR protein from *Torpedo* appears not to exhibit any preference for Lo domains in vitro (Bermudez, Antollini, Fernandez Nievas, Aveldano, & Barrantes, 2010). However, inclusion of some sphingomyelin molecular species (brain sphingomyelins, 16:0, 18:0, or 24:1 sphingomyelins) that generate bilayer asymmetry by enriching the sphingolipid content of the outer leaflet of the lipid bilayer can favor the partitioning of the nAChR in Lo domains (Perillo, Penalva, Vitale, Barrantes, & Antollini, 2016). This can be correlated with the observation that Lo domains in the outer leaflet of a bilayer can induce liquid order in the inner leaflet by a coupling mechanism involving in-register Lo domains in the two halves of the bilayer (Lin & London, 2015).

6. EVOLUTIONARY CONSERVATION OF STEROL-RECOGNITION MOTIFS

The occurrence of consensus cholesterol-recognition motifs covers a wide evolutionary span, from *H. sapiens* back to the bacterial pentameric channels, structural homologs of the nAChR found in prokaryotes, i.e., the cyanobacterium *Gloeobacter violaceus* and its orthologue from *Erwinia chrysanthemi*. Cyanobacteria possess hopanoids, which are structurally and functionally similar to sterols. The remarkable preservation of the CCMs through millions of years in the evolutionary scale has led us to suggest that this domain has important structural and/or functional roles (Barrantes, 2015; Barrantes & Fantini, 2016). In support of this hypothesis is the extensive experimental work showing that mutations in amino acid residues in the TM regions of the nAChR alter channel gating (see review in Barrantes, 2007). Some of these functionally relevant mutations are very close to or within CARC/CARC-like domains. The CCMs may have had other functional roles in prokaryotic AChR-like and other channel-forming proteins, but upon appearance of cholesterol in the course of phylogeny, this lipid probably acquired protagonism in eukaryotes for transducing regulatory signals from the plasma membrane to the protein moiety and, concomitantly, cholesterol-recognizing sequences became integrated into the genes coding for many hormone and neurotransmitter receptors as well as channel proteins.

As discussed in preceding sections, the vectorial topography of the CARC sequence makes this motif ideally suited for interaction with cholesterol in the outer leaflet of biological membranes. This is indeed the case with type I membrane proteins, in which the carbon chain enters the membrane bilayer from the extracellular, outer leaflet in the N- to C-terminus direction, as depicted in Fig. 2. The same vectoriality applies to the TM segments 1, 3, 5, and 7 in the GPCRs (Fig. 4), with the corresponding outer leaflet topography for CARC. Since the cholesterol/sphingolipid-enriched domains ("rafts") stem essentially from the tight contacts between these two lipid molecules in the outer membrane leaflet, it is tempting to speculate that one of the possible evolutionary forces leading to the establishment of the CARC sequence was the need to optimize ordered lipid domains in eukaryotic biomembranes.

ACKNOWLEDGMENTS

Experimental work quoted in this chapter was supported by Grants PICT 2011-0604 from FONCYT, Ministry of Science and Technology and PIP No. N° 112-201101-01023 from the National Scientific and Technical Research Council of Argentina (CONICET).

REFERENCES

Anderson, R. G., & Jacobson, K. (2002). A role for lipid shells in targeting proteins to caveolae, rafts, and other lipid domains. *Science, 296*, 1821–1825.

Baginski, M., Tempczyk, A., & Borowski, E. (1989). Comparative conformational analysis of cholesterol and ergosterol by molecular mechanics. *European Biophysics Journal: EBJ, 17*, 159–166.

Baier, C. J., Fantini, J., & Barrantes, F. J. (2011). Disclosure of cholesterol recognition motifs in transmembrane domains of the human nicotinic acetylcholine receptor. *Scientific Reports, 1*, 69.

Barrantes, F. J. (1989). The lipid environment of the nicotinic acetylcholine receptor in native and reconstituted membranes. *Critical Reviews in Biochemistry and Molecular Biology, 24*, 437–478.

Barrantes, F. J. (2004). Structural basis for lipid modulation of nicotinic acetylcholine receptor function. *Brain Research. Brain Research Reviews, 47*, 71–95.

Barrantes, F. J. (2007). Cholesterol effects on nicotinic acetylcholine receptor. *Journal of Neurochemistry, 103*(Suppl. 1), 72–80.

Barrantes, F. J. (2010). Cholesterol effects on nicotinic acetylcholine receptor: Cellular aspects. In: R. Harris (Ed.), *Cholesterol binding proteins and cholesterol transport. Chapter 17* (pp. 467–487). ISSN: 0306-0225. Springer Verlag. Subcellular Biochemistry 51.

Barrantes, F. J. (2012). Regulation of the nicotinic acetylcholine receptor by cholesterol as a boundary lipid. In: I. Levitan & F. J. Barrantes (Eds.), *Cholesterol regulation of ion channels and receptors* (pp. 183–204): Hoboken, NJ: John Wiley & Sons.

Barrantes, F. J. (2015). Phylogenetic conservation of protein–lipid motifs in pentameric ligand-gated ion channels. *Biochimica et Biophysica Acta, 1848*, 1796–1805.

Barrantes, F. J., & Fantini, J. (2016). From hopanoids to cholesterol: Molecular clocks of pentameric ligand-gated ion channels. *Progress in Lipid Research, 63*, 1–13.

Bermudez, V., Antollini, S. S., Fernandez Nievas, G. A., Aveldano, M. I., & Barrantes, F. J. (2010). Partition profile of the nicotinic acetylcholine receptor in lipid domains upon reconstitution. *Journal of Lipid Research, 51*, 2629–2641.

Bruses, J., Chauvet, N., & Rutishauser, U. (2001). Membrane lipid rafts are necessary for the maintenance of the (alpha)7 nicotinic acetylcholine receptor in somatic spines of ciliary neurons. *The Journal of Neuroscience, 21*, 504–512.

Campagna, J. A., & Fallon, J. (2006). Lipid rafts are involved in C95 (4,8) agrin fragment-induced acetylcholine receptor clustering. *Neuroscience, 138*, 123–132.

Cherezov, V., Rosenbaum, D. M., Hanson, M. A., Rasmussen, S. G., Thian, F. S., Kobilka, T. S., et al. (2007). High-resolution crystal structure of an engineered human beta2-adrenergic G protein-coupled receptor. *Science, 318*, 1258–1265.

Clay, A. T., Lu, P., & Sharom, F. J. (2015). Interaction of the P-glycoprotein multidrug transporter with sterols. *Biochemistry, 54*, 6586–6597.

Criado, M., Eibl, H., & Barrantes, F. J. (1982). Effects of lipids on acetylcholine receptor. Essential need of cholesterol for maintenance of agonist-induced state transitions in lipid vesicles. *Biochemistry, 21*, 3622–3629.

Czub, J., & Baginski, M. (2006). Comparative molecular dynamics study of lipid membranes containing cholesterol and ergosterol. *Biophysical Journal, 90*, 2368–2382.

de Almeida, R. F. M., Loura, L. M. S., Prieto, M., Watts, A., Fedorov, A., & Barrantes, F. J. (2004). Cholesterol modulates the organization of the gamma M4 transmembrane domain of the muscle nicotinic acetylcholine receptor. *Biophysical Journal, 86*, 2261–2272.

Di Scala, C., Chahinian, H., Yahi, N., Garmy, N., & Fantini, J. (2014). Interaction of Alzheimer's beta-amyloid peptides with cholesterol: Mechanistic insights into amyloid pore formation. *Biochemistry, 53*, 4489–4502.

Dopico, A. M., & Bukiya, A. N. (2014). Lipid regulation of BK channel function. *Frontiers in Physiology, 5*, 312.

el Battari, A., Ah-Kye, E., Muller, J. M., Sari, H., & Marvaldi, J. (1985). Modification of HT 29 cell response to the vasoactive intestinal peptide (VIP) by membrane fluidization. *Biochimie, 67*, 1217–1223.

Epand, R. M. (2006). Cholesterol and the interaction of proteins with membrane domains. *Progress in Lipid Research, 45*, 279–294.

Epand, R. M. (2008). Proteins and cholesterol-rich domains. *Biochimica et Biophysica Acta, 1778*, 1576–1582.

Epand, R. M., Thomas, A., Brasseur, R., & Epand, R. F. (2010). Cholesterol interaction with proteins that partition into membrane domains: An overview. *Sub-Cellular Biochemistry, 51*, 253–278.

Epand, R. F., Thomas, A., Brasseur, R., Vishwanathan, S. A., Hunter, E., & Epand, R. M. (2006). Juxtamembrane protein segments that contribute to recruitment of cholesterol into domains. *Biochemistry, 45*, 6105–6114.

Fantini, J., & Barrantes, F. J. (2009). Sphingolipid/cholesterol regulation of neurotransmitter receptor conformation and function. *Biochimica et Biophysica Acta, 1788*, 2345–2361.

Fantini, J., & Barrantes, F. J. (2013). How cholesterol interacts with membrane proteins: An exploration of cholesterol-binding sites including CRAC, CARC, and tilted domains. *Frontiers in Physiology, 4*, 31.

Fantini, J., Di Scala, C., Baier, C. J., & Barrantes, F. J. (2016). Molecular mechanisms of protein-cholesterol interactions in plasma membranes: Functional distinction between topological (tilted) and consensus (CARC/CRAC) domains. *Chemistry and Physics of Lipids*, *199*, 52–60.

Fantini, J., Di Scala, C., Evans, L. S., Williamson, P. T., & Barrantes, F. (2016). A mirror code for protein-cholesterol interactions in the two leaflets of biological membranes. *Scientific Reports*, *6*, 21907.

Fantini, J., & Yahi, N. (2015). *Brain lipids in synaptic function and neurological disease. Clues to innovative therapeutic strategies for brain disorders*. New York: Academic Press.

Gal, Z., Hegedus, C., Szakacs, G., Varadi, A., Sarkadi, B., & Ozvegy-Laczka, C. (2015). Mutations of the central tyrosines of putative cholesterol recognition amino acid consensus (CRAC) sequences modify folding, activity, and sterol-sensing of the human ABCG2 multidrug transporter. *Biochimica et Biophysica Acta*, *1848*, 477–487.

Gimpl, G. (2016). Interaction of G protein coupled receptors and cholesterol. *Chemistry and Physics of Lipids*, *199*, 61–73.

Griffie, J., Burn, G., & Owen, D. M. (2015). The nanoscale organization of signaling domains at the plasma membrane. *Current Topics in Membranes*, *75*, 125–165.

Hamouda, A. K., Chiara, D. C., Sauls, D., Cohen, J. B., & Blanton, M. P. (2006). Cholesterol interacts with transmembrane alpha-helices M1, M3, and M4 of the Torpedo nicotinic acetylcholine receptor: Photolabeling studies using [3H]azicholesterol. *Biochemistry*, *45*, 976–986.

Hanson, M. A., Cherezov, V., Griffith, M. T., Roth, C. B., Jaakola, V. P., Chien, E. Y., et al. (2008). A specific cholesterol binding site is established by the 2. Å structure of the human beta2-adrenergic receptor. *Structure*, *16*, 897–905.

Jafurulla, M., Tiwari, S., & Chattopadhyay, A. (2011). Identification of cholesterol recognition amino acid consensus (CRAC) motif in G-protein coupled receptors. *Biochemical and Biophysical Research Communications*, *404*, 569–573.

Jamin, N., Neumann, J. M., Ostuni, M. A., Vu, T. K., Yao, Z. X., Murail, S., et al. (2005). Characterization of the cholesterol recognition amino acid consensus sequence of the peripheral-type benzodiazepine receptor. *Molecular Endocrinology (Baltimore, Md.)*, *19*, 588–594.

Jaremko, L., Jaremko, M., Giller, K., Becker, S., & Zweckstetter, M. (2014). Structure of the mitochondrial translocator protein in complex with a diagnostic ligand. *Science*, *343*, 1363–1366.

Lagane, B., Gaibelet, G., Meilhoc, E., Masson, J. M., Cezanne, L., & Lopez, A. (2000). Role of sterols in modulating the human mu-opioid receptor function in *Saccharomyces cerevisiae*. *The Journal of Biological Chemistry*, *275*, 33197–33200.

Lazar, D. F., & Medzihradsky, F. (1992). Altered microviscosity at brain membrane surface induces distinct and reversible inhibition of opioid receptor binding. *Journal of Neurochemistry*, *59*, 1233–1240.

Lee, A. G. (2004). How lipids affect the activities of integral membrane proteins. *Biochimica et Biophysica Acta*, *1666*, 62–87.

Leibel, W. S., Firestone, L. L., Legler, D. C., Braswell, L., & Miller, K. W. (1987). Two pools of cholesterol in acetylcholine receptor-rich membranes from Torpedo. *Biochimica et Biophysica Acta*, *8987*, 249–260.

Levitan, I., Singh, D. K., & Rosenhouse-Dantsker, A. (2014). Cholesterol binding to ion channels. *Frontiers in Physiology*, *5*, 65. http://dx.doi.org/10.3389/fphys.2014.00065. eCollection 2014.

Li, H., & Papadopoulos, V. (1998). Peripheral-type benzodiazepine receptor function in cholesterol transport. Identification of a putative cholesterol recognition/interaction amino acid sequence and consensus pattern. *Endocrinology*, *139*, 4991–4997.

Lin, Q., & London, E. (2015). Ordered raft domains induced by outer leaflet sphingomyelin in cholesterol-rich asymmetric vesicles. *Biophysical Journal, 108*, 2212–2222.

Lingwood, D., & Simons, K. (2010). Lipid rafts as a membrane-organizing principle. *Science, 327*, 46–50.

Maguire, P. A., & Druse, M. J. (1989). The influence of cholesterol on synaptic fluidity, dopamine D1 binding and dopamine-stimulated adenylate cyclase. *Brain Research Bulletin, 23*, 69–74.

Marchand, S., Devillers-Thiery, A., Pons, S., Changeux, J. P., & Cartaud, J. (2002). Rapsyn escorts the nicotinic acetylcholine receptor along the exocytic pathway via association with lipid rafts. *The Journal of Neuroscience, 22*, 8891–8901.

Marsh, D. (1991). General features of phospholipid phase transitions. *Chemistry and Physics of Lipids, 57*, 109–120.

Marsh, D., & Barrantes, F. J. (1978). Immobilized lipid in acetylcholine receptor-rich membranes from *Torpedo marmorata. Proceedings of the National Academy of Sciences of the United States of America, 75*, 4329–4333.

Maxfield, F. R. (2002). Plasma membrane microdomains. *Current Opinion in Cell Biology, 14*, 483–487.

Morrill, G. A., & Kostellow, A. B. (2016). Molecular properties of globin channels and pores: Role of cholesterol in ligand binding and movement. *Frontiers in Physiology, 7*, 360. http://dx.doi.org/10.3389/fphys.2016.00360. eCollection 2016.

Morrill, G. A., Kostellow, A. B., & Gupta, R. K. (2014). The pore-lining regions in cytochrome c oxidases: A computational analysis of caveolin, cholesterol and transmembrane helix contributions to proton movement. *Biochimica et Biophysica Acta, 1838*, 2838–2851.

Morrill, G. A., Kostellow, A. B., & Gupta, R. K. (2016). The role of receptor topology in the vitamin D3 uptake and Ca2+ response systems. *Biochemical and Biophysical Research Communications, 477*, 834–840.

Nishio, M., Umezawa, Y., Fantini, J., Weiss, M. S., & Chakrabarti, P. (2014). CH–π hydrogen bonds in biological macromolecules. *Physical Chemistry Chemical Physics, 16*, 12648–12683.

Opekarova, M., & Tanner, W. (2003). Specific lipid requirements of membrane proteins—A putative bottleneck in heterologous expression. *Biochimica et Biophysica Acta, 1610*, 11–22.

Palmer, M. (2004). Cholesterol and the activity of bacterial toxins. *FEMS Microbiology Letters, 238*, 281–289.

Pathak, P., & London, E. (2015). The effect of membrane lipid composition on the formation of lipid ultrananodomains. *Biophysical Journal, 109*, 1630–1638.

Perillo, V. L., Penalva, D. A., Vitale, A. J., Barrantes, F. J., & Antollini, S. S. (2016). Transbilayer asymmetry and sphingomyelin composition modulate the preferential membrane partitioning of the nicotinic acetylcholine receptor in Lo domains. *Archives of Biochemistry and Biophysics, 591*, 76–86.

Picazo-Juarez, G., Romero-Suarez, S., Nieto-Posadas, A., Llorente, I., Jara-Oseguera, A., Briggs, M., et al. (2011). Identification of a binding motif in the S5 helix that confers cholesterol sensitivity to the TRPV1 ion channel. *The Journal of Biological Chemistry, 286*, 24966–24976.

Popot, J. L., Demel, R. A., Sobel, A., van Deenen, L. L., & Changeux, J. P. (1977). Preferential affinity of acetylcholine receptor protein for certain lipids studied using monolayer cultures. *Comptes Rendus Hebdomadaires des Séances de l'Académie des Sciences Série D: Sciences Naturelles, 285*, 1005–1008.

Posada, I. M., Fantini, J., Contreras, F. X., Barrantes, F., Alonso, A., & Goni, F. M. (2014). A cholesterol recognition motif in human phospholipid scramblase 1. *Biophysical Journal, 107*, 1383–1392.

Rao, M., & Mayor, S. (2014). Active organization of membrane constituents in living cells. *Current Opinion in Cell Biology*, *29*, 126–132.

Rosenbaum, D. M., Cherezov, V., Hanson, M. A., Rasmussen, S. G., Thian, F. S., Kobilka, T. S., et al. (2007). GPCR engineering yields high-resolution structural insights into beta2-adrenergic receptor function. *Science*, *318*, 1266–1273.

Rosenhouse-Dantsker, A., Noskov, S., Durdagi, S., Logothetis, D. E., & Levitan, I. (2013). Identification of novel cholesterol-binding regions in Kir2 channels. *The Journal of Biological Chemistry*, *288*, 31154–31164.

Sharpe, L. J., Rao, G., Jones, P. M., Glancey, E., Aleidi, S. M., George, A. M., et al. (2015). Cholesterol sensing by the ABCG1 lipid transporter: Requirement of a CRAC motif in the final transmembrane domain. *Biochimica et Biophysica Acta*, *1851*, 956–964.

Simons, K., & Ikonen, E. (1997). Functional rafts in cell membranes. *Nature*, *387*, 569–572.

Simons, K., & van Meer, G. (1988). Lipid sorting in epithelial cells. *Biochemistry*, *27*, 6198–6202.

Singh, A. K., McMillan, J., Bukiya, A. N., Burton, B., Parrill, A. L., & Dopico, A. M. (2012). Multiple cholesterol recognition/interaction amino acid consensus (CRAC) motifs in cytosolic C tail of Slo1 subunit determine cholesterol sensitivity of Ca2+- and voltage-gated K+ (BK) channels. *The Journal of Biological Chemistry*, *287*, 20509–20521.

Singh, D. K., Shentu, T. P., Enkvetchakul, D., & Levitan, I. (2011). Cholesterol regulates prokaryotic Kir channel by direct binding to channel protein. *Biochimica et Biophysica Acta*, *1808*, 2527–2533.

Stetzkowski-Marden, F., Gaus, K., Recouvreur, M., Cartaud, A., & Cartaud, J. (2006). Agrin elicits membrane lipid condensation at sites of acetylcholine receptor clusters in C2C12 myotubes. *Journal of Lipid Research*, *47*, 2121–2133.

Vaidehi, N., Bhattacharya, S., & Larsen, A. B. (2014). Structure and dynamics of G-protein coupled receptors. *Advances in Experimental Medicine and Biology*, *796*, 37–54.

van Zanten, T. S., & Mayor, S. (2015). Current approaches to studying membrane organization. *F1000Research*, *4*, pii: F1000 Faculty Rev-1380. http://dx.doi.org/10.12688/f1000research.6868.1. eCollection 2015. Review.

Veatch, S. L., & Keller, S. L. (2002). Organization in lipid membranes containing cholesterol. *Physical Review Letters*, *89*, 268101.

Veatch, S. L., & Keller, S. L. (2003a). Separation of liquid phases in giant vesicles of ternary mixtures of phospholipids and cholesterol. *Biophysical Journal*, *85*, 3074–3083.

Veatch, S. L., & Keller, S. L. (2003b). A closer look at the canonical 'Raft Mixture' in model membrane studies. *Biophysical Journal*, *84*, 725–726.

Veatch, S. L., Polozov, I. V., Gawrisch, K., & Keller, S. L. (2004). Liquid domains in vesicles investigated by NMR and fluorescence microscopy. *Biophysical Journal*, *86*, 2910–2922.

Warne, T., Serrano-Vega, M. J., Baker, J. G., Moukhametzianov, R., Edwards, P. C., Henderson, R., et al. (2008). Structure of a beta1-adrenergic G-protein-coupled receptor. *Nature*, *454*, 486–491.

Willmann, R., Pun, S., Stallmach, L., Sadasivam, G., Santos, A. F., Caroni, P., et al. (2006). Cholesterol and lipid microdomains stabilize the postsynapse at the neuromuscular junction. *The EMBO Journal*, *25*, 4050–4060.

Yeagle, P. L. (1989). Lipid regulation of cell membrane structure and function. *The FASEB Journal*, *3*, 1833–1842.

Yesylevskyy, S. O., Demchenko, A. P., Kraszewski, S., & Ramseyer, C. (2013). Cholesterol induces uneven curvature of asymmetric lipid bilayers. *The Scientific World Journal*, *2013*, 965230.

Strandberg, E., & Killian, J. A. (2003). Snorkeling of lysine side chains in transmembrane helices: How easy can it get? *FEBS Letters*, *544*, 69–73.

Zhu, D., Xiong, W. C., & Mei, L. (2006). Lipid rafts serve as a signaling platform for nicotinic acetylcholine receptor clustering. *The Journal of Neuroscience*, *26*, 4841–4851.

CHAPTER TWO

Differential Effects of Sterols on Ion Channels: Stereospecific Binding vs Stereospecific Response

Nicolas Barbera*, Manuela A.A. Ayee*, Belinda S. Akpa†, Irena Levitan*,[1]

*University of Illinois at Chicago, Chicago, IL, United States
†North Carolina State University, Raleigh, NC, United States
[1]Corresponding author: e-mail address: levitan@uic.edu

Contents

Abstract

Numerous ion channels have been shown to be regulated by the level of membrane cholesterol, but the mechanisms responsible for these effects are still not well understood. The key question in the field is how to discriminate between the contributions of the two central mechanisms that might be responsible for the sensitivity of ion channels to cholesterol: specific sterol–protein interactions or regulation of channels by the

Current Topics in Membranes, Volume 80
ISSN 1063-5823
http://dx.doi.org/10.1016/bs.ctm.2017.06.001

bilayer physical properties. Comparative analysis of cholesterol and its isomers on the function of an ion channel is a powerful tool to achieve this goal. An increasing number of studies show that cholesterol regulates several types of ion channels in a stereospecific manner, suggesting an involvement of specific sterol–protein interactions. However in this chapter, we present evidence that the stereospecificity of cholesterol–ion channel interactions may be mediated, not by a lack of binding, as has been generally assumed, but by the specificity of the interaction, which results in a functional effect, in the case of native cholesterol, and a lack of functional effect, in the case of a cholesterol isomer. In other words, accumulating evidence suggests that the structural requirements of ion channel cholesterol-binding sites are lax, allowing chiral isomers of cholesterol to bind to the same site in a nonstereospecific way, but the ability of a sterol to confer a functional effect on the channel activity can still be stereospecific. This is an important distinction both conceptually and methodologically. Indeed, our analysis shows that the orientations of cholesterol and its chiral isomer *ent*-cholesterol within a hydrophobic binding pocket of Kir2.2 are significantly different, and we propose that this difference may underlie distinct functional outcomes.

1. INTRODUCTION

Cholesterol is a major lipid component of the cell membrane of eukaryotic cells, constituting anywhere between 10 and 45 mol% with respect to other lipids (Yeagle, 1985), and it plays an important role in regulating the physical properties of the membrane and the function of membrane proteins (Yeagle, 1985, 1991). Numerous studies show that changes in membrane cholesterol content play a major role in regulation of a variety of ion channels (Levitan, Singh, & Rosenhouse-Dantsker, 2014; Rosenhouse-Dantsker, Mehta, & Levitan, 2012). The predominant effect of cholesterol on ion channels is a decrease in the channel activity that typically results from a decrease in the open probability or the number of active channels in the membrane, as described for multiple types of K^+ channels, including inwardly rectifying K^+ channels (Romanenko, Fang, et al., 2004; Romanenko, Rothblat, & Levitan, 2002), Ca^{2+}-sensitive K^+ channels (Bukiya, Belani, Rychnovsky, & Dopico, 2011; Crowley, Treistman, & Dopico, 2003), and voltage-gated K^+ channels (Abi-Char et al., 2007; Hajdú, Varga, Pieri, Panyi, & Gáspár, 2003), as well as voltage-gated Na^+ and Ca^{2+} channels (Lundbaek et al., 2004) and volume-regulated anion channels (Romanenko, Rothblat, & Levitan, 2004). Cholesterol may also be required for activating ion channels and stabilizing them in an open state, as was shown for the nicotinic acetylcholine receptors (Baenziger, Henault, Therien, & Sun, 2015; Barrantes, 2007; Fantini & Barrantes, 2013).

In addition, these receptors may be regulated by the thickness of the bilayers, highlighting the complex nature of their cholesterol regulation (Baenziger et al., 2015; daCosta, Dey, Therien, & Baenziger, 2013). Cholesterol was also shown to have both positive and negative effects on different types of TRP channels (Picazo-Juarez et al., 2011; Rosenhouse-Dantsker, Mehta, et al., 2012). The central question regarding the mechanism responsible for cholesterol-dependent regulation of ion channel function is whether these effects are mediated by changes in the physical properties of the lipid bilayer that surrounds the channel or by specific direct interactions between the cholesterol molecule and the channel protein. In the first scenario, cholesterol has an indirect effect on channel function by modulating the stiffness, curvature, or thickness of the bilayer. Alternatively, cholesterol may act as a specific ligand stabilizing the channel in a particular conformation. Discriminating between these possibilities, however, is difficult because of the multiple interactions that cholesterol has with different components of the membrane, as well as the limitations of tools used to probe for the specificity of the effects. An increasing number of studies show that cholesterol regulation of multiple types of ion channels is stereospecific, which indicates that specific cholesterol–protein interactions play the central role in these interactions (Fig. 1). It is also important to

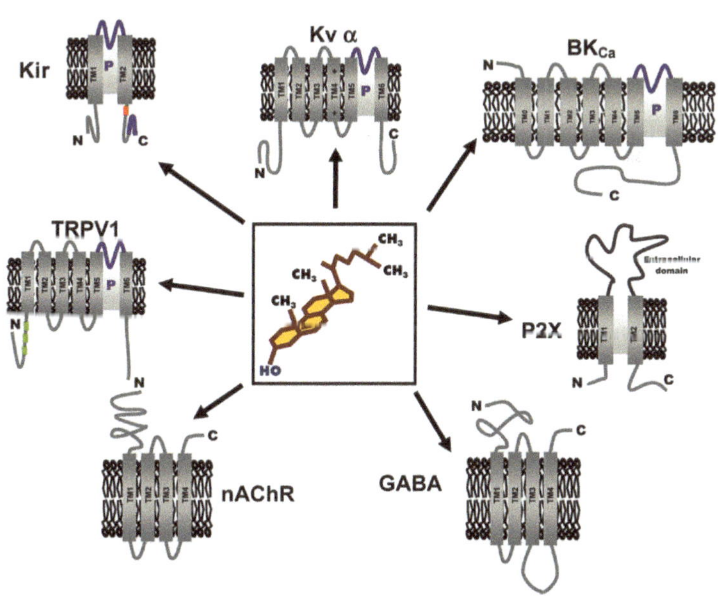

Fig. 1 Overview of structures of ion channels known to be regulated by cholesterol via specific sterol–protein interactions.

recognize that the two mechanisms are not mutually exclusive. The goal of this chapter is to discuss different approaches to discriminate between these mechanisms, highlighting alternative interpretations and discussing possible misconceptions.

2. PHYSICAL PROPERTIES OF MEMBRANE BILAYER: STEROL ANALYSIS

Cholesterol (3β-hydroxy-5-cholestene) is well known to have a major impact on the physical properties of the membrane bilayer, including an increase in lipid order, a decrease in membrane fluidity, and an increase in bilayer thickness and stiffness (Needham & Nunn, 1990; Yeagle, 1985, 1991). It is logical to expect, therefore, that a decrease in membrane fluidity or a decrease in the deformability of the bilayer could significantly affect the ability of ion channels to undergo the conformation changes between the closed and the open states. More specifically, it was proposed that since channel proteins have to undergo conformation changes within the medium of the lipid bilayer, an increase in membrane stiffness or a decrease in membrane deformability may result in higher membrane deformation energy required for this process, which in turn would inhibit the activity of the channels (Lundbaek, Birn, Hansen, & Andersen, 1996). A hydrophobic mismatch between the lipid bilayer and the transmembrane domains of the channel protein may also play a major role in the regulation of channel function, because transitions between the channel conformation states may depend on the thickness of the membrane hydrophobic core (daCosta et al., 2013; Lundbaek et al., 2004). Altering membrane cholesterol levels by depleting or enriching cells with cholesterol cannot discriminate between these mechanisms and specific interactions between cholesterol and channel proteins via direct binding. A powerful tool to achieve this goal is to substitute membrane cholesterol with other sterols that have differential effects on the physical properties of the lipid bilayer. The justification for this approach is that if different sterols have similar effects on the characteristics of the membrane, but different effects on the activity of the ion channel in question, the argument can be made that cholesterol regulates the channel via a specific interaction rather than by altering the physical properties of the bilayer.

Two synthetic chiral cholesterol analogues are typically used: epicholesterol (3α-hydroxy-5-cholestene) and *ent*-cholesterol (enantiomeric cholesterol). In epicholesterol, the only structural difference with cholesterol

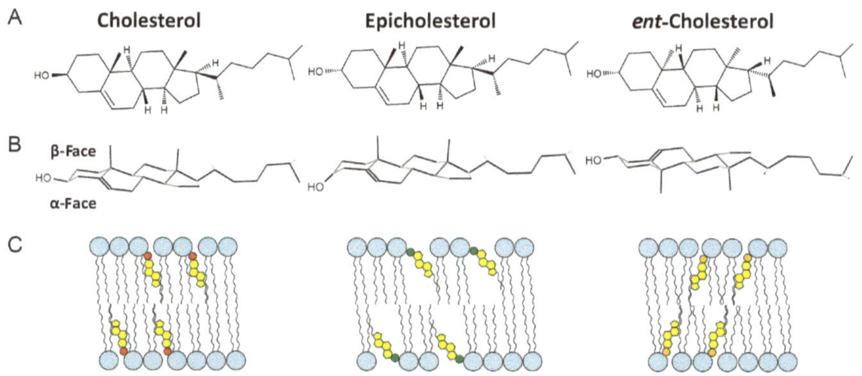

Fig. 2 Cholesterol isomers: structures and position in the lipid bilayers. (A) Chemical structures of cholesterol, epicholesterol, and *ent*-cholesterol with *solid wedges* indicating bonds pointing above the plane of the page and *dashed wedges* indicating bonds pointing below the plane of the page. (B) Side views of the 3-dimensional structures of the three isomers of cholesterol, showing the positions of substituents on the α- and β-faces. (C) Schematic of the localization of the three cholesterol isomers within phospholipid bilayers and their effects on membrane packing.

is the placement of the hydroxyl group (3α instead of 3β). The *ent*-cholesterol is a mirror image of cholesterol, with an opposite configuration for each of the eight stereocenters of cholesterol (Fig. 2A) (Westover & Covey, 2004; Xu & London, 2000). It is important to note, however, that while cholesterol chiral analogues were shown to have effects similar to those of cholesterol on the physical properties of the lipid bilayer, such as membrane fluidity and formation of ordered membrane domains (Gimpl, Burger, & Fahrenholz, 1997; Westover & Covey, 2004; Xu & London, 2000), there are also some notable differences, particularly between cholesterol and epicholesterol.

2.1 3D Sterol Structure of Cholesterol Chiral Analogues

The difference between altering the orientation of one hydroxyl group (epicholesterol) vs altering the configuration at all eight stereocenters (*ent*-cholesterol) results in a major difference in the 3-dimensional structure of the molecule, with *ent*-cholesterol actually being structurally closer to cholesterol than epicholesterol. The cholesterol molecule is composed of an asymmetric ringed system with a smooth α-face and a rough β-face possessing two methyl group substituents. The hydroxyl group on the third carbon is also oriented toward the β-face (Róg, Pasenkiewicz-Gierula, Vattulainen, & Karttunen, 2007). The smooth surface of the epicholesterol isomer is interrupted by the α-orientation of its hydroxyl group, as compared to the β-orientation in natural cholesterol (Mannock, Lee, Lewis, & McElhaney, 2008). In contrast, the enantiomeric, mirror-image form of cholesterol possesses both rough and smooth surfaces just as the natural form does. However, in *ent*-cholesterol, the α-face is rough while the β-face is smooth, and the hydroxyl group is oriented toward the rough α-face (Westover & Covey, 2004). Furthermore, the difference in the 3-dimensional structures of cholesterol, epicholesterol, and *ent*-cholesterol results in significant differences in their interactions with phospholipids and, consequently, differential effects on the physical properties of membrane bilayers. See the 3D structures of all three sterols in Fig. 2B.

2.2 Position Within the Bilayer

Both cholesterol and the two chiral analogues direct their hydroxyl groups toward the hydrophilic regions of the bilayer. For cholesterol and for *ent*-cholesterol, this results in an orientation of the molecule parallel to the surrounding lipids, whereas epicholesterol assumes a tilted orientation in the bilayer due to the α-conformation of its hydroxyl group (Fig. 2C), as was demonstrated by deuterium nuclear magnetic resonance (NMR) (Dufourc, Parish, Chitrakorn, & Smith, 1984; Murari, Murari, & Baumann, 1986). In contrast, due to the almost identical (but mirrored) chemical structures and amphiphilic nature of cholesterol and *ent*-cholesterol, it is predicted that they assume the same orientation in a membrane. It has been shown that the β-hydroxyl group of cholesterol forms hydrogen bonds with phospholipid backbone carbonyl groups (Pasenkiewicz-Gierula, Róg, Kitamura, & Kusumi, 2000). Similarly, it is predicted that *ent*-cholesterol also forms hydrogen bonds between its hydroxyl group and the phospholipid carbonyl groups due to the orientation of the hydroxyl group on its rough face. Although epicholesterol also forms hydrogen bonds with carbonyl oxygen atoms, it additionally forms more direct hydrogen bonds with phosphate oxygen atoms in phospholipid headgroups than does cholesterol, causing epicholesterol to localize at a position in the bilayer about 2 Å higher than that of cholesterol (Róg & Pasenkiewicz-Gierula, 2003). This upward shifting and tilting of epicholesterol molecules impacts the biophysical properties of membranes.

2.3 Lipid Order and Packing

From experimental NMR studies as well as atomistic MD simulations, it was shown that both cholesterol and epicholesterol have an ordering effect on the bilayer, though epicholesterol increases phospholipid tail ordering and packing to a lesser extent than cholesterol (Dufourc et al., 1984; Róg & Pasenkiewicz-Gierula, 2003). This can again be explained by the α-conformation of the hydroxyl group in epicholesterol, which disrupts the smooth α-face of the molecule, in addition to causing a tilted sterol conformation and a geometric shift toward the polar region of the bilayer. It is well established that the presence of cholesterol has a condensing effect on lipid monolayers and leads to reduced area per lipid and increased lipid tail order (e.g., Bonn et al., 2004; Demel, Bruckdorfer, & van Deenen, 1972). Specifically, the addition of a 1:1 molar ratio of cholesterol condensed the mean molecular area of monolayers from 82 $Å^2$ for a pure phospholipid monolayer to 49 $Å^2$ at 37 °C, whereas addition of an equivalent amount of epicholesterol decreased the molecular area to 60 $Å^2$ (Demel et al., 1972). Similar experiments showed that lipid monolayers containing either cholesterol or *ent*-cholesterol condensed to the same extent in the presence of either isomer as indicated by identical surface pressure–molecular area isotherms (Westover, Covey, Brockman, Brown, & Pike, 2003). Thus, these isomers are expected to cause similar changes in the packing and ordering of phospholipid tails in bilayers.

An increase in lipid packing results in changes in the permeability of the membrane to small solutes including water, sugars, or ions (Bittman & Blau, 1972; Demel et al., 1972; Róg & Pasenkiewicz-Gierula, 2003). For example, addition of cholesterol to egg lecithin liposomes resulted in a 42% decrease in the permeability of the liposomes to water, whereas addition of an equivalent amount of epicholesterol decreased membrane permeability by 27% (Bittman & Blau, 1972). It was also shown, however, that while cholesterol decreased the permeability of lecithin liposomes to glucose, glycerol, and Rb^+, epicholesterol caused no significant change in permeability of liposomes to these solutes (Demel et al., 1972). The latter observation could be the result of a decreased level of incorporation of epicholesterol into the bilayer, as it was reported that epicholesterol was incorporated into liposomes 3.5 times less than cholesterol. With regard to *ent*-cholesterol, due to its similar structure to cholesterol and similar membrane ordering capacity, it is expected to reduce membrane permeability to the same extent as cholesterol. Experimental studies are still needed to validate this prediction.

Taken together, these studies suggest that *if cholesterol regulates a channel by altering the physical properties of the bilayer only, then substitution with ent-cholesterol should have the same effect as cholesterol and substitution with epicholesterol may have similar but diminished effects*. However, it is important to note that a similarity between the effects of cholesterol and its isomers on a channel function *does not* constitute proof that the effect is mediated by the physical properties of the bilayer because, as described in more detail below, it is entirely possible that cholesterol isomers also bind to the channel protein. Further comparison of multiple sterols and comparative analysis of their effects on the channel function and on the lipid properties of the membrane might help to clarify this issue. In contrast, *if the effect of cholesterol is lost or antagonized by the substitution with either of these analogues, it does indicate the involvement of a specific sterol–protein interaction.* This conclusion also needs to be made with caution, because, in the complex environment of a cellular membrane, this approach does not discriminate between direct interaction of cholesterol with the channel protein and an indirect effect via cholesterol interaction with another regulatory protein or lipid, which in turn regulates the activity of the channel.

3. DIFFERENTIAL EFFECTS OF STEROLS ON THE ION CHANNEL FUNCTION: PREVALENCE OF STEREOSPECIFIC EFFECTS

Several studies employed the approach of substituting cholesterol with its chiral analogues to elucidate the mechanism of cholesterol regulation of ion channels. These experiments were performed for several types of ion channels that are suppressed by the elevation of membrane cholesterol: *inwardly rectifying K^+ channels (Kir)* (D'Avanzo, Hyrc, Enkvetchakul, Covey, & Nichols, 2011; Romanenko et al., 2002; Singh, Rosenhouse-Dantsker, Nichols, Enkvetchakul, & Levitan, 2009), *large-conductance Ca^{2+}-sensitive voltage-gated K^+ channels (BK)* (Bukiya et al., 2011), *transient receptor potential vanilloid 1 (TRPV1) channels* (Picazo-Juarez et al., 2011), *and volume-regulated anion channels (VRACs)* (Romanenko, Rothblat, et al., 2004) and for two types of ion channels that require cholesterol to support agonist-induced activation, ligand-gated *nicotinic acetylcholine receptor (nAChR)* (Addona, Sandermann, Kloczewiak, & Miller, 2003) *and $GABA_A$ receptor* (Sooksawate & Simmonds, 2001). Of these channels, with the exception of VRAC and nAChR, the effects of cholesterol were shown to be stereospecific, indicating that cholesterol regulates these channels via specific interactions. Further studies, however, are needed to determine whether the specificity of the cholesterol effects should be attributed to the direct interactions of cholesterol with the channels or with other proteins or regulatory lipids, which in turn may regulate channel function. Interestingly, while cholesterol regulation of nAChR was found not to be stereospecific, the propensity of evidence indicates that nAChR is also regulated by specific interactions with cholesterol, as described in more detail below.

3.1 Stereospecific Effects of Cholesterol on Ion Channels

3.1.1 Inwardly Rectifying K^+ (Kir) Channels

Our studies demonstrated that while an increase in cellular cholesterol strongly suppresses Kir2.1 channels in endothelial cells (Romanenko et al., 2002) and in a heterologous expression system (Romanenko, Fang, et al., 2004), loading the cells with epicholesterol has the opposite effect (Fig. 3). Specifically, removing ~50% of endogenous cholesterol and substituting it with a similar amount of epicholesterol resulted in a strong increase in Kir current density that was much higher than in cells depleted of cholesterol to the same degree (Romanenko et al., 2002). Our interpretation of these observations was that epicholesterol is not just "invisible" to Kir channels but that it competes with cholesterol for a binding site and therefore antagonizes cholesterol function. This is an important point,

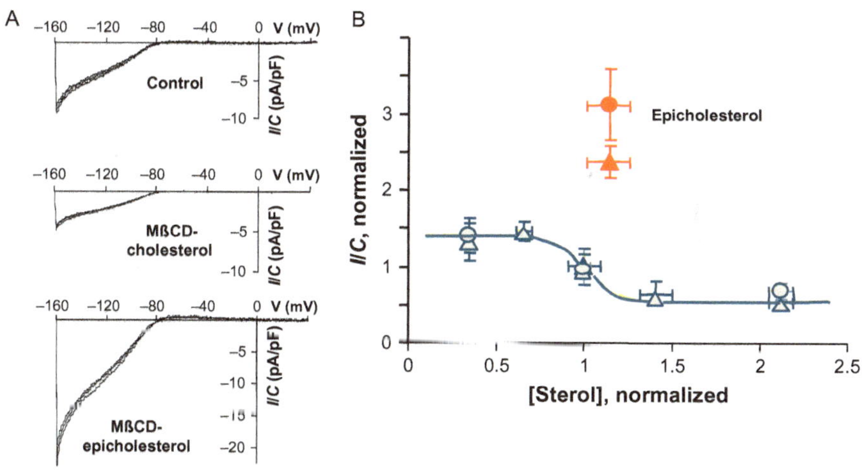

Fig. 3 Stereospecific effect of cholesterol on Kir currents. (A) Typical current traces recorded from three individual cells: a control cell, a cell enriched with cholesterol, and a cell enriched with epicholesterol. Three superimposing traces are shown for each cell. (B) The functional dependence of peak Kir currents on the level of membrane cholesterol in the absence and in the presence of epicholesterol. *Modified from Romanenko, V. G., Rothblat, G. H., & Levitan, I. (2002). Modulation of endothelial inward rectifier K+ current by optical isomers of cholesterol.* Biophysical Journal, 83, *3211–3222.*

because it suggests that epicholesterol can also bind to the Kir channel protein, occupying the same or a similar site to that of cholesterol, but having no functional effect. Another interpretation could be that cholesterol has both specific and bilayer-mediated effects, whereas epicholesterol may regulate the channels only through altering the physical properties of the bilayer.

However, as we emphasized previously, these data alone do not discriminate between cholesterol binding to the channel protein and cholesterol binding to an intermediate signaling molecule which might regulate channel activity. Indeed, it is important to recognize that in the complex environment of a cellular membrane, cholesterol may interact with multiple regulatory molecules, such as other regulatory lipids or proteins. The most straightforward approach to discriminate between these possibilities is to explore cholesterol regulation of purified ion channels incorporated into artificial lipid bilayers that contain no regulatory intermediates. To achieve this goal, we took advantage of a purified bacterial homologue of Kir, KirBac1.1, that could be incorporated into liposomes that contained none of the possible regulatory proteins or lipids (Enkvetchakul et al., 2004; Singh et al., 2009). Mammalian Kir2 channels were not available at that time in the purified form. Using this approach, we showed that, in the absence of any signaling intermediates, purified Kir channels are suppressed by an increase in cholesterol level of the bilayer. We further demonstrated that that there is no correlation between the activity of these channels and membrane fluidity as modulated by an array of different sterols (Singh et al., 2009). A later study also showed that purified mammalian Kir2.1 is likewise suppressed by cholesterol when incorporated into liposomes, whereas *ent*-cholesterol has no effect (D'Avanzo et al., 2011), further supporting our conclusion that cholesterol regulates Kir channels by specific direct interactions.

3.1.2 Large-Conductance Ca^{2+}-Sensitive Voltage-Gated K^{+} Channels (BK)

Similar to Kir channels, BK channels are suppressed by the elevation of membrane cholesterol (Bolotina, Omelyanenko, Heyes, Ryan, & Bregestovski, 1989; Bukiya et al., 2011; Bukiya, Vaithianathan, Toro, & Dopico, 2008; Crowley et al., 2003). Based on the correlation between the impact of cholesterol on the physical properties of the bilayer and on the BK activity, it was initially proposed that BK channels are suppressed by decrease in membrane fluidity (Bolotina et al., 1989) or increase in bilayer lateral stress (Chang, Reitstetter, Mason, & Gruener, 1995). However, more recent studies by Dopico and colleagues demonstrated that the effect of cholesterol on BK channels is strongly stereospecific. Substitution of cholesterol with either epicholesterol or *ent*-cholesterol resulted in a total loss of the inhibitory effect (Bukiya et al., 2011). Interestingly, however, the inhibitory effect of sterols on BK channels is not unique to cholesterol. Two other sterols, coprastanol and cholestanol, have similar effects on BK activity to that of cholesterol, whereas a chiral analogue of coprastanol, epicoprastanol, has no effect. These observations indicate that cholesterol regulates BK channels by specific sterol–protein interactions and not by altering the physical properties of the bilayer. Notably, these data also suggest that BK cholesterol-binding site can accommodate not only cholesterol but also other sterols, such as coprastanol and cholestanol. It is not clear yet, however, whether the chiral analogues of cholesterol and of coprastanol have no functional effects on BK channels because they do not bind to the channels or because they bind but do not elicit a functional response. In any case, the ability of the BK-binding site to accommodate several different sterols suggests significant flexibility of the structural requirements for sterol binding.

3.1.3 Transient Receptor Potential Vanilloid 1 (TRPV1) Channels

TRPV1 channels were also shown to be suppressed by cholesterol enrichment, whereas substitution of cholesterol by epicholesterol was shown to abrogate the inhibitory effect, indicating that this effect is stereoselective (Picazo-Juarez et al., 2011). It was concluded that TRPV1 channels are also regulated by a direct specific interaction of cholesterol with the channel protein.

3.1.4 GABA$_A$ Receptors

GABA$_A$ receptors are regulated by cholesterol in a biphasic way, with cholesterol depletion and cholesterol enrichment resulting in the inhibition of channel function. Substitution of cholesterol by epicholesterol abrogates the positive, but not the negative, effects of cholesterol, suggesting that the channels are regulated by a combination of specific and nonspecific mechanisms (Sooksawate & Simmonds, 2001).

3.2 Nonstereospecific Effects of Cholesterol on Ion Channels

3.2.1 Nicotinic Acetylcholine Receptor (nAChR)

The most prominent example of an ion channel that is regulated by cholesterol in a nonstereospecific way is the nicotinic acetylcholine receptor. The first study that discovered the lack of the stereospecific effect of nAchR was the study of Popot, Demel, Sobel, Van Deenen, and Changeux (1978), who tested the efficiency of nAchR incorporation into lipid monolayers. The study showed that there was no difference in nAchR incorporation between the monolayers containing cholesterol and epicholesterol, but there was a significant difference between cholesterol and ergosterol. Since cholesterol and ergosterol have similar effects on the physical properties of the membrane (Demel et al., 1972), Popot et al. concluded that there is a direct interaction between cholesterol and nAchR. Importantly, this early study suggested that a lack of stereospecificity of cholesterol effects on an ion channel does not necessarily mean a lack of a direct interaction. The nonstereospecific nature of cholesterol regulation of nAchR was confirmed by demonstrating that both epicholesterol and *ent*-cholesterol act similarly to cholesterol in supporting the functional activity of nAchR channels (Addona et al., 2003). These observations were not interpreted, however, as evidence to suggest that nAChR is regulated by the physical properties of the membrane, because Addona et al. found that nAChR functional activity can also be supported by coprastanol, a sterol that lacks the planar structure that is critical for the typical effects of cholesterol on lipid bilayer. In contrast, it was concluded that nAChR is regulated by the direct interaction with the cholesterol molecule and that this interaction is not stereospecific. Notably, direct interaction of cholesterol with nAChR was further verified using a photoactivatable cholesterol analogue that cross-links with the protein upon UV illumination (Corbin, Wang, & Blanton, 1998; Hamouda, Chiara, Sauls, Cohen, & Blanton, 2006). It is also important to note, however, that the study of Baenziger and colleagues showed recently that nAChR is also regulated by the thickness of the lipid bilayer, indicating that both specific sterol–protein interactions and physical properties of the membrane are important for activating nAChR (daCosta et al., 2013). Thus, in designing and interpreting experiments, it is critical to note that comparing cholesterol only with its chiral isomers might be insufficient to discriminate between direct and indirect effects of cholesterol. The structural differences between the chiral isomers might be insufficient to significantly affect the ability of the sterol to bind and regulate a channel. If no stereospecific effect is observed, it is important to test multiple sterols for more robust interpretation of the results.

3.2.2 Volume-Regulated Anion Channels (VRACs)

Another example of nonstereospecific cholesterol regulation of an ion channel is VRAC. Similar to Kir, BK, and TRPV1 and in contrast to nAChR, VRACs are also suppressed by cholesterol enrichment and enhanced by cholesterol depletion (Klausen, Hougaard, Hoffmann, & Pedersen, 2006; Levitan, Christian, Tulenko, & Rothblat, 2000). Similarly to nAChR, however, substitution of cholesterol with epicholesterol has no impact on VRAC activity, indicating that cholesterol effect on VRAC is not stereoselective (Romanenko, Rothblat, et al., 2004). Furthermore, extending the study to additional sterols showed an effect consistent with the sensitivity of VRACs to the lipid packing of the lipid bilayer. More studies are needed to explore the nature of this interaction.

In summary, two general conclusions emerge from comparative analysis of the differential effects of sterols on channel function: First, the majority of the ion channels that have been studied so far, nAChR, $GABA_A$, Kir, BK, and TRPV1, are regulated by direct sterol–ion channel interactions. It is also important to note that the same channel can be regulated by both direct interaction with a sterol and by sterol-induced changes in the bilayer properties, as was demonstrated for nAChR (daCosta et al., 2013) and for $GABA_A$ receptor (Sooksawate & Simmonds, 2001). The second conclusion is that it appears that cholesterol-binding sites of several types of ion channels might accommodate not only cholesterol but also its chiral analogues or other sterols, which makes it more difficult to discriminate between specific and nonspecific interactions. The overall picture that emerges from these studies is that the major mechanism by which cholesterol regulates ion channel function is direct interaction with the channel protein with different degrees of specificity between different sterols.

4. EVIDENCE FOR DIRECT BINDING AND STEROL COMPETITION FOR THE BINDING SITES: KIR CHANNELS

4.1 Challenges of the Approach

Direct binding studies are typically performed by mixing a protein with a radiolabeled ligand and then separating the bound and free ligand by different centrifugation or filtration techniques. A major constraint in using this approach for cholesterol binding to channels is that both cholesterol and the channel proteins need to be solubilized in organic solvents which is likely to denature the protein and thus might interfere with the binding and also create a background of nonspecific interactions mediated via the solvent molecules (Gimpl, 2010; Radhakrishnan, Sun, Kwon, Brown, &

Goldstein, 2004). Only a few studies, therefore, provided direct evidence for cholesterol binding to proteins, with the two most known examples being cholesterol-binding proteins, SCAP1 and NPC1 (Gimpl, 2010; Infante, Abi-Mosleh, et al., 2008; Infante, Wang, et al., 2008; Radhakrishnan et al., 2004). It is essential, however, to test directly whether specific ion channels bind cholesterol and *to establish whether the binding is stereospecific.*

4.2 Cholesterol Binding to KirBac1.1 Channels Is Not Stereospecific

Following the methodology developed in earlier studies of cholesterol-binding proteins, we established that, as we expected, cholesterol binds to the purified KirBac1.1 protein (Singh, Shentu, Enkvetchakul, & Levitan, 2011). No cholesterol binding was observed for the denatured protein. The specificity of the binding was further verified by *saturability* and *competition* with the cold ligand (Singh et al., 2011). The existence of a binding site in a bacterial channel may represent an existence of a hydrophobic pocket that is a general structural feature of Kir channels. We also found that, as expected, nonradioactive cholesterol can fully displace the radioactive ligand from the channel protein. Most significantly, epicholesterol was found to compete with the radioactive cholesterol for the KirBac1.1 binding as efficiently as native cholesterol, indicating that both *cholesterol and epicholesterol bind to KirBac1.1 channels to the same sites* (Fig. 4). This conclusion is consistent with the hypothesis we proposed for mammalian Kir channels based on the increase in Kir activity in cells loaded with epicholesterol. Notably, two other sterols tested in this study, 25-hydroxycholesterol and 5-androsten 3β-17β-diol, competed only slightly or not at all, showing that the binding of cholesterol is specific. A lack of the stereospecificity of cholesterol binding, with cholesterol and epicholesterol competing for the same sites, is not unique to Kir channels, as it was also reported for a sterol-sensing domain of SCAP (Radhakrishnan et al., 2004). As described above, epicholesterol and *ent*-cholesterol were also found to interact with nAChR (Addona et al., 2003). The conclusion of these studies is that, while both cholesterol and epicholesterol appear to be capable of binding to the channels, only cholesterol exerts the modulatory effect. Our next goal, therefore, was to explore whether cholesterol-binding sites can indeed accommodate different cholesterol isomers based on energetic considerations.

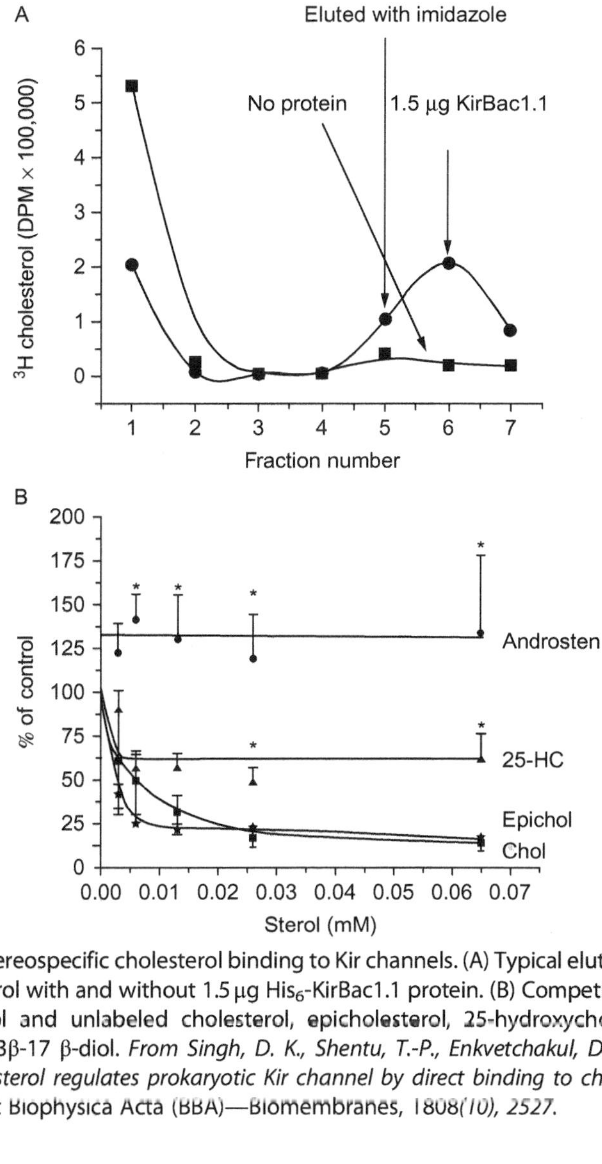

Fig. 4 Nonstereospecific cholesterol binding to Kir channels. (A) Typical elution profiles of [^{3}H]-cholesterol with and without 1.5 μg His$_6$-KirBac1.1 protein. (B) Competition between ^{3}H-cholesterol and unlabeled cholesterol, epicholesterol, 25-hydroxycholesterol, and 5-androsten 3β-17 β-diol. *From Singh, D. K., Shentu, T.-P., Enkvetchakul, D., & Levitan, I. (2011). Cholesterol regulates prokaryotic Kir channel by direct binding to channel protein.* Biochimica et Biophysica Acta (BBA)—Biomembranes, *1808(10), 2527.*

5. COMPUTATIONAL ANALYSIS OF THE STEREOSPECIFICITY OF CHOLESTEROL BINDING TO ION CHANNELS

Identification and characterization of a binding site is a major step in elucidating the mechanism of the interaction between a ligand and a protein.

There is still a paucity of information about cholesterol-binding sites in ion channels, and several aspects of this topic are discussed in this volume. Briefly, the most well studied are cholesterol-binding motifs, CRAC and its mirror image CARC, that have been identified in multiple ion channels (Fantini & Barrantes, 2013). However, while multiple copies of these motifs have been found in a variety of ion channels, both in the transmembrane and in the cytosolic domains, in many cases it is not clear whether a specific CRAC or CARC motif is really a part of a cholesterol-binding site or not. Two approaches are usually used to address this question: energetic considerations that are based on the computational analysis of the physical forces generated between the cholesterol and the protein molecules and site-directed mutations of the residues of the CRAC/CARC motifs to test whether a specific residue is required for the regulatory effect. A limitation of the first approach, however, is that if computational analysis is applied only to the regions containing the known motifs, it will not identify any alternative sites that do not include these motifs. The second approach, verification of the cholesterol-binding motifs as part of a binding site using site-directed mutagenesis, should also be used with caution, because mutations that result in the loss of cholesterol sensitivity are not necessarily part of a cholesterol-binding site. For example, we previously identified an array of residues in Kir2.1 channels that are critical for cholesterol sensitivity of the channels but are located in the regions that energetically cannot accommodate a cholesterol-binding site (Rosenhouse-Dantsker, Logothetis, & Levitan, 2011; Rosenhouse-Dantsker, Noskov, Durdagi, Logothetis, & Levitan, 2013; Rosenhouse-Dantsker, Noskov, et al., 2012; Rosenhouse-Dantsker, Noskov, Logothetis, & Levitan, 2013). When using this approach, therefore, it is important to show that mutations of multiple residues of the specific CRAC/CARC motifs alter the sensitivity of the channels to cholesterol, whereas mutations of the residues that are not predicted to interact with cholesterol have no effect.

A stronger approach is a combined computational/experimental analysis that involves an unbiased docking of a cholesterol molecule to a channel protein followed by molecular dynamics simulations and validation by site-directed mutagenesis. Briefly, docking analyses, which probe the energetic landscape between a protein and ligand, provide an initial prediction for ligand-binding sites, helping to narrow the search down to a reasonable number of potential interacting residues. These predicted binding pockets are then probed with computational molecular dynamic simulations to investigate the dynamic interactions between the ligand and predicted binding site, validating or invalidating the docking prediction. The final predictions from the simulations are further tested and verified by mutating the residues predicted to interact with the cholesterol molecule and testing the functional effect. Indeed, two independent studies of Kir channels that used a more general search for cholesterol-binding sites, unbiased for the presence of the known motifs, identified putative cholesterol bindings at the nonannular hydrophobic pockets of the channels that do not contain any known motifs (Fürst, Nichols, Lamoureux, & D'Avanzo, 2014; Rosenhouse-Dantsker, Noskov, Durdagi, et al., 2013; Rosenhouse-Dantsker, Noskov, Logothetis, et al., 2013). Multiple mutations of the putative cholesterol-interacting residues were tested to verify the validity of these predictions. Notably, even though there are minor differences in the predictions of the two studies for the cholesterol-binding site in Kir2 channels, which are most likely based on differences in the initial structure taken for the analysis, overall a very similar nonannular site was found in both studies. Moreover, this computational approach was used to identify potential cholesterol-binding pockets on nAChR (Brannigan, Hénin, Law, Eckenhoff, & Klein, 2008), the $GABA_A$ receptor (Henin, Salari, Murlidaran, & Brannigan, 2014), and on mitochondrial VDAC (Weiser, Salari, Eckenhoff, & Brannigan, 2014), and in none of the three cases was it noted by the authors that these binding sites are based on the CRAC or CARC motifs. These observations underscore the importance of energetic considerations in identifying and characterizing cholesterol-binding sites.

The next major question that we address here is whether the predicted cholesterol-binding sites can accommodate the chiral isomers of cholesterol. As discussed above, in previous studies, cholesterol stereospecificity typically leads to the conclusion that the *binding* of cholesterol is stereospecific meaning that while cholesterol binds to the channel, an analogue of cholesterol does not. However, the stereospecificity of cholesterol regulation of Kir channels appears to be not due to the specificity of the binding but to the ability of a sterol molecule to stabilize the channels in a closed state. Here, using a computational approach via docking analyses, we present further evidence to support this conclusion by exploring the relative ability of cholesterol and its analogues, epicholesterol and *ent*-cholesterol, to bind to Kir channels. As mentioned previously, docking analyses allow one to make predictions about protein–ligand binding by exploring a ligand's conformational space and the energetic landscape governing protein–ligand interactions. To that end, we performed three sets of docking analyses on each of cholesterol, epicholesterol, and *ent*-cholesterol to the previously identified crystal structure of a Kir2.2 channel. The expectation was that if binding is indeed *not* stereospecific, then the results for the three sterols would be similar in predicted energy, conformation, and binding location. Our analysis shows that this is indeed the case.

Specifically, in comparing the results of the docking analyses, we show there is *considerable overlap among all three sterols in binding energy, conformation, and location*. The most favorable binding conformations of all three cholesterol isomers tested in our analysis, cholesterol, epicholesterol, and *ent*-cholesterol were found to overlap in a binding pocket between the transmembrane domains of two subunits, above the slide helix, similar to the site that has been identified previously as the cholesterol-binding pocket (Fürst et al., 2014; Rosenhouse-Dantsker, Noskov, Durdagi, et al., 2013; Rosenhouse-Dantsker, Noskov, Logothetis, et al., 2013). Similarly to the previous studies, cholesterol-binding pocket is found at a nonannular site between the transmembrane helices of the adjacent subunits in the proximity of the slide helix (top view Fig. 5). The location of this cholesterol-binding pocket corresponds to a type of binding site that is formed within the protein and which is expected to regulate protein function by altering the free energy difference between the open and the closed conformations of the channels (Bruno, Rusinova, Gleason, Koeppe, & Andersen, 2013). Further analysis is needed though to characterize more fully different conformations of cholesterol binding to these channels. Furthermore, as was also described previously, cholesterol molecules can assume multiple poses within the binding pocket of the Kir protein, which can be represented as a "cloud" of cholesterol molecules interacting with the channel protein (Fig. 6A). Both cholesterol isomers assume multiple poses in the same pocket. The "cloud" of *ent*-cholesterol molecules in the same pocket is shown in Fig. 6B and the overlap between the two in Fig. 6C. The overlap of all three isomers in the binding pocket is shown in Fig. 6D. In terms of binding energy, the results for the three sterols were also comparable. For cholesterol, the predicted binding energy for the most favorable conformation was −9.75 kcal/mol, while for *ent*-cholesterol it was −9.9 kcal/mol, and for epicholesterol's two conformations it was −9.8 and −10.4 kcal/mol (all of these values are within the standard error for AutoDock's calculation of binding energy, viz., −2.5 kcal/mol). These results further support our hypothesis regarding the difference between stereospecificity in binding and in the regulatory effect of cholesterol on ion channel function. For all three sterols, docking analyses yielded very similar predictions for interaction

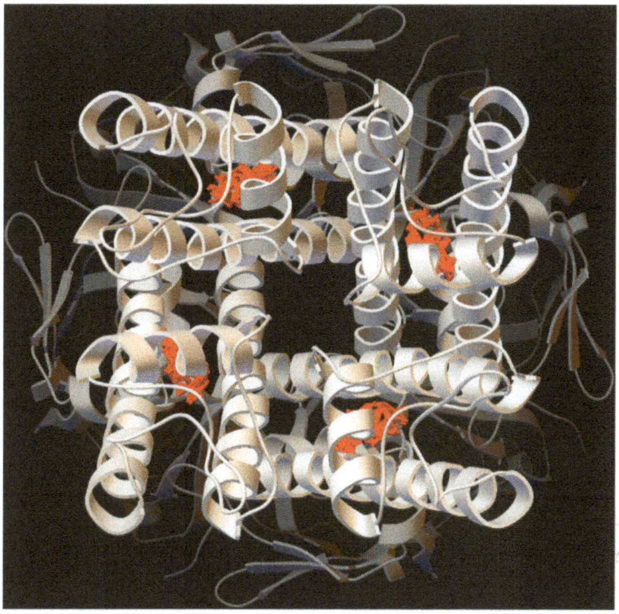

Fig. 5 Nonannular hydrophobic cholesterol-binding pocket in Kir2.2 channels. A top view of the Kir2.2 tetramer, highlighting the location of cholesterol binding (*red*), as predicted by molecular docking carried out using AutoDockTools (ADT) 1.5.6. The crystal structure for the Kir2.2 channel was obtained from the Protein Data Bank (PDB ID: 3SPI). The 3-dimensional structure for cholesterol was obtained from the ChemSpider database. A PDBQT structure file was generated for the channel beginning with protonation, followed by the assignment of Gasteiger charges and the merging of nonpolar hydrogen atoms to their bonded carbon, as well as the assignment of atom types using ADT. The search box was centered on the transmembrane domain of one subunit in the tetramer, and grid size of 0.375 Å was used. For each docking analysis, 250 independent docking calculations were made, with a population size of 150 for every calculation.

energy and conformation, implying that the pocket in which these different ligands bind is promiscuous enough to accommodate all three isomers.

Interestingly, however, there are some notable differences between the orientations of the three isomers in the binding pocket: cholesterol is found to be oriented with the "rough face" of cholesterol's methyl groups directed toward the protein, which was shown previously in MD simulations to be a dynamically stable location and orientation for cholesterol (Fürst et al., 2014). In contrast, *ent*-cholesterol which enters the same binding pocket is oriented in a different direction so that the tetracyclic rings of cholesterol and *ent*-cholesterol overlap when compared, while their methyl groups face opposite directions (see Fig. 7A and B for the comparison of the relative

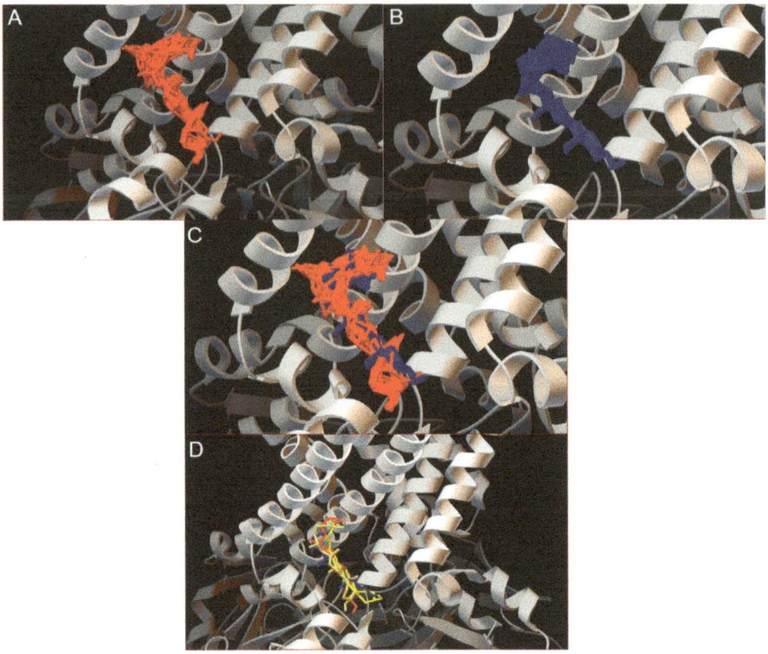

Fig. 6 Kir-binding site is predicted to accommodate all three cholesterol isomers: cholesterol, *ent*-cholesterol, and epicholesterol. A "cloud" of representative poses of (A) cholesterol (*red*) and (B) *ent*-cholesterol (*blue*) in the most energetically favorable conformations, as determined by docking analyses. (C) Overlay of the two clouds. (D) A single representative pose for the most energetically favorable conformation for cholesterol (*red*), *ent*-cholesterol (*blue*), and epicholesterol (*yellow*). Docking analyses were performed three times for cholesterol, epicholesterol, and *ent*-cholesterol. The 3-dimensional structures for all three sterols were obtained from the ChemSpider database. In each docking run, results were organized by ligand pose and orientation. Ligand poses within a root mean square deviation of 2 Å were grouped into clusters, which represented potential binding sites. These clusters were assigned a single value for binding energy. Potential binding sites were identified by comparing across all three docking analyses the sites where the ligand bound with the highest frequency (defined as the cluster with the greatest percentage of poses).

orientations of the two molecules). As described above, epicholesterol enters the same binding pocket as cholesterol and *ent*-cholesterol but, uniquely, it can assume both orientations with half of the poses corresponding to an epicholesterol conformation wherein the methyl groups faced the same direction as cholesterol, while the other half corresponded to methyl groups aligning with *ent*-cholesterol (Fig. 7C and D). The overlap between all four positions is shown in Fig. 7E. These differences might be key to the understanding of the stereospecificity of the cholesterol functional impact of Kir channels.

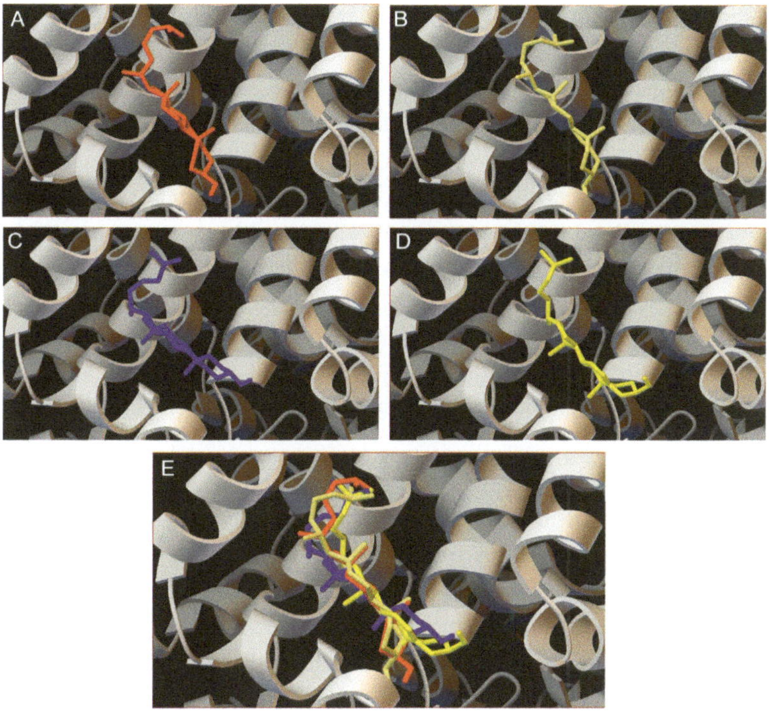

Fig. 7 Differential orientations of cholesterol, *ent*-cholesterol, and epicholesterol within the Kir2.2 predicted binding site. Close-up views of the representative pose of the most energetically favorable conformation of (A) cholesterol (*red*), (B) *ent*-cholesterol (*blue*), and (C,D) epicholesterol (*yellow*). (E) A close-up view of all four representative poses overlaid atop one another. Note the opposite orientations of the methyl groups for cholesterol and its analogues within the binding site.

6. CONCLUSIONS AND FUTURE DIRECTIONS

Structural sterol analysis is a powerful tool to elucidate the mechanisms of cholesterol regulation of ion channels and other membrane proteins. In general, the stereospecificity of the functional response is evidence that cholesterol regulates the protein by specific sterol–protein interactions and not by altering the physical properties of the membrane bilayer. It is also possible that the stereospecificity can be due to differential abilities of cholesterol and its isomers to couple between the channel proteins and the lipid bilayer. The opposite, however, is not necessarily true because a lack of stereospecificity does not preclude direct binding of cholesterol to the protein. Multiple sterols exerting differential effects on the physical properties of the membrane should be tested to discriminate between these possibilities.

It should be further noted that the stereospecificity of the cholesterol functional effect does not necessarily mean that a cholesterol isomer does not bind to the channel. In general, in the case of a ligand binding-mediated structural change of a protein, the underlying assumption is that upon binding, the ligand exerts a thermodynamic or steric effect, which causes a shift in the target protein's secondary or tertiary structure, effecting a conformational change. However, it is also entirely possible that a modified ligand can bind to a given target channel without acting as an agent of change. Therefore, in examining the effects of cholesterol analogues on the regulation of ion channels, it should be noted that the absence of a physiological effect on the ion channel does not necessarily preclude the existence of ligand-channel-binding events. Rather, it simply shows that the analogue or ligand in question does not have a regulatory effect on the channel, and further experiments are needed to clarify this issue. This is a critical distinction that has major implications both structurally, in terms of identifying cholesterol-binding sites, and mechanistically, in terms determining how cholesterol binding exerts its effect on the channel function. First, it indicates that the standard interpretation that the stereospecificity of cholesterol effects means a lack of binding of the isomer is not always correct and, therefore, using this comparison to identify cholesterol-binding sites might be incorrect and misleading. More importantly, differential effects of sterol isomers that bind to the channel protein but have significantly different impact on the channel activity may provide the most powerful and unique insights into the mechanistic basis of the regulatory effects: what are the differences in the way the sterols interact with the channel that lead to the functional effects vs no effect. Discriminating these two scenarios and obtaining appropriate structural mechanistic insights is the major goal of future studies in this area.

ACKNOWLEDGMENTS

We are very grateful to Dr. Olaf Andersen for the critical reading of this manuscript. We also thank Mr. Gregory Kowalsky for his help in designing Fig. 1. The work was supported by the National Institutes of Health grants HL073965 and HL083298 (to I.L.).

REFERENCES

Abi-Char, J., Maguy, A., Coulombe, A., Balse, E., Ratajczak, P., Samuel, J.-L., et al. (2007). Membrane cholesterol modulates Kv1.5 potassium channel distribution and function in rat cardiomyocytes. *The Journal of Physiology*, *582*(3), 1205.

Addona, G. H., Sandermann, H. J., Kloczewiak, M. A., & Miller, K. W. (2003). Low chemical specificity of the nicotinic acetylcholine receptor sterol activation site. *Biochimica et Biophysica Acta*, *1609*, 177–182.

Baenziger, J. E., Henault, C. M., Therien, J. P., & Sun, J. (2015). Nicotinic acetylcholine receptor-lipid interactions: Mechanistic insight and biological function. *Biochimica et Biophysica Acta, 1848*(9), 1806–1817.

Barrantes, F. J. (2007). Cholesterol effects on nicotinic acetylcholine receptor. *Journal of Neurochemistry, 103*, 72.

Bittman, R., & Blau, L. (1972). The phospholipid-cholesterol interaction. Kinetics of water permeability in liposomes. *Biochemistry, 11*, 4821–4839.

Bolotina, V., Omelyanenko, V., Heyes, B., Ryan, U., & Bregestovski, P. (1989). Variations of membrane cholesterol alter the kinetics of Ca2+-dependent K+ channels and membrane fluidity in vascular smooth muscle cells. *Pflügers Archiv, 415*, 262–268.

Bonn, M., Roke, S., Berg, O., Juurlink, L. B., Stamouli, A., & Müller, M. (2004). A molecular view of cholesterol-induced condensation in a lipid monolayer. *The Journal of Physical Chemistry B, 108*(50), 19083–19085.

Brannigan, G., Hénin, J., Law, R., Eckenhoff, R., & Klein, M. L. (2008). Embedded cholesterol in the nicotinic acetylcholine receptor. *Proceedings of the National Academy of Sciences of the United States of America, 105*(38), 14418–14423.

Bruno, M. J., Rusinova, R., Gleason, N. J., Koeppe, R. E., 2nd, & Andersen, O. S. (2013). Interactions of drugs and amphiphiles with membranes: Modulation of lipid bilayer elastic properties by changes in acyl chain unsaturation and protonation. *Faraday Discussions, 161*, 461–480. discussion 563–589.

Bukiya, A. N., Belani, J. D., Rychnovsky, S., & Dopico, A. M. (2011). Specificity of cholesterol and analogs to modulate BK channels points to direct sterol-channel protein interactions. *The Journal of General Physiology, 137*(1), 93–110.

Bukiya, A. N., Vaithianathan, T., Toro, L., & Dopico, A. M. (2008). The second transmembrane domain of the large conductance, voltage- and calcium-gated potassium channel beta(1) subunit is a lithocholate sensor. *FEBS Letters, 582*(5), 673.

Chang, H. M., Reitstetter, R., Mason, R. P., & Gruener, R. (1995). Attenuation of channel kinetics and conductance by cholesterol: An interpretation using structural stress as a unifying concept. *Journal of Membrane Biology, 143*, 51–63.

Corbin, J., Wang, H. H., & Blanton, M. P. (1998). Identifying the cholesterol binding domain in the nicotinic acetylcholine receptor with [125I]azido-cholesterol. *Biochimica et Biophysica Acta, 1414*, 65–74.

Crowley, J. J., Treistman, S. N., & Dopico, A. M. (2003). Cholesterol antagonizes ethanol potentiation of human brain BKCa channels reconstituted into phospholipid bilayers. *Molecular Pharmacology, 64*(2), 365–372.

daCosta, C. J. B., Dey, L., Therien, J. P. D., & Baenziger, J. E. (2013). A distinct mechanism for activating uncoupled nicotinic acetylcholine receptors. *Nature Chemical Biology, 9*(11), 701.

D'Avanzo, N., Hyrc, K., Enkvetchakul, D., Covey, D. F., & Nichols, C. G. (2011). Enantioselective protein-sterol interactions mediate regulation of both prokaryotic and eukaryotic inward rectifier K^+ channels by cholesterol. *PLoS One, 6*(4). e19393.

Demel, R. A., Bruckdorfer, K. R., & van Deenen, L. L. M. (1972). The effect of sterol structure on the permeability of liposomes to glucose, glycerol and Rb+. *Biochemica et Biophysica Acta, 255*, 321–330.

Dufourc, E. J., Parish, E. J., Chitrakorn, S., & Smith, I. C. (1984). Structural and dynamical details of cholesterol-lipid interaction as revealed by deuterium NMR. *Biochemistry, 23*(25), 6062–6071.

Enkvetchakul, D., Bhattacharyya, J., Jeliazkova, I., Groesbeck, D. K., Cukras, C. A., & Nichols, C. G. (2004). Functional characterization of a prokaryotic Kir channel. *The Journal of Biological Chemistry, 279*(45), 47076–47080.

Fantini, J., & Barrantes, F. J. (2013). How cholesterol interacts with membrane proteins: An exploration of cholesterol-binding sites including CRAC, CARC and tilted domains. *Frontiers in Physiology, 4*, 31.

Fürst, O., Nichols, C. G., Lamoureux, G., & D'Avanzo, N. (2014). Identification of a cholesterol-binding pocket in inward rectifier K(+) (Kir) channels. *Biophysical Journal, 107*, 2786–2796.

Gimpl, G. (2010). Cholesterol-protein interaction: Methods and cholesterol reporter molecules. In J. R. Harris (Ed.), *Cholesterol binding and cholesterol transport proteins: 51* (pp. 1–45). Heidelberg: Springer.

Gimpl, G., Burger, K., & Fahrenholz, F. (1997). Cholesterol as modulator of receptor function. *Biochemistry, 36*, 10959–10974.

Hajdú, P., Varga, Z., Pieri, C., Panyi, G., & Gáspár, R. J. (2003). Cholesterol modifies the gating of Kv1.3 in human T lymphocytes. *Pflügers Archiv, 445*, 674–682.

Hamouda, A. K., Chiara, D. C., Sauls, D., Cohen, J. B., & Blanton, M. P. (2006). Cholesterol interacts with transmembrane alpha-helices M1, M3, and M4 of the Torpedo nicotinic acetylcholine receptor: Photolabeling studies using [3H]azicholesterol. *Biochemistry, 45*(3), 976.

Henin, J., Salari, R., Murlidaran, S., & Brannigan, G. (2014). A predicted binding site for cholesterol on the GABAA receptor. *Biophysical Journal, 106*(9), 1938–1949.

Infante, R. E., Abi-Mosleh, L., Radhakrishnan, A., Dale, J. D., Brown, M. S., & Goldstein, J. L. (2008). Purified NPC1 protein. *Journal of Biological Chemistry, 283*(2), 1052–1063.

Infante, R. E., Wang, M. L., Radhakrishnan, A., Kwon, H. J., Brown, M. S., & Goldstein, J. L. (2008). NPC2 facilitates bidirectional transfer of cholesterol between NPC1 and lipid bilayers, a step in cholesterol egress from lysosomes. *Proceedings of the National Academy of Sciences of the United States of America, 105*(40), 15287–15292.

Klausen, T. K., Hougaard, C., Hoffmann, E. K., & Pedersen, S. F. (2006). Cholesterol modulates the volume-regulated anion current in Ehrlich-Lettre ascites cells via effects on Rho and F-actin. *American Journal of Physiology. Cell Physiology, 291*(4), C757–C771.

Levitan, I., Christian, A. E., Tulenko, T. N., & Rothblat, G. H. (2000). Membrane cholesterol content modulates activation of volume-regulated anion current in bovine endothelial cells. *The Journal of General Physiology, 115*(4), 405–416.

Levitan, I., Singh, D. K., & Rosenhouse-Dantsker, A. (2014). Cholesterol binding to ion channels. *Frontiers in Physiology, 5*, 65.

Lundbaek, J. A., Birn, P., Hansen, A. J., & Andersen, O. S. (1996). Membrane stiffness and channel function. *Biochemistry, 35*, 3825–3830.

Lundbaek, J. A., Birn, P., Hansen, A. J., Sogaard, R., Nielsen, C., Girshman, J., et al. (2004). Regulation of sodium channel function by bilayer elasticity: The importance of hydrophobic coupling. Effects of micelle-forming amphiphiles and cholesterol. *The Journal of General Physiology, 123*(5), 599–621.

Mannock, D. A., Lee, M. Y., Lewis, R. N., & McElhaney, R. N. (2008). Comparative calorimetric and spectroscopic studies of the effects of cholesterol and epicholesterol on the thermotropic phase behaviour of dipalmitoylphosphatidylcholine bilayer membranes. *Biochimica et Biophysica Acta (BBA)—Biomembranes, 1778*(10), 2191–2202.

Murari, R., Murari, M. P., & Baumann, W. J. (1986). Sterol orientations in phosphatidylcholine liposomes as determined by deuterium NMR. *Biochemistry, 25*, 1062–1067.

Needham, D., & Nunn, R. S. (1990). Elastic deformation and failure of lipid bilayer membranes containing cholesterol. *Biophysical Journal, 58*, 997–1009.

Pasenkiewicz-Gierula, M., Róg, T., Kitamura, K., & Kusumi, A. (2000). Cholesterol effects on the phosphatidylcholine bilayer polar region: A molecular simulation study. *Biophysical Journal, 78*(3), 1376–1389.

Picazo-Juarez, G., Romero-Suarez, S., Nieto-Posadas, A., Llorente, I., Jara-Oseguera, A., Briggs, M., et al. (2011). Identification of a binding motif in the S5 helix that confers cholesterol-sensitivity to TRPV1. *Journal of Biological Chemistry, 286*(28), 24966–24976.

Popot, J. L., Demel, R. A., Sobel, A., Van Deenen, L. L., & Changeux, J. P. (1978). Interaction of the acetylcholine (nicotinic) receptor protein from Torpedo marmorata electric organ with monolayers of pure lipids. *European Journal of Biochemistry*, *85*, 27–42.

Radhakrishnan, A., Sun, L.-P., Kwon, H. J., Brown, M. S., & Goldstein, J. L. (2004). Direct binding of cholesterol to the purified membrane region of SCAP: Mechanism for a sterol-sensing domain. *Molecular Cell*, *15*(2), 259.

Róg, T., & Pasenkiewicz-Gierula, M. (2003). Effects of epicholesterol on the phosphatidylcholine bilayer: A molecular simulation study. *Biophysical Journal*, *84*(3), 1818–1826.

Róg, T., Pasenkiewicz-Gierula, M., Vattulainen, I., & Karttunen, M. (2007). What happens if cholesterol is made smoother: Importance of methyl substituents in cholesterol ring structure on phosphatidylcholine–sterol interaction. *Biophysical Journal*, *92*(10), 3346–3357.

Romanenko, V. G., Fang, Y., Byfield, F., Travis, A. J., Vandenberg, C. A., Rothblat, G. H., et al. (2004). Cholesterol sensitivity and lipid raft targeting of Kir2.1 channels. *Biophysical Journal*, *87*, 3850–3861.

Romanenko, V. G., Rothblat, G. H., & Levitan, I. (2002). Modulation of endothelial inward rectifier K+ current by optical isomers of cholesterol. *Biophysical Journal*, *83*, 3211–3222.

Romanenko, V. G., Rothblat, G. H., & Levitan, I. (2004). Sensitivity of volume-regulated anion current to cholesterol structural analogues. *The Journal of General Physiology*, *123*(1), 77–87.

Rosenhouse-Dantsker, A., Logothetis, D. E., & Levitan, I. (2011). Cholesterol sensitivity of KIR2.1 is controlled by a belt of residues around the cytosolic pore. *Biophysical Journal*, *100*(2), 381.

Rosenhouse-Dantsker, A., Mehta, D., & Levitan, I. (2012). Regulation of ion channels by membrane lipids. *Comprehensive Physiology*, *2*, 31–68.

Rosenhouse-Dantsker, A., Noskov, S., Durdagi, S., Logothetis, D. E., & Levitan, I. (2013). Identification of novel cholesterol-binding regions in Kir2 channels. *Journal of Biological Chemistry*, *288*(43), 31154–31164.

Rosenhouse-Dantsker, A., Noskov, S., Han, H., Adney, S. K., Tang, Q.-Y., RodrÃguez-Menchaca, A. A., et al. (2012). Distant cytosolic residues mediate a two-way molecular switch that controls the modulation of inwardly rectifying potassium (Kir) channels by cholesterol and phosphatidylinositol 4,5-bisphosphate (PI(4,5)P2). *Journal of Biological Chemistry*, *287*(48), 40266–40278.

Rosenhouse-Dantsker, A., Noskov, S., Logothetis, D. E., & Levitan, I. (2013). Cholesterol sensitivity of KIR2.1 depends on functional inter-links between the N and C termini. *Channels*, *7*, 303–312.

Singh, D. K., Rosenhouse-Dantsker, A., Nichols, C. G., Enkvetchakul, D., & Levitan, I. (2009). Direct regulation of prokaryotic Kir channel by cholesterol. *Journal of Biological Chemistry*, *284*(44), 30727–30736.

Singh, D. K., Shentu, T.-P., Enkvetchakul, D., & Levitan, I. (2011). Cholesterol regulates prokaryotic Kir channel by direct binding to channel protein. *Biochimica et Biophysica Acta (BBA)—Biomembranes*, *1808*(10), 2527.

Sooksawate, T., & Simmonds, M. A. (2001). Effects of membrane cholesterol on the sensitivity of the GABA(A) receptor to GABA in acutely dissociated rat hippocampal neurones. *Neuropharmacology*, *40*, 178–184.

Weiser, B. P., Salari, R., Eckenhoff, R. G., & Brannigan, G. (2014). Computational investigation of cholesterol binding sites on mitochondrial VDAC. *The Journal of Physical Chemistry. B*, *118*(33), 9852–9860.

Westover, E. J., & Covey, D. F. (2004). The enantiomer of cholesterol. *Journal of Membrane Biology*, *202*(2), 61.

Westover, E. J., Covey, D. F., Brockman, H. L., Brown, R. E., & Pike, L. J. (2003). Cholesterol depletion results in site-specific increases in epidermal growth factor receptor phosphorylation due to membrane level effects studies with cholesterol enantiomers. *Journal of Biological Chemistry*, *278*(51), 51125–51133.

Xu, X., & London, E. (2000). The effect of sterol structure on membrane lipid domains reveals how cholesterol can induce lipid domain formation. *Biochemistry*, *39*, 843–849.

Yeagle, P. L. (1985). Cholesterol and the cell membrane. *Biochimica et Biophysica Acta*, *822*, 267–287.

Yeagle, P. L. (1991). Modulation of membrane function by cholesterol. *Biochimie*, *73*, 1303.

SECTION 2

Structural Determinants of Cholesterol-Ion Channels Interactions

CHAPTER THREE

Regulation of Ca^{2+}-Sensitive K^+ Channels by Cholesterol and Bile Acids via Distinct Channel Subunits and Sites

Alex M. Dopico[1], Anna N. Bukiya
College of Medicine, The University of Tennessee Health Science Center, Memphis, TN, United States
[1]Corresponding author: e-mail address: adopico@uthsc.edu

Contents

Abstract

Cholesterol (CLR) conversion into bile acids (BAs) in the liver constitutes the major pathway for CLR elimination from the body. Moreover, these steroids regulate each other's metabolism. While the roles of CLR and BAs in regulating metabolism and tissue function are well known, research of the last two decades revealed the existence of specific

Current Topics in Membranes, Volume 80
ISSN 1063-5823
http://dx.doi.org/10.1016/bs.ctm.2017.07.001

protein receptors for CLR or BAs in tissues with minor contribution to lipid metabolism, raising the possibility that these lipids serve as signaling molecules throughout the body. Among other lipids, CLR and BAs regulate ionic current mediated by the activity of voltage- and Ca^{2+}-gated, K^+ channels of large conductance (BK channels) and, thus, modulate cell physiology and participate in tissue pathophysiology. Initial work attributed modification of BK channel function by CLR or BAs to the capability of these steroids to directly interact with bilayer lipids and thus alter the physicochemical properties of the bilayer with eventual modification of BK channel function. Based on our own work and that of others, we now review evidence that supports direct interactions between CLR or BA and specific BK protein subunits, and the consequence of such interactions on channel activity and organ function, with a particular emphasis on arterial smooth muscle. For each steroid type, we will also briefly discuss several mechanisms that may underlie modification of channel steady-state activity. Finally, we will present novel computational data that provide a chemical basis for differential recognition of CLR vs lithocholic acid by distinct BK channel subunits and recognition sites.

1. BIOLOGY AND CHEMISTRY OF CHOLESTEROL AND BILE ACIDS

1.1 Cholesterol in Physiology and Disease

Cholesterol (CLR) is one of the major building blocks of life: in animal cells, CLR serves as a structural lipid in the plasma membrane and, to a lesser extent, in membranes of organelles, including the nuclear envelope, and as a common metabolic precursor of steroids (Dopico & Tigyi, 2007; Nes, 2011; Pikuleva, 2008; Simons & Sampaio, 2011). In cell membranes, CLR molecules usually arrange in vertical domains that span the whole lipid bilayer leading to the formation of the so-called lipid rafts; these domains are also enriched in sphingolipids and cluster membrane proteins receptors and associated signaling molecules. Thus, lipid rafts are critical for cell signaling, protein sorting, and molecular trafficking (Brown & London, 1998; Gimpl, Burger, & Fahrenholz, 1997; Sonnino & Prinetti, 2013). The availability of CLR in central nervous system (CNS) is vital for myelin formation, brain maturation, and synapse formation and function (Pfrieger, 2003; Saher et al., 2005). Cholesterol regulates cell motility and organization of the actin cytoskeleton (Ramprasad et al., 2007), interactions between bacterial or viral pathogens and mammalian cells (Goluszko & Nowicki, 2005; Schroeder, 2010), and expression and function of protein receptors, including ion channels (Gimpl et al., 1997; Levitan, Fang, Rosenhouse-Dantsker, & Romanenko, 2010; Tillman & Cascio, 2003).

Both CLR deficiency and excess result in a multitude of pathological conditions. The consequences of CLR deficiency (e.g., in Smith–Lemli–Opitz syndrome) span from physical malformations to developmental delay and mental retardation (Nowaczyk & Irons, 2012), the latter being consistent with the central role that CLR has in myelin formation and CNS development (Saher et al., 2005). Abnormally low levels of CLR in systemic circulation (i.e., hypocholesterolemia) are often observed after major surgery or trauma, severe infections, and prolonged hypovolemic shock, and very low CLR levels are predictors of mortality in critically ill patients (Dunham, Fealk, & Sever, 2003; Guimarães, Lima, Cipullo, Lobo, & Burdmann, 2008; Stachon et al., 2000; Vyroubal et al., 2008). In the general population, low CLR levels positively correlate with increased mortality due to respiratory, digestive disease, or trauma (Jacobs et al., 1992). In turn, 30% of the US population is at risk of pathological conditions triggered by hypercholesterolemia (http://www.cdc.gov/cholesterol/facts.htm). These include atherosclerosis, increased incidence of stroke, cardiovascular morbidity/mortality, accelerated progression of Alzheimer disease and ocular disorders (Bouhairie & Goldberg, 2016; El-Sayyad, Elmansi, & Bakr, 2015; Meng et al., 2014; Navi & Segal, 2009; Saini, Arneja, & Dhalla, 2004). The ubiquitous presence of CLR in animal cells, the prevalence of pathophysiological conditions associated with abnormal CLR levels, and the wide distribution and key roles of ion channels in both excitable and nonexcitable tissues has led to an expanding area of investigation aimed at addressing the regulation of ion channel function by manipulating CLR levels, and the consequences of such regulation on cell and organ function (Dopico, Bukiya, & Singh, 2012b; Gimpl et al., 1997; Levitan et al., 2010; Tillman & Cascio, 2003).

1.2 Bile Acids in Physiology and Disease

In most species, including humans, bile acids (BAs) are synthesized mainly in the liver where they are conjugated with glycine and, to a lesser extent, taurine. BA synthesis is tightly regulated by negative feedback through the nuclear farnesoid X receptor (FXR) (Sinal et al., 2000; Teodoro, Rolo, & Palmeira, 2011). Conjugates of the so-called primary BAs, cholic acid (CA), and chenodeoxycholic acid (CDCA), and their secondary metabolites are actively secreted from the hepatocyte to the canaliculus. BAs are stored in the gallbladder and released into the duodenum after a meal (reviewed in Boyer, 2013).

BAs are best known for forming micelles in water media and thus emulsify dietary fat, lipophilic vitamins, and CLR itself. The resulting mixed micelles greatly facilitate the absorption of dietary lipid by the intestine (reviewed in Wahlström, Sayin, Marschall, & Bäckhed, 2016). Bile acid's detergent properties (referred to the property of BAs to insert into a lipid monolayer with consequent introduction of monolayer-positive curvature and membrane destabilization) have also been claimed primary responsible for the antibacterial effects of BA on gut microbiota (Copple & Li, 2016; Wahlström et al., 2016).

Primary BAs are deconjugated and partially converted into secondary BAs by dehydroxylation, which increases the hydrophobicity of the molecule. In humans, the most abundant secondary BAs are lithocholic acid (LCA) and deoxycholic acid (DCA) (Ridlon, Kang, & Hylemon, 2006; Setchell, Lawson, Tanida, & Sjövall, 1983). After reaching the lumen of the terminal ileum, BAs are reabsorbed and returned to the liver through the portal circulation. This enterohepatic circulation preserves more than 95% of the BA pool (reviewed in Porez, Prawitt, Gross, & Staels, 2012; de Aguiar Vallim, Tarling, & Edwards, 2013).

In addition to BA actions in the regulation of metabolism, the widespread location of BA receptors revealed by work conducted mainly during the last two decades raised the concept that BAs could serve as signaling molecules throughout the body. For example, FXR contributes to a plethora of vascular events: its activation favors apoptosis (Bishop-Bailey, Walsh, & Warner, 2004) and inhibits inflammation and vascular smooth muscle cell migration (Li, Swales, Thomas, Warner, & Bishop-Bailey, 2007), which may hamper vascular remodeling and atherosclerosis. FXR promotes angiotensin type 2 receptor transcription (Zhang et al., 2008) and participates in the hypotensive effects of angiotensin type 1 receptor blockers (Nakajima et al., 1995). FXR activation decreases myogenic tone by increasing the expression of endothelial nitric oxide synthase (eNOS) with eventual production of nitrogen oxide ($NO^{\bullet}$) and by promoting catabolism of symmetric dimethyl-arginine (ADMA), an endogenous eNOS inhibitor (Hu et al., 2006; Kida, Murata, Hori, & Ozaki, 2009), while decreasing expression of the vasoconstrictor endothelin-1 (He et al., 2006; Li et al., 2008; Rask-Madsen & King, 2007). Under physiological conditions, however, systemic BA levels during fasting barely reach 3 μM, which is six times lower than BA levels in the portal circulation (Ceryak, Bouscarel, & Fromm, 1993; Güldütuna, You, Kurts, & Leuschner, 1993; Scharschmidt et al., 1982) and low to significantly activate the vast majority of BA-recognizing receptors, including ion channels (see below). In contrast, several human pathological conditions such as mild liver cirrhosis (Jönsson, Hedenborg, Wisén, & Norman, 1992; Lindblad, Lundholm, & Scherstén, 1977), chronic active hepatitis (Korman, Hofmann, & Summerskill, 1974), extrahepatic cholestasis (Lindblad et al., 1977), and the stagnant-loop syndrome (Lewis, Tabaqchali, Panveliwalla, & Wootton, 1969) are associated with abnormally high BA levels in the systemic circulation due to BA spillover from the portal circulation (Greco & Mingrone, 1993; Hamdan & Stacey, 1993; Lewis et al., 1969; Tarantino et al., 1989). This spillover may render a 10–100-fold increase in BA systemic concentrations (Ceryak et al., 1993; Greco & Mingrone, 1993; Ostrow, 1993), which may be responsible for the systemic hypotension observed in patients with liver damage and/or significant portosystemic circulatory shunt (Bomzon, 1994; Bomzon & Ljubuncic, 1995). Under physiological conditions, local high levels of BAs have been hypothesized as responsible for the local hyperemia associated with fat absorption by the ileum (Bomzon & Ljubuncic, 1995). Using in vitro and in vivo experimental models, it has been consistently demonstrated that BAs reduce vascular reactivity, responses to vasoconstrictors, and reduce systemic blood pressure (Bomzon & Ljubuncic, 1995; Pak & Lee, 1993). Experiments conducted in rat cerebral arteries clearly indicated that in this model BA-induced vasodilation was endothelium-independent and mediated by targets located in the vascular smooth muscle itself (Bukiya, Liu, Toro, & Dopico, 2007).

In addition to their vascular actions, BAs have been found to function as pheromones in sea lamprey (Li et al., 2002), central olfactory stimuli in the salmon (Lo et al., 1994), and regulators of immunity (Copple & Li, 2016) and of several neurological functions in mammals, including memory and motor coordination, blood–brain barrier permeability, and regulation of the hypothalamic–pituitary axis (for a comprehensive review on BA regulation of neurological function and disease, see McMillin & DeMorrow, 2016).

1.3 Connections in the Chemistry and Metabolism of Cholesterol and Bile Acids

The chemistry and metabolic pathways of CLR and BAs are heavily intertwined. Both CLR and BAs are *steroids*, i.e., lipids that share a polycyclic, tetra-ring (A–D) system of 17 carbon atoms (C) termed cyclopentaneperhydrophenanthrene (Pine, 1987). Cholesterol, however, consists of a basic structure of 27C, which defines all steroids encompassed under the term *cholestanes* while for most BAs the basic structure consists of

24C, which defines the steroids termed *cholanes*. In CLR, the steroidal ring system features a double bond in ring B between C5 and C6, which confers some rigidity to the steroidal core of the molecule. At C17, CLR includes a freely movable, nonpolar iso-octyl chain. Overall, the only hydrophilic group in cholesterol is a single β-hydroxyl group at C3, and the molecule is asymmetric: the so-called α-face of cholesterol is a smooth, rather planar surface while the β-face has rough edges given by the methyl groups at C18 and C19 (Fig. 1A). In turn, BAs are characterized as planar amphiphiles: in

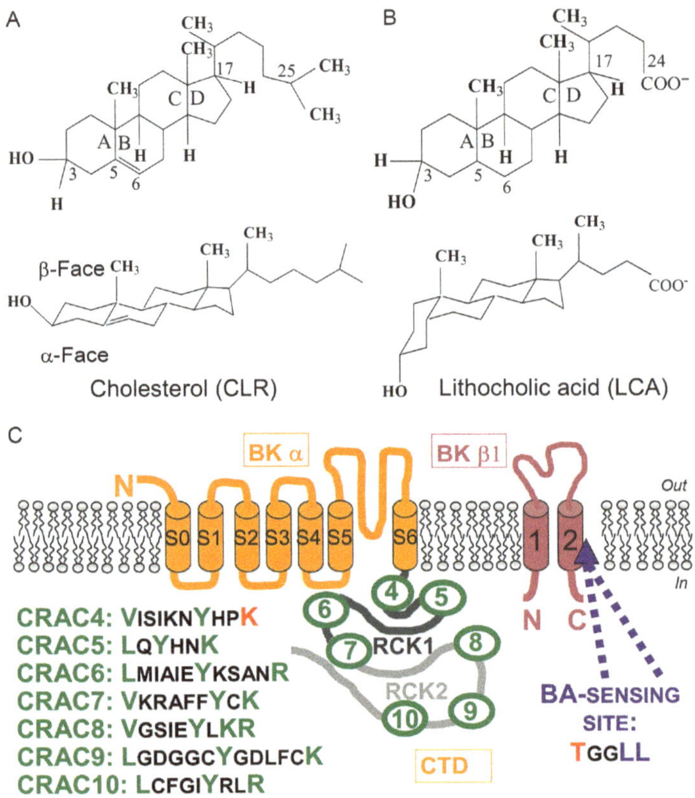

Fig. 1 Structural bases of cholesterol and bile acid modulation of BK channel activity. (A) Chemical structure of cholesterol. (B) Chemical structure of lithocholic (bile) acid. (C) Schematic structure of a BK channel heterodimer in vascular smooth muscle. The BK channel-forming α subunit is depicted in *orange* while the smooth muscle abundant, accessory β1 subunit is shown in *purple*. Cholesterol recognition amino acid consensus (CRAC) motifs on BK α subunits cloned from rat cerebral artery smooth muscle (cbv1 proteins; GenBank AY330293.2) are shown in *green*. CRAC4 exerts a profound effect on the overall cholesterol sensitivity of cbv1 channels (Singh et al., 2012). Location of the bile acid (BA)-sensing site on BK β1 subunits is shown in *blue*. In both CRAC4 and BA-sensing site, the amino acids that form hydrogen bonds with hydroxyl groups of cholesterol or bile acid are in *red font*. CTD: cytosolic tail domain.

this case, the movable chain attached to C17 is shorter than that of CLR and ends in a carboxyl group. The pK_a of the free (i.e., unconjugated BA) carboxyl in aqueous solution has been reported to have apparent pK_as ranging from 4.98 (for cholate at concentrations below critical micellar concentrations) to 6.2 (for chenodeoxycholate in soluble micelles), or up to ~7 when BAs are embedded within a proteolipid environment (Cabral, Hamilton, & Small, 1986; Carey & Small, 1972; Ko et al., 1994; Scharschmidt et al., 1982). Therefore, the carboxyl end remains ionized in a significant number of BA molecules at physiological pH (6.8–7.4) and thus contribute to the polar fraction of the BA molecule. BA polarity, however, is primarily determined by the hydroxyl groups, most in α-configuration for natural BAs, and *usually* attached to C3, C7, and/or C12 (Carey, 1985). In humans, BAs may be hydroxylated at C1–7, 12, 19, 22, and 23 (Sjövall, 2004).

The vast majority of natural BAs presents a *cis* joint between rings A and B, which renders an overall bean-shaped rigid amphiphile defined by (a) a concave, polar α *side* containing the hydroxyls and, under some circumstances, the carboxylate in the movable side chain and (b) a convex, lipophilic β *side* defined by the protruding methyls at C18 and C19 (Carey, 1985). The bean shape of natural BAs contrasts with the rather flat structure of the CLR ring structure (Fig. 1A and B).

Cholesterol and BA metabolisms are uniquely related. BAs are synthesized from CLR in the liver (Hofmann & Borgstroem, 1964) and their synthesis constitutes a significant way to eliminate CLR from the body (Taoka et al., 2016). Bile acid synthesis recognizes two pathways: under normal physiology, 75% of BA production starts with 7α-hydroxylation of CLR by CLR 7α-hydroxylase (CYP7A1) (Thomas, Pellicciari, Pruzanski, Auwerx, & Schoonjans, 2008). This is the rate-limiting enzyme and determines the amount of BAs produced. An alternative pathway starts with hydroxylation of CLR by sterol-27-hydroxylase (CYP27A1) (Russell, 2003). BAs are strong transcriptional inhibitors of CYP7A1, exerting a *negative feedback regulation on their own synthesis* (Russell & Setchell, 1992). Moreover, BA targeting of CYP7A1 enables these cholanes to regulate CLR metabolism itself (Staels & Fonseca, 2009). In addition, BAs can decrease the expression of the proprotein convertase subtilisin/kexin 9 gene. This gene product inhibits low-density lipoprotein (LDL) receptor activity and thus promotes intracellular degradation of this receptor (Langhi et al., 2008).

In FXR knockout (KO) mice, expression of scavenger receptor B1 (SRB1), which mediates liver uptake of high-density lipoprotein (HDL), is decreased whereas HDL–CLR is increased (Sinal et al., 2000). Consistently, FXR ligands increase hepatic SRB1 expression and decrease

HDL–CLR levels (Zhang et al., 2010). Moreover, FXR activation seems to be beneficial for treating hypertriglyceridemia, yet increases plasma CLR levels, whereas *reduction of FXR activity by BAs* lowers CLR levels while increasing triglyceridemia (reviewed in Porez et al., 2012).

Finally, the relationship between CLR and BA metabolism has been exploited in human therapy: as an alternative to statins, hypercholesterolemia has been treated with BA-binding resins (BABR) (Staels & Kuipers, 2007). BABR absorbs BAs in the lumen of the intestine, thereby preventing their uptake by the ileum, and facilitating their elimination within feces. This leads to a reduced BA pool size, repression of FXR-mediated signaling, and induction of CYP7A1 expression and synthesis of BAs from CLR to replenish the BA pool. A decrease in intrahepatic CLR activates sterol regulatory element-binding transcription factor 2, which induces the expression of LDL receptor to enhance CLR uptake by the liver, with eventual reduction of serum CLR levels (Staels & Kuipers, 2007; Taoka et al., 2016). Ongoing clinical studies are aimed at determining the efficacy of selective agonists of FXR or other BA-recognizing receptors (e.g., TGR5) in the treatment of metabolic syndrome, which often includes hypercholesterolemia (Taoka et al., 2016). The ability of CDCA to inhibit the rate-limiting step in the synthesis of CLR led to the therapeutic use of this BA in the dissolution of CLR stones and in cerebrotendinous xanthomatosis, which features abnormally high levels of cholestanol in the brain and hypercholesterolemia (Broughton, 1994).

1.4 Evidence for Direct Interactions Between Cholesterol and Ion Channel Proteins

Biological actions of CLR and BAs were initially attributed to their capacity to self-associate and directly interact with other lipids. Cholesterol insertion into lipid membranes was early recognized to introduce the liquid-order (i.o.) phase, which is vital for mammalian membranes, modify a wide variety of physicochemical properties of the lipid bilayer, and lead to the formation of vertical lipid domains, also enriched in sphingolipids, which clusters protein receptors and other signaling molecules (Brown & London, 1998).

Work of several decades, however, has clearly documented that CLR can directly bind to proteins and, in some cases, modulate their function. To begin with, CLR is specifically recognized by key enzymes that participate in the steroid synthesis or CLR catabolism, such as the CLR side chain cleaving enzyme CYP11A1, which enables CLR conversion to pregnenolone, and cytochrome P450 46A1, the principal CLR hydroxylase in brain

(Mast et al., 2008; Pikuleva, Mackman, Kagawa, Waterman, & Ortiz de Montellano, 1995). Moreover, CLR binds to plasma carriers or apolipoproteins, including apolipoprotein B and apolipoprotein A1, which contribute to carry CLR in plasma as LDL and HDL, respectively (Clay & Barter, 1996; Yu, Zhang, & Xu, 2015). Data from biochemical binding, structural biology, high-resolution imaging, and computational dynamics, however, have provided evidence that the CLR molecule may be directly recognized by target proteins and participate in signaling process during physiology or disease. Some widely recognized examples are beta-cryptogein, HIV-1 envelope protein gp41, steroid acute regulatory (StARD) proteins, Niemann-Pick C (NPC) 1, NPC2, and retinoic acid-related orphan, sigma, translocator protein (TSPO), or peripheral benzodiazepine receptor, cannabinoid receptor (CB1), galanin, and opioid delta receptors (Davies & Ioannou, 2000; Kallen, Schlaeppi, Bitsch, Delhon, & Fournier, 2004; Kwon et al., 2009; Lascombe et al., 2002; Li, Liu, Valls, Hiser, & Ferguson-Miller, 2015; Li, Yao, Degenhardt, Teper, & Papadopoulos, 2001; Oddi et al., 2011; Palmer, Mahen, Schnell, Djamgoz, & Aydar, 2007; Pang, Graziano, & Wang, 1999; Schroeder, 2010; Ujčíková et al., 2014; van der Kant, Zondervan, Janssen, & Neefjes, 2013; Xu, Benoff, Liou, Lobel, & Stock, 2007).

Complex protein ion channels that fully expand the membrane bilayer are not exception to direct CLR–protein recognition; the techniques listed in the previous paragraph have been used to document direct recognition of CLR molecules by a heterogeneous set of nonconserved ion channel proteins that often differ not only in primary and secondary structure but also in topology and quaternary structure: nicotinic acetylcholine receptor (nAch), NMDA, $GABA_A$, inwardly rectifying K^+ channels (Kir), and BK channels themselves (Barrantes, 2010; Hénin, Salari, Murlidaran, & Brannigan, 2014; Korinek et al., 2015; Levitan, Singh, & Rosenhouse-Dantsker, 2014).

1.5 Evidence for Direct Interactions Between Bile Acids and Ion Channel Proteins

The detergent properties of BAs were considered responsible for both dietary fat emulsification and antibacterial actions that regulate gut microbiota (see above). Moreover, BA–bilayer lipid interactions have been traditionally considered to explain BA modulation of ion channel function: thus, BA insertion into membranes introduces positive monolayer curvature and changes in the bilayer physical properties with eventual change in protein function (e.g., Dopico, Walsh, & Singer, 2002).

Evidence for direct protein–BA interactions is revealed all along the BA metabolic pathway: (1) primary BAs are selectively conjugated by BA–CoA synthetase and BA–CoA amino acid N-acyltransferase (Falany, Johnson, Barnes, & Diasio, 1994; Kwakye, Barnes, & Diasio, 1993); (2) the whole enterohepatic circulation of BAs involves a wide variety of protein transporters including the Na^+-taurocholate cotransporting polypeptide and BA export pump (also known as ABCB11) expressed in the sinusoidal and canalicular hepatocyte membranes, respectively, and the apical Na^+-dependent BA transporter and heterodimeric organic solute transporters expressed in the apical brush border and basolateral membranes, respectively, of the ileal enterocyte (Boyer, 2013; Dawson & Karpen, 2015); (3) deconjugation and dihydroxylation of BA in the ileum lumen are catalyzed by specific enzymes of the microbial flora (Inagaki et al., 2006); (4) the highly toxic LCA, involved in bowel inflammation and colon carcinogenesis, is efficiently metabolized by CYP3A enzymes so only LCA trace amounts reach the liver under normal conditions (Copple & Li, 2016; Degirolamo, Modica, Palasciano, & Moschetta, 2011); and (5) even BA's antimicrobial effects in the gut have been linked to steroid interaction with FXR and eventual induction of transcription of antimicrobial agents (e.g., iNOS and IL-18) (Inagaki et al., 2006).

As mentioned above, studies over the past two decades further revealed that BAs may function as signaling molecules through a variety of specific BA-recognizing receptors. These include FXR, vitamin D receptor (VDR), pregnane X receptor (PXR), and two members of the G protein-coupled receptor superfamily: TGR5 and sphingosine-1-phosphate receptor 2. These receptors are expressed in tissues within *and beyond the enterohepatic circulation*, where they lead to systemic actions of BAs. For example, TGR5 is expressed in not only sinusoidal endothelial cells, bile duct epithelial cells, Kupffer cells, enteroendocrine cells, and enteric neurons but also adipocytes, myocytes, immune cells, macrophages, astrocytes, and brain neurons (McMillin & DeMorrow, 2016; Pols, Noriega, Nomura, Auwerx, & Schoonjans, 2011). In turn, while having the highest expression levels in liver and ileum, FXR is robustly expressed in the kidney and, at lower levels, in heart, ovary, thymus, eye, spleen, and testes (Lefebvre, Cariou, Lien, Kuipers, & Staels, 2009; Teodoro et al., 2011), vascular endothelium (He et al., 2006), and smooth muscle cells of coronary arteries and aorta (Bishop-Bailey et al., 2004).

Bile acid protein receptors, nevertheless, are largely involved in regulating CLR and BA metabolism (see above). FXR in particular modulates CLR homeostasis through several interrelated mechanisms, which have been reviewed elsewhere (Copple & Li, 2016; Porez et al., 2012). Thus, FXR and TGR5 have emerged as promising targets for treating disorders associated with metabolic syndrome (Porez et al., 2012). In another example, activation of VDR by LCA and 3-keto-LCA in the intestine induces the BA metabolizing cytochrome p450 CYP3A4 that detoxifies BAs and protects the intestine from toxicity by these hydrophobic BAs (Copple & Li, 2016; Makishima et al., 2002). Collectively, BA receptors have been demonstrated to regulate lipid, glucose, and energy metabolism (Copple & Li, 2016; Forman et al., 1995; Li & Chiang, 2015; Makishima et al., 1999; Maruyama et al., 2002; Parks et al., 1999; Schaap, Trauner, & Jansen, 2014).

2. MODULATION OF BK CHANNELS BY CHOLESTEROL OR BILE ACIDS

2.1 Voltage- and Ca^{2+}-Gated, Large Conductance K^+ (BK) Channels

Recent comprehensive reviews on BK channel phenotypes and their structural bases, subunit composition, tissue distribution, and the genetic, epigenetic, posttranslational, and pharmacological regulation of these channels can be found elsewhere (Dopico et al., 2012b; Geng & Magleby, 2015; Hermann, Sitdikova, & Weiger, 2015; Hoshi & Heinemann, 2016; Kyle & Braun, 2014; Toro et al., 2014; Torres, Granados, & Latorre, 2014; Yang, Zhang, & Cui, 2015; Zhang & Yan, 2014). Succinctly, BK channels result from tetrameric association of channel-forming α (also termed slo or slo1) subunits, which are encoded by the *Slo1 or KCNMA1* in mammals (see reviews above) (Fig. 1C). *Slo1* splicing significantly contributes to BK structural and ionic current phenotype diversity (Fodor & Aldrich, 2009). However, all slo1 subunits share a common design and present significant homology with voltage-gated K^+ channels of the six transmembrane domain (TM6) family (K_V channels); S1–S4 contributes to voltage-sensing and S5–S6 conforms the activation gate-pore region (Latorre & Brauchi, 2006; Salkoff, Butler, Ferreira, Santi, & Wei, 2006: Lee & Cui, 2010). In addition to the TM6 core, slo1 subunits contain S0, which leads to an exofacial N-end, and a large cytosolic tail domain (CTD) (Wallner, Meera, & Toro, 1996; Wang & Sigworth, 2009; Yuan, Leonetti, Pico, Hsiung, & MacKinnon, 2010). The CTD includes two regulator of conductance of K^+ (RCK) domains. RCKs contain distinct sites for sensing Ca^{2+}, which allows BK channels to increase channel open probability (Po) in response to increase in Ca^{2+}_i within physiological levels (Bao & Cox, 2005; Wei, Solaro, Lingle, & Salkoff, 1994). The crystal structure of the CTD shows that the two RCKs from each slo1 subunit form an octameric Ca^{2+}-gating ring (Yang et al., 2015; Yuan et al., 2010). Ca^{2+}-binding, voltage sensor movement, and pore opening are conceptualized as independent equilibria that interact allosterically with each other (Horrigan & Aldrich, 2002; Latorre, Morera, & Zaelzer, 2010). The CTD also contains regions subject to phosphorylation/dephosphorylation (Braun, 2009; Schubert & Nelson, 2001; Tian, Coghill, MacDonald, Armstrong, & Shipston, 2003; Tian et al., 2001; Yan et al., 2008; Zhou et al., 2010) as well as defined sequences for specific interactions with physiological signals, such as carbon monoxide and heme (Hou, Xu, Heinemann, & Hoshi, 2008; Jaggar et al., 2005), caveolin-1 (Alioua et al., 2008), carbon monoxide (Hou, Heinemann, & Hoshi, 2009), small amphiphilic drugs such as ethanol (Bukiya, Kuntamallappanavar, et al., 2014), and membrane lipids, including phosphatidylinositol 4,5-bisphosphate (Vaithianathan et al., 2008) and CLR itself (see below).

In cell membranes from vertebrates, native BK channels usually result from the association of slo1 proteins with auxiliary proteins termed BK β subunits (Lu et al., 2006; Orio, Rojas, Ferreira, & Latorre, 2002). Four types of BK β subunits have been identified, each being encoded by a different gene: *KCNMB1–4* (Orio et al., 2002; Torres, Morera, Carvacho, & Latorre, 2007). All BK βs have two TM segments joined by an extracellular loop, plus two short intracellular N and C ends (Fig. 1C) (Orio et al., 2002). Association of BK β proteins with slo1 subunits leads to modification of channel ion current kinetics, pharmacology (Orio et al., 2002; Wang & Brenner, 2006) and regulation by lipids (Dopico & Bukiya, 2014), including BAs (see below). In the plasma membrane of nonexcitable LNCaP prostate cancer cells, slo1 subunits associate with the leucine-rich repeat-containing protein LRRC26, which has been termed BK γ subunit (Yan & Aldrich, 2010). The Ca^{2+}-sensitive slo1 channels are associated with many other membrane proteins, including ion channels that control Ca^{2+} signaling, such as voltage-gated Ca^{2+} (Ca_V) channels (Dai, Hall, & Hell, 2009; Fakler & Adelman, 2008; Kato, Tanaka, Ishii, Yin, & Sakuma, 2009; Marrion & Tavalin, 1998) and *N*-methyl-D-aspartate (NMDA) receptors (Isaacson & Murphy, 2001), as well as with chaperone proteins (e.g., AKAPs; Berkefeld, Fakler, & Schulte, 2010).

BK channels are widely distributed and play critical roles in the physiology of excitable and nonexcitable tissues. In neurons, BK channels regulate action potential (AP) repolarization, sharpening individual somatic APs (Bock & Stuart, 2016), and even dendritic spikes (Golding, Jung, Mickus, & Spruston, 1999). In spite of these generalizations, functional diversity and distinct regulatory roles of BK channels in specific CNS regions and peripheral tissues arise from different current phenotypes due to differential BK subunit expression, β subunits in particular (Lorenz, Heils, Kasper, & Sander, 2007; Orio et al., 2002; Pape, Munsch, & Budde, 2004; Shao, Halvorsrud, Borg-Graham, & Storm, 1999; Wang, Ding, Xia, & Lingle, 2002; Xia, Ding, & Lingle, 1999). Other processes in which BK play a fundamental role are regulation of neurotransmitter and hormone release (Yazejian, Sun, & Grinnell, 2000), suppression of high-frequency cycles in the auditory system (Fettiplace & Fuchs, 1999; Ramanathan & Fuchs, 2002), relaxation of vascular (Brayden & Nelson, 1992; Brenner, Jegla, Wickenden, Liu, & Aldrich, 2000; Jaggar, Porter, Lederer, & Nelson, 2000), and nonvascular smooth muscle (Choudhury, Garg, Singh, & Mishra, 2011; France, Bhattarai, Galligan, & Xu, 2012; Heppner, Herrera, Bonev, Hill-Eubanks, & Nelson, 2003; Kotlikoff, 1993; Semenov, Wang, Herlihy, & Brenner, 2006), flow-induced K^+ secretion in the nephron distal segment (Grimm, Foutz, Brenner, & Sansom, 2007), K^+ secretion in the distal colon (Sausbier et al., 2006), and cancer proliferation and metastasis (Bloch et al., 2007; Khaitan et al., 2009; Sontheimer, 2008).

2.2 Cholesterol and BK Channel-Mediated Currents

Manipulation of CLR content at the organ level can be achieved by a high CLR diet which effectively evokes hypercholesterolemia in animal models, including rat and mouse (reviewed by Bukiya & Rosnehouse-Dantsker, 2015). Rabbits, however, represent one of the species that is most susceptible to dietary manipulation with CLR level: a high CLR diet may result in up to 30-fold increase in blood CLR levels (Bukiya & Rosnehouse-Dantsker, 2015). Considering that both CLR and BK channels play a critical role in cardiovascular physiology and disease, it comes at no surprise that the majority of work devoted to studying the effects of high CLR diet on BK currents has been conducted in vascular preparations. An early study reported that rabbits subjected to a high CLR diet presented an increased contribution of charybdotoxin-sensitive currents (likely mediated by BK channel activity) to in vitro dilation of deendothelialized carotid artery evoked by nitric oxide ($NO^•$) (Najibi & Cohen, 1995). Also in rabbits, a high CLR diet has been shown to reduce BK channel-mediated, in vivo vasodilation in the hindlimb evoked by acetylcholine or bradykinin, two $NO^•$-dependent vasodilators (Jeremy & McCarron, 2000). The discrepancy between these studies could be explained by tissue-specific (carotid artery vs hindlimb perfusion) actions of high CLR diet on vascular BK currents or the loss of endothelial modulation in the in vitro preparation. Interestingly, the $NO^•$-independent component of BK current was actually increased in rabbits subjected to high CLR diet (Jeremy & McCarron, 2000). In rats, a high CLR diet results in decreased in vitro constriction of deendothelialized, pressurized middle cerebral arteries (MCA) evoked by the highly selective BK channel blocker paxilline (Bukiya, Dopico, Leffler, & Fedinec, 2014). Thus, both studies seem to indicate that a high CLR diet reduces the contribution of BK currents to myogenic tone and artery diameter. Constriction of rat MCA secondary to BK channel inhibition in response to intoxicating levels of ethanol (50 mM) is also significantly decreased by the high CLR diet. Moreover, high CLR diet-induced attenuation of ethanol action was also detected in arteries with intact endothelium and in cerebral arterioles in vivo (Bukiya, Dopico, et al., 2014). In synthesis, the effect of high CLR diet on BK current-mediated vascular physiology and pharmacology is rather complex: both $NO^•$-dependent and $NO^•$-independent mechanisms may play a role, with regional (i.e., arterial territory) and species differences likely playing a part.

As alternative to high CLR diets, manipulation of CLR content at the organ/tissue/cell levels can be achieved by sample incubation with high-density lipoproteins (HDL) and LDLs, which renders CLR depletion and enrichment, respectively (Bisen et al., 2016; Wu, Wang, Deng, Sun, & Li, 2013). The efficiency of these approaches, however, relies on the status of reverse CLR transport (e.g. CLR efflux) and LDL endocytosis (Maxfield, 2014; Rosenson et al., 2016). Cholesterol-lowering can be also achieved by tissue treatment with statins, such as lovastatin (mevinolin), a highly potent competitive inhibitor of hydroxymethylglutaryl-coenzyme A reductase, a key enzyme that enables cholesterol synthesis (Alberts et al., 1980). A more straightforward approach to manipulate CLR levels, however, results from incubation of the biological sample in solutions containing the CLR carrier methyl-beta-cyclodextrin (MβCD) per se and MβCD that had been saturated with CLR, which renders CLR depletion and enrichment, respectively. Moreover, manipulation of MβCD/MβCD–CLR concentration and time of exposure enables fair control on the degree of change in CLR content in the biological sample (Bukiya, Vaithianathan, Kuntamallappanavar, Asuncion-Chin, & Dopico, 2011; Zidovetzki & Levitan, 2007). This technique has been used to explore CLR modulation of BK channels and associated physiology: MβCD-mediated CLR depletion resulted in inhibition of phasic, but not tonic contractions in rat ureters. Normal contractile behavior could be restored by CLR replenishment or pharmacological inhibition of BK channels (Babiychuk et al., 2004), leading to the suggestion that CLR depletion with MβCD evokes BK channel activation.

Direct modulation of native BK currents by treatments that modify membrane CLR levels was first reported in cultured myocytes from human and rabbit aortas: tissue incubation in CLR-enriching medium resulted in approximately a twofold decrease in BK open channel probability (Po), while CLR depletion with mevinolin led to an opposite effect (Bolotina, Omelyanenko, Heyes, Ryan, & Bregestovski, 1989; Bregestovski & Bolotina, 1989). Similar findings have been reported in human coronary artery smooth muscle cells in culture (Wu et al., 2013) and melanoma IGR39 cells (Tajima, Itokazu, Korpi, Somerharju, & Käkelä, 2011). These findings are mirrored by reports that described BK channel activation upon CLR depletion in a wide array of species and tissues: mouse colonic epithelial cells, mouse and rat cerebrovascular myocytes, rat coronary artery myocytes, clonal rat pituitary GH3 cells, bovine aortic endothelial cells, human myometrial myocytes in culture, and a glioma cell line (reviewed in Dopico et al., 2012b). Noteworthy, studies using CLR depletion to study BK channel function outnumber those that used CLR enrichment. This fact is probably due to the high level of CLR already present in the plasma membranes of native cells: it is estimated that CLR accounts for 30–50mol% of membrane lipids (Gimpl et al., 1997). Within this concentration range, CLR induced BK channel inhibition reaches a plateau (Fig. 2A and B), making it difficult to observe further inhibition with CLR enrichment. Whether this treatment leads to some adaptive response in BK channels to very high CLR levels remains to be determined.

In contrast to previously described findings, BK channel *inhibition* upon CLR depletion has been reported in uterine myocytes freshly isolated from rats (Shmygol, Noble, & Wray, 2007) and in chicken hair cells (Purcell, Liu, Thomas, & Duncan, 2011). In agreement with these reports, BK channel activity in atherosclerotic plaques (that are presumably rich in CLR) was increased when compared to nonatherosclerotic sections of human coronary

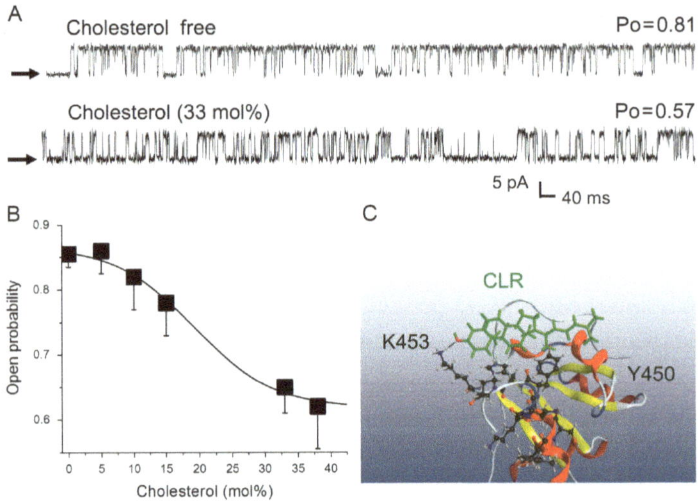

Fig. 2 Cholesterol induces slo1 channel inhibition. (A) Records showing cholesterol-induced decrease in the activity (Po) of cbv1 channels reconstituted into POPE:POPS (3:1, w/w) planar bilayers. Channel openings are shown as upward deflections; *arrows* point at the baseline (all channels closed). (B) Concentration response curve for cholesterol-induced BK channel inhibition of reconstituted cbv1 channels (Bukiya, Vaithianathan, Toro, & Dopico, 2008). (C) Molecular dynamics snapshot showing the interaction of cholesterol with Lys and Tyr residues (K453 and Y450) within the cbv1 CRAC4 motif (Singh et al., 2012). Cholesterol is depicted in *green*; hydrogen bond is shown as a *gray line* between K453 and the cholesterol's hydroxyl group. Part of the protein molecule is hidden to improve visual clarity.

arteries (Wiecha et al., 1997). Bases for discrepancy between studies remain undetermined. Several nonmutually exclusive (and possibly interrelated) possibilities should be briefly considered (for an expanded discussion of these topics, see Dopico et al., 2012b). First, CLR may sort BK channel subunits into membrane rafts (Lam, Shaw, & Duszyk, 2004; Purcell et al., 2011). Cholesterol enrichment of human melanoma IGR39 cells has been shown to increase the proportion of BK channels in rafts, which decreases their activity (Tajima et al., 2011). Conversely, CLR depletion resulted in a location shift of BK channels from detergent-insoluble (compatible with "rafts") to detergent-soluble domains (Lam et al., 2004). This shift was accompanied by an increase in BK channel function as caveolin-1, a protein highly abundant in most of rafts, interacted with BK channels and thus exerted a negative regulatory effect on channel function (Wang et al., 2005). More often than not, proteins anchored to lipid rafts are associated with cytoskeletal elements.

Thus, differential raft formation and associated BK protein sorting across different cell types could contribute to the differential CLR effect on BK channel function reported in different preparations. Second, these differences may obey local differences in the regulation of BK activity by closely associated ion channel and receptors. Manipulation with CLR level could decouple BK channels from voltage-gated Ca^{2+} channels, ryanodine, IP3, or other receptors that may provide Ca^{2+} fluxes for BK channel activation (Löhn et al., 2000; Purcell et al., 2011; Weaver, Olsen, McFerrin, & Sontheimer, 2007). A third mechanism recognizes perturbation of BK channel mobility and communication with protein partners secondary to cytoskeletal disruption; indeed, disruption of actin in the cytoskeleton has been associated with a great increase in BK Po (Brainard, Miller, Martens, & England, 2005). This mechanism may actually connect the two aforementioned mechanisms, as cytoskeletal proteins are inextricably linked to raft proteins such as caveolin-1. A fourth mechanism of regional differences in the CLR response of BK channels may recognize variations in the BK channel subunit composition and/or stoichiometry. Indeed, CLR application failed to modulate BK channels resulting from β2 or β4 subunit assembly with hslo1 (cloned from human substantia nigra) and coexpressed in HEK-293 cells (King et al., 2006). In the same expression system, β1 subunit assembly with slo1 resulted in channel inhibition upon CLR enrichment with either MβCD–CLR complex or LDL whereas homomeric slo1 channels were inhibited upon CLR depletion (Wu et al., 2013). Thus, there seems to be dissociation between CLR effects on homomeric slo1 channels vs heteromeric slo1 + different β subunits. Conceivably, this dissociation is the result of CLR regulation of BK subunit trafficking from the plasma membrane to cell cytosol and vice versa. Following coexpression of slo1 with β1 subunits in HEK-293 cells, CLR enrichment via MβCD–CLR treatment downregulated slo1 expression in the plasma membrane. Noteworthy, lactacystin and bafilomycin A2 (e.g., inhibitors of lysosomal transport) rescued plasma levels of slo1 subunit in presence of the MβCD–CLR complex (Wu et al., 2013). Whether similar CLR-driven mechanisms apply to slo1 or other BK subunits in native channels remain to be established. Last but not least, regional differences in the modifications of the membrane physicochemical properties introduced by CLR insertion in the lipid bilayer (see below "Lipid-mediated effects of cholesterol on BK channels") could also contribute to the different regional responses of BK channels to manipulation of CLR levels.

In contrast to the variability of outcomes from work on CLR levels and function of BK channels that include regulatory β subunits, BK channel homomers seem to be consistently activated and inhibited by CLR depletion and enrichment, respectively. Indeed, current resulting from homomeric BK channels expressed in HEK-293 was downregulated by CLR enrichment (Wu et al., 2013). Moreover, channel inhibition by CLR was reported when hslo1 channels (cloned from human brain) were reconstituted into planar, artificial lipid bilayers, a system that allows not only titration of CLR levels in the lipid mixture that is used to form the bilayer but also evaluation of steroid action in an extremely simplified proteolipid system. This study documented for the first time that CLR inhibited hslo1 channel activity in a bilayer exclusively composed of 1-palmitoyl-2-oleoyl-*sn*-glycero-3-phosphoethanolamine (POPE) (Crowley, Treistman, & Dopico, 2003). As found with hslo1, CLR inhibition of cbv1 channels (cloned from rat cerebral artery myocytes, Liu, Liu, Asuncion-Chin, & Dopico, 2005) in planar binary bilayers (i.e., made of two phospholipid species) was concentration-dependent up to 40 mol% CLR (Fig. 2A and B) (Bukiya, McMillan, Parrill, & Dopico, 2008). Overall, CLR inhibition of slo1 or native BK channel function in artificial or native membranes occurs in absence of major changes in the channel's ion permeation properties (Dopico et al., 2012b). Thus, disregarding the specific mechanism, CLR action on each functional channel seems largely limited to that of a gating modifier leading to reduced steady-state activity (Po). As corollary of CLR-induced modification of equilibria among the BK channel functional states, the presence of the steroid leads to modification of BK channel responses to other modulators and drugs, alcohol in particular (Yuan, Chen, Covey, Johnston, & Treistman, 2011; Bukiya, Dopico, et al., 2014; for a review on the topic, see Dopico et al., 2012b).

2.3 Mechanisms Underlying Cholesterol Modulation of BK Channel Activity

The mechanisms underlying CLR modulation of ion channel function in general and of BK channel activity in particular have been categorized in two major groups: (i) lipid-mediated effects: according to this view, CLR insertion into the bilayer (usually authors refer to the bulk bilayer, not the bilayer–BK channel interface) leads to modifications in one or several physicochemical properties of the *bulk lipid bilayer* eventually leading to change in BK Po and (ii) protein-mediated effects: in this case, change in BK Po is secondary to initial and direct contact (binding) between CLR and channel proteins. Whether these contacts may lead to changes in lipid bilayer properties, which in turn, lead to the actual change in Po has been usually disregarded. The two types of mechanisms are briefly summarized as follows.

2.3.1 Mechanisms Driven by Changes in the Physicochemical Properties of the Lipid Bilayer

Early interpretations of CLR-induced inhibition of BK channel activity have been attributed to the ability of CLR to modify critical physical features of the lipid membranes, including those that make CLR indispensable for maintaining membrane organization in living cells (Quinn & Wolf, 2009; Sonnino & Prinetti, 2013). Cholesterol molecules insert into membranes with their steroid hydrophobic area being embedded deep into the bilayer hydrophobic core and their hydroxyl group facing the lipid–aqueous interface, usually at the "alcohol" (glycerol/sphingol)–polar head–group interface (Bittman, 1997; Ohvo-Rekilä, Ramstedt, Leppimäki, & Slotte, 2002). This insertion introduces the liquid-ordered (l.o.) lipid phase, which prevails in most native membranes and is essential for life as we know it; this phase allows for both translation of signaling proteins at ease (hindered in the solid-ordered phase) and makes ion transbilayer leak difficult (facilitated in the liquid-disordered phase) (Clarke, Heron, Seddon, & Law, 2006; Quinn & Wolf, 2009). Cholesterol insertion is associated with the so-called condensation effect, i.e., the area of the membrane per mole of phospholipid is reduced when compared to that in the absence of CLR. Cholesterol insertion into bilayers is also associated with bilayer thickening, increased order parameters, increased stiffness, and change in lateral pressure profile, any of these modifications per se being proposed to modify ion channel behavior (Cantor, 1999; de Meyer & Smit, 2009; Kaiser et al., 2009; Lundbaek, Andersen, Werge, & Nielsen, 2003; Quinn & Wolf, 2009). Cholesterol-induced modulation of membrane "fluidity" (measured as the rotational diffusion coefficient of 1,6-diphenyl-1,3,5-hexatriene) was detected along with CLR regulation of Ca^{2+}-dependent K^+ currents in cultured smooth muscle rabbit aortic cells; while CLR depletion increased membrane "fluidity" and Po, CLR enrichment had opposite effect on both parameters (Bolotina et al., 1989). The idea that CLR-driven changes in membrane physical properties may underlie CLR-induced inhibition of BK channels in artificial bilayers was further developed by Chang, Reitstetter, Mason, and Gruener (1995). In this work, BK channels from rat brain homogenates were incorporated into POPE:1-palmitoyl-2-oleoyl-*sn*-glycero-3-phosphoserine (POPS) bilayers, with increasing amounts of CLR (up to 35% of total lipid weight) progressively decreasing

Po. This decrease was paralleled by progressive reduction in channel mean open time and increase in mean closed time. Similar effect of CLR on BK channel dwell times in artificial POPE:POPS bilayers has been reported with recombinant proteins: hslo1 (Crowley et al., 2003) and cbv1 (Bukiya, Belani, Rychnovsky, & Dopico, 2011). It has been proposed that the progressive decrease in mean open time reflects a CLR-driven increase in bilayer lateral stress secondary to CLR insertion as a wedge between phospholipid molecules (Chang et al., 1995). This action and its connection to CLR's condensing effect are discussed elsewhere (Dopico et al., 2012b).

More recent work, however, disputed a significant involvement of increased lateral stress in CLR-induced inhibition of BK channel activity: enantiomeric CLR (*ent*-CLR), i.e., the mirror image of the natural CLR molecule failed to inhibit cbv1 channels in POPE:POPS bilayers where CLR was effective (Bukiya, Belani, et al., 2011). Natural and *ent*-CLR only differ slightly in their effect on the shapes of ordered domains and have similar abilities to interact with sphingomyelins, phosphatidylcholine, and to induce lipid condensation in a monolayer system (Alakoskela et al., 2008; Westover, Covey, Brockman, Brown, & Pike, 2003). Thus, any drastic difference in biological effects of CLR vs *ent*-CLR has been widely used as a good predictor of the existence of specific CLR-recognition site(s) in a protein (Westover et al., 2003). In a structure–activity relationship (SAR) study, Bukiya, Belani, et al. (2011) used not only *ent*-CLR but also a variety of steroids that differed in capability to alter bilayer physicochemical properties (e.g., lipid condensation). From this detailed SAR study, it was hypothesized that the most likely explanation for CLR action on BK channel-forming properties would be the existence of specific CLR-recognition domains in the slo1 subunit, a hypothesis proven correct in a later study (see next subheading).

2.3.2 Direct Cholesterol–BK Protein Interactions

Since BK regulatory β subunits are not needed for CLR-induced BK channel inhibition (Bukiya, Singh, Parrill, & Dopico, 2011; Bukiya, Vaithianathan, et al., 2008; Chang et al., 1995; Crowley et al., 2003), the efforts to map CLR-sensing sites in BK proteins were focused on the channel-forming slo1 subunit. Considering that truncated (CTD-free) BK channel homomers were insensitive to CLR (Singh et al., 2012), CLR-sensing motifs were expected to be located within CTD. The CTD sequence includes seven cholesterol-recognition amino acid consensus (CRAC) motifs, e.g., motifs with the following sequence:

(Leu/Val)–X_{1-5}–(Tyr)–X_{1-5}–(Lys/Arg), where X represents any amino acid (Fig. 1C), the positively charged residue enables hydrogen bonding with CLR's hydroxyl and the central Tyr is key for CLR to interact with the overall domain (Epand, 2006; Li et al., 2001). Indeed, membrane CLR sequestration by CRAC motifs located in protein transmembrane areas or juxtamembrane (i.e., adjacent to the bilayer) segments has been well documented (Epand, 2006; Fantini & Barrantes, 2013). Consistently, cumulative Tyr to Phe substitutions within the BK CTD CRAC domains progressively blunted CLR-induced BK channel inhibition (Singh et al., 2012). The contribution of CRAC4 to CLR action on BK channels was examined in detail, as CRAC4 is the BK CTD CRAC motif nearest to the bilayer (Lee & Cui, 2010; Yuan et al., 2010): computational molecular dynamics revealed hydrogen bond formation between CLR hydroxyl group and Lys453, with Tyr450 retaining the steroid polycyclic ring system within CRAC4 (Fig. 2), which obeys the overall design of CRAC–CLR model docking.

It should be noted that with the exception of CRAC7, which has a juxtamembrane location, the remaining CRACs in the slo1 CTD are expected to reside away from BK protein transmembrane segments. Cholesterol sequestering ability of CRAC motifs that are present in the sequence of nontransmembrane protein segments has been questioned repeatedly (Fantini & Barrantes, 2013; Palmer, 2004). Whether the CLR-sensing CRAC motifs in the BK channel CTD serve as true binding sites for CLR or link the CLR binding event onto another protein sites with changes in channel activity, remains to be explored. Thus, binding or imaging studies showing actual binding of CLR to slo1 CTD CRACs are needed to answer this question.

2.4 Bile Acids and BK Channel-Mediated Currents

Bile acid concentrations in the aqueous phase (e.g., serum, experimental saline solution) reported to evoke vasodilation range from sub-μM to 100 μM (Bomzon & Ljubuncic, 1995; Dopico, Bukiya, & Singh, 2012a). These levels are well below the critical micellar concentration (CMC) of physiological BAs, which suggests that BA-induced vasodilation involves BA monomers in solution. Conceivably, BA monomers could interact with membrane lipids and thus form mixed micelles leading to modification of ion channel function or, alternatively, could target discrete cell targets (whether distinct lipid species or proteins associated with the functional ion channel complex) to evoke vasodilation. The first report of BA activation of BK channels dates back to 1994, in which LCA and other BAs (100 μM) were reported to activate BK channels in rabbit myocytes freshly isolated from the superior mesenteric artery, an artery that contributes to peripheral resistance (Dopico, Kirber, Singer, & Walsh, 1994). Initial data were extended to a wide BA concentration range and to myocytes isolated from the nonresistive circulation (main pulmonary artery), and nonvascular SMC isolated from the gallbladder. Collectively, data demonstrated that naturally occurring BAs (3–1000 μM in the aqueous phase) reversibly activated BK channel activity, an action that could explain or, at least, contribute to, BA-induced SM relaxation and even vasodilation (Dopico et al., 2002). LCA results on SM BK channels were extended to SM cells isolated from resistance size (120–250 μm), rat cerebral arteries that control cerebral blood flow. In this system, full concentration–responses curves to LCA activation of BK channels (i.e., increase in Po) rendered $EC_{50} \asymp 45\,\mu M$ and $E_{max} = 150–300\,\mu M$, the latter raising BK activity to ~300% of pre-LCA values in a reversible fashion (Bukiya et al., 2007). This quantitative evaluation of LCA action on BK channels clearly indicated that the BA effectively increases channel activity at concentrations well below CMC. The fact that LCA activates BK currents at several tens of μM is consistent with the low affinity of protein receptors for BAs: LCA activates human VDR with an EC_{50} of ~8 μM (Makishima et al., 2002). In a competitive binding assay for LCA on PXR, the steroid showed an IC_{50} of ~10 μM (Staudinger, Liu, Madan, Habeebu, & Klaassen, 2001). Moreover, the EC_{50} for BA action of the human bile acid-sensitive ion channel (BASIC) is ~2 mM (Lefèvre et al., 2014).

The rank efficacy order for BA activation of native BK channels in rabbit mesentery SM cells (LCA ≫ DCA ≫ CA; Dopico et al., 2012b), however, does not fully match the efficacies/potency for BA interaction with known BA receptors: LCA ≍ DCA > CA for rat FXR activation (Makishima et al., 1999); LCA ≥ DCA > CDCA > CA for activation of human TGR5 expressed in Chinese hamster ovary cells (Copple & Li, 2016; Kawamata et al., 2003). It is noteworthy that most of our electrophysiological recordings of BA action on BK channels were conducted in isolated membrane patches, no less than 10 min after patch excision from the SM cell under the continuous perfusion of solutions without nucleotides (e.g., ATP) or other "regenerating" systems, and under conditions of heavy Ca^{2+} buffering (Bukiya et al., 2007; Dopico et al., 2002). These conditions ruled out involvement of freely diffusible cytosolic signals secondary to possible BA receptor activation, endothelial activators of SM BK channels (e.g., $NO^{\bullet}$), intact intracellular organelles, and BA metabolism by the cell. Such conditions make it difficult for BA activation of BK channels to involve TRG5 receptors: these receptors activate cystic fibrosis transmembrane conductance regulator (the ion channel known as CFTR) in biliary epithelial cells (Keitel et al., 2009) and ATP-gated K^+ channels in gallbladder SM cells due to increase in freely diffusible cAMP (Li et al., 2011) secondary to TRG5–BA direct interaction. Moreover, LCA activation of SM BK channels was evoked following reconstitution of recombinant α (cbv1, that is, an isoform cloned from SM myocytes freshly isolated from rat cerebral arteries, Liu et al., 2005) + smooth muscle-abundant β1 subunits into planar 3:1 POPE:POPS (w/w) bilayers (Bukiya, Vaithianathan, et al., 2008), indicating that complex lipid domains and intact membrane cytoarchitecture are not necessary for LCA to activate BK channels. Collectively, these studies led authors to postulate a *direct* interaction between LCA and a discrete target(s) in the membrane, these targets being limited to the BK α or β1 subunits and their immediate proteolipid microenvironment.

2.5 Mechanisms Underlying Bile Acid Modulation of BK Channel Activity: Evidence for Direct, Bile Acid–BK Subunit-Specific Recognition

Results obtained with naturally occurring BAs and synthetic analogs documented that BA-induced increase in BK Po was an exponential function of the overall hydrophobicity of the BA, with LCA being the most effective among the cholanes tested (Bukiya, McMillan, et al., 2008; Dopico et al., 2002). These findings pointed at a hydrophobic microdomain in the lipid or protein components of the system which LCA interacts with to increase channel activity. The unlikely possibility that BA could alter BK function by perturbing membrane lipids is fully discussed elsewhere (Dopico et al., 2002). BK channel activation by BAs, however, required that at least one hydroxyl (e.g., in LCA at C3) was present (Dopico et al., 2002), which suggested that a polar partner for hydroxyl bonding was also required. Moreover, a structural–activity relationship for LCA and related monohydroxylated BA revealed that a hydroxyl group facing the α plane of the BA molecule (as found for LCA's hydroxyl at C3) and a *cis* configuration in the joint between steroidal rings A and B are both needed for maximal activation of BK channels (Bukiya, McMillan, et al., 2008), two features that, together with overall hydrophobicity, explain the high efficacy of LCA and the need for a bean-shaped, rigid amphiphile to active BK channels (Bukiya, McMillan, et al., 2008; Dopico et al., 2002).

In parallel, data obtained after heterologous expression of BK channel-forming subunits cloned from rat cerebral artery myocytes (Liu et al., 2005) and expressed in heterologous systems in absence and presence of BK β1–4 subunits demonstrated that LCA-induced BK channel activation required β1 but not other BK β subunits (Bukiya, Vaithianathan, Toro, & Dopico, 2009; Bukiya et al., 2007). Moreover, homomeric cbv1 channels reconstituted into planar 3:1 POPE:POPS (w/w) bilayers consistently remain insensitive to 33 mol% of LCA in the lipid mixture. This LCA content is close to E_{max} for LCA activation of BK $\alpha + \beta 1$ heteromers (Bukiya et al., 2007). The strict chemical requirements for monohydroxylated BAs to activate BK channels (see paragraph above) and results from recombinant BK proteins strongly suggested the possibility of a direct interaction between LCA and the BK β1 subunit itself. It is noteworthy that parallel studies on rat vascular SM cells (which highly express BK β1; Behrens et al., 2000; Brenner, Jegla, et al., 2000, Brenner, Peréz, et al., 2000) vs skeletal muscle fibers (which poorly express BK β1; Behrens et al., 2000; Brenner, Jegla, et al., 2000) matched the results from recombinant BK proteins, suggesting a significant role for LCA–BK β1 interaction and eventual BK channel activation in the intact cell.

The use of chimeric β proteins formed by swapping extracellular (EC) loop, TM, and intracellular (IC) domains between β1, which emboldens BK channels with sensitivity to <100 μM LCA, and β4, which does not, identified TM2 as the β1 region that confers LCA sensitivity to BK channels (Bukiya, McMillan, et al., 2008). Based on this finding and the LCA structure–activity relationship (SAR) studies described above, computational modeling for LCA docking onto TM2 was developed and validated by electrophysiological studies and Ala-scanning of the BK β1 subunit: the LCA C3 hydroxyl hydrogen bonds with Thr169, a residue that is found in BK β1 but not in BK β2–4, while Leu172 and Leu173 interact via van der Waals forces with the hydrophobic surface of the LCA concave hemisphere (Fig. 3) (Bukiya, McMillan, et al., 2008; Bukiya, Singh, et al., 2011).

The contribution of smooth muscle β1-containing BK channels to LCA-induced vasodilation has been determined using rodent models both in vitro and in vivo. Data clearly demonstrate that genetic ablation of *KCNBM1* in mouse blunts artery dilation in response to LCA (Bukiya et al., 2007). Data from rat cerebral arteries demonstrate that LCA-induced dilation is identical whether tested in intact or deendothelized arteries, ruling out involvement of $NO^{\bullet}$ or other endothelial factors in LCA action on BK channels. In both

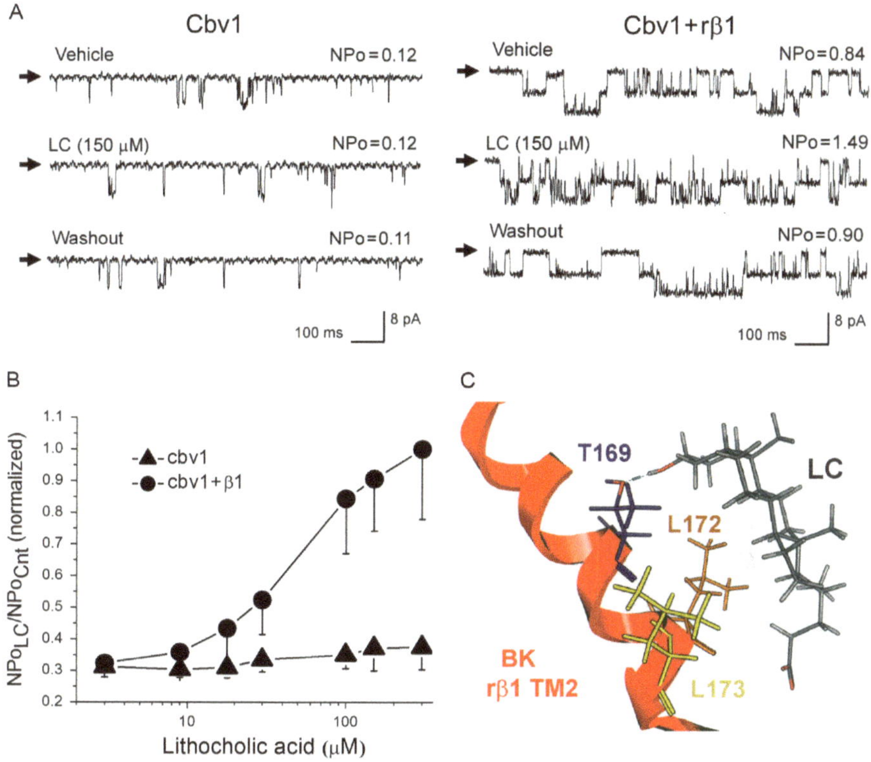

Fig. 3 Bile acid-induced BK channel activation. (A) Records showing cbv1 channel activity before, during, and after application of 150 μM (lithocholic acid) LCA in cell-free patches from *Xenopus* oocytes. Channel openings are shown as downward deflections; *arrows* point at the baseline (all channels closed) (Bukiya, Singh, et al., 2011). (B) Concentration response curves for LCA-induced BK channel inhibition obtained from homomeric cbv1 vs heteromeric cbv1 + β1 channels in cell-free patches from *Xenopus* oocytes (Bukiya et al., 2007). (C) Molecular dynamics snapshot showing LCA interaction with BK β1 TM2 domain. Hydrogen bond between TM2 T169 and the C3-hydroxyl of LCA is indicated as a *blue dashed line* (Bukiya, Singh, et al., 2011).

mouse and rat, 45 μM LCA evokes a ~9% increase in artery diameter which, according to Poiseuille's law, would render an increase in blood flow of ~41%.

On a practical note, LCA may constitute a unique pharmacological tool to selectively target β1-containing BK channels whether in heterologous expression systems, planar lipid bilayers, intact cells, or whole tissues. However, the many other biological actions of LCA (e.g., transcriptional activator; see above) would preclude the use of this BA as tool for chronic vasodilating therapy. The LCA site in BK β1 TM2, however, accommodates nonsteroidal compounds, whether endogenous, such as leukotriene B4 (Bukiya, McMillan, et al., 2014) or exogenous, such as sodium 3-hydroxyolean-12-en-30-oate (HENA) (Bukiya et al., 2013). Recognition of the latter ligand by the LCA site renders both BK channel activation and endothelium independent, artery dilation (Bukiya et al., 2013). Because (1) BK β1 is abundant in all types of smooth muscle and highly conserved among species, (2) the LCA site was identified using rat and human β1, the BK β1 LCA-sensing site represents an attractive target for nonsteroidal, LCA-structural analogs (Bukiya, Patil, Li, Miller, & Dopico, 2012; Bukiya et al., 2013; McMillan et al., 2014; Patil, Bukiya, Li, Dopico, & Miller, 2008) that could serve not only as vasodilators but also to counteract SM enhanced contraction in nonvascular tissues, as found in asthma, erectile, bladder, and uterine dysfunction.

2.6 On the Possibility that BK Protein Sites that Recognize Cholesterol or Lithocholate are Common to Both Ligands

As described above, CLR and LCA molecules share a considerable number of chemical features: an amphiphilic character (albeit CLR is much more hydrophobic than hydrophobic BA such as LCA); the presence of a hydroxyl group at C3 (albeit in β and α configuration in CLR and natural BA, respectively), the large hydrophobic steroidal nucleus (cholestane and cholane for CLR and BAs), and a freely movable lateral chain at C17 (much more hydrophobic in CLR). When minimum energy conformations of CLR and LCA molecules were flexibly aligned, the alignment showed high degree of overlap between heavy atoms of both compounds (Fig. 4A). The alignment became possible due to the relative flexibility of LCA structure that lost its characteristic bean-like shape and adopted the rather planar configuration characteristic of CLR.

The possibility that LCA might successfully occupy CLR-sensing CRAC4 region on BK protein have been addressed using computational docking. Docking showed that the most energetically favorable placement of LCA molecule on CRAC4 would be perpendicular to the axis of CRAC4 domain (Fig. 4B). This orientation contrast with the positioning of the CLR molecule, that is, *along* the CRAC4 axis (Fig. 2C). LCA docking also made formation of hydrogen bond with K453 quite difficult. Overall, although LCA could be docked onto the CRAC4 structure, this docking is strikingly different from that of CLR, and will inevitably result in functional consequences that differ from CLR effect on BK channel function.

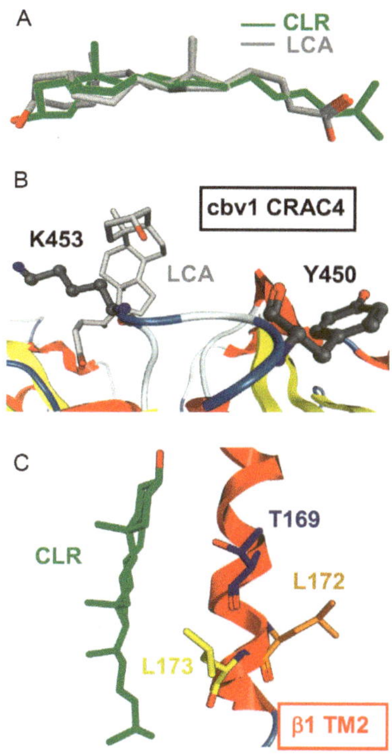

Fig. 4 Cholesterol and lithocholic acid (LCA) are unlikely to occupy each other's recognition site in the BK protein. (A) Flexible alignment of cholesterol and LCA molecules. (B) Preferred docking of LCA onto the cbv1 CRAC4 domain shows lack of hydrogen bonding between LCA and K453 due to the large distance between relevant chemical moieties. Moreover, LCA steroidal ring does not seem to interact with Tyr that is central to CRAC structure. (C) Cholesterol docking onto the BK β1 TM2 region containing the LCA-sensing site. Cholesterol's preferred docking shows that the sterol cannot accommodate its planar steroidal rings to the convexity provided by L172 and L173 (a structural requirement for LCA bonding; Bukiya, Singh, et al., 2011). As a result, cholesterol molecule is positioned away from the steroid-sensing site, with hydrogen bonding between cholesterol hydroxyl group and T169 unable to be formed due to the large distance between relevant chemical moieties.

In contrast to LCA docking on CRAC4, CLR docking on BK β1 TM2 steroid-sensing site was unsuccessful. In particular, due to the planar shape and longer dimensions of CLR molecule, accommodation by the T169, L172, and L173 trio became impossible (Fig. 4C). Although CLR molecule could be docked in the vicinity of the LCA-sensing site, this docking lacked the critical hydrogen bonding that was proven crucial for LCA effect on BK channel function (Bukiya, Singh, et al., 2011).

3. CONCLUSION

Functional studies demonstrate that BK channels represent one of the many proteins targeted by CLR or BAs, leading to modulation of BK channel function and thus, cell physiology. Modulation of BK channels by exposure to CLR or BAs is sustained in a very simple lipid environment (a binary phospholipid bilayer or an excised membrane patch) and does not require complex cellular cytoarchitecture, suggesting that BK channel proteins themselves provide recognition site(s) for these steroids. In spite of these common features, drastic differences in CLR vs BA interaction with BK channels exist. Regarding CLR, (1) *at molar fractions found in natural membranes*, the steroid inhibits the activity of homomeric slo1 channels, whether cloned from human brain or from rat cerebral artery smooth muscle. This modulation involves several CRAC motifs located in the slo1 CTD. Whether these motifs constitute actual binding sites for CLR remains unknown. In turn, accessory BK β1 subunits, which are abundant in smooth muscle and a few brain areas, may modify CLR effect on slo1 channels. (2) CLR inhibition of slo1 function results from a decrease in Po, with some minor reduction in unitary conductance. In contrast, LCA and related cholane steroids *at aqueous concentrations found in circulation during pathological conditions and associated with BA-induced vasodilation*: (1) potentiate BK currents via BA (LCA) docking onto a distinct "LCA site" mapped to the BK β1 subunit and (2) LCA potentiation of current results solely from increased Po without modification of unitary conductance. New computational models presented in this review make it unlikely for CLR to dock onto the LCA site and for LCA to dock onto CLR CRAC(4) motifs. Thus, these two related steroids are directly sensed by BK channels and thus modify channel function by interacting with distinct BK subunits and sites. The distinct gating mechanisms targeted by CLR and LCA leading to these steroids' opposite actions on slo1 Po (decrease and increased, respectively) are under current investigation.

ACKNOWLEDGMENT

This work was supported by grants R01HL10634, R3711560 (A.M.D.) and R01 AA023764 (A.N.B.).

REFERENCES

Alakoskela, J. M., Sabatini, K., Jiang, X., Laitala, V., Covey, D. F., & Kinnunen, P. K. (2008). Enantiospecific interactions between cholesterol and phospholipids. *Langmuir, 24*, 830–836.

Alberts, A. W., Chen, J., Kuron, G., Hunt, V., Huff, J., Hoffman, C., et al. (1980). Mevinolin: A highly potent competitive inhibitor of hydroxymethylglutaryl-coenzyme A reductase and a cholesterol-lowering agent. *Proceedings of the National Academy of Sciences of the United States of America*, *77*, 3957–3961.

Alioua, A., Lu, R., Kumar, Y., Eghbali, M., Kundu, P., Toro, L., et al. (2008). Slo1 caveolin-binding motif, a mechanism of caveolin-1-Slo1 interaction regulating Slo1 surface expression. *Journal of Biological Chemistry*, *283*, 4808–4817.

Babiychuk, E. B., Smith, R. D., Burdyga, T., Babiychuk, V. S., Wray, S., & Draeger, A. (2004). Membrane cholesterol regulates smooth muscle phasic contraction. *Journal of Membrane Biology*, *198*, 95–101.

Bao, L., & Cox, D. H. (2005). Gating and ionic currents reveal how the BK_{Ca} channel's Ca^{2+} sensitivity is enhanced by its beta1 subunit. *Journal of General Physiology*, *126*, 393–412.

Barrantes, F. J. (2010). Cholesterol effects on nicotinic acetylcholine receptor: Cellular aspects. *Subcellular Biochemistry*, *51*, 467–487.

Behrens, R., Nolting, A., Reimann, F., Schwarz, M., Waldschütz, R., & Pongs, O. (2000). hKCNMB3 and hKCNMB4, cloning and characterization of two members of the large-conductance calcium-activated potassium channel beta subunit family. *FEBS Letters*, *474*, 99–106.

Berkefeld, H., Fakler, B., & Schulte, U. (2010). Ca^{2+}-activated K^+ channels: From protein complexes to function. *Physiological Reviews*, *90*, 1437–1459.

Bisen, S., Seleverstov, O., Belani, J., Rychnovsky, S., Dopico, A. M., & Bukiya, A. N. (2016). Distinct mechanisms underlying cholesterol protection against alcohol-induced BK channel inhibition and resulting vasoconstriction. *Biochimica et Biophysica Acta*, *1861*, 1756–1766.

Bishop-Bailey, D., Walsh, D. T., & Warner, T. D. (2004). Expression and activation of the farnesoid X receptor in the vasculature. *Proceedings of the National Academy of Sciences of the United States of America*, *101*, 3668–3673.

Bittman, R. (1997). Has nature designed the cholesterol side chain for optimal interaction with phospholipids? *Subcellular Biochemistry*, *28*, 145–171.

Bloch, M., Ousingsawat, J., Simon, R., Schraml, P., Gasser, T. C., Mihatsch, M. J., et al. (2007). *KCNMA1* gene amplification promotes tumor cell proliferation in human prostate cancer. *Oncogene*, *26*, 2525–2534.

Bock, T., & Stuart, G. J. (2016). The impact of BK channels on cellular excitability depends on their subcellular location. *Frontiers in Cellular Neuroscience*, *10*, 206.

Bolotina, V., Omelyanenko, V., Heyes, B., Ryan, U., & Bregestovski, P. (1989). Variations of membrane cholesterol alter the kinetics of $Ca^{2(+)}$-dependent K^+ channels and membrane fluidity in vascular smooth muscle cells. *Pflügers Archiv*, *415*, 262–268.

Bomzon, A. (1994). Impaired responsiveness to angiotensin II in experimental cirrhosis. *Hepatology*, *20*, 266–268.

Bomzon, A., & Ljubuncic, P. (1995). Bile acids as endogenous vasodilators? *Biochemical Pharmacology*, *49*(5), 581–589.

Bouhairie, V. E., & Goldberg, A. C. (2016). Familial hypercholesterolemia. *Endocrinology and Metabolism Clinics of North America*, *45*, 1–16.

Boyer, J. L. (2013). Bile formation and secretion. *Comprehensive Physiology*, *3*, 1035–1078.

Brainard, A. M., Miller, A. J., Martens, J. R., & England, S. K. (2005). Maxi-K channels localize to caveolae in human myometrium: A role for an actin–channel–caveolin complex in the regulation of myometrial smooth muscle K+ current. *American Journal of Physiology. Cell Physiology*, *289*, C49–C57.

Braun, A. P. (2009). Distinct post-translational modifications regulate BK channel activity: The interplay between protein palmitoylation and phosphorylation. *Channels (Austin, TX)*, *3*, 144–145.

Brayden, J. E., & Nelson, M. T. (1992). Regulation of arterial tone by activation of calcium-dependent potassium channels. *Science*, *256*, 532–535.

Bregestovski, P. D., & Bolotina, V. N. (1989). Membrane fluidity and kinetics of Ca^{2+}-dependent potassium channels. *Biomedica Biochimica Acta, 48*, S382–S387.
Brenner, R., Jegla, T. J., Wickenden, A., Liu, Y., & Aldrich, R. W. (2000). Cloning and functional characterization of novel large conductance calcium-activated potassium channel beta subunits, hKCNMB3 and hKCNMB4. *Journal of Biological Chemistry, 275*, 6453–6461.
Brenner, R., Peréz, G. J., Bonev, A. D., Eckman, D. M., Kosek, J. C., Wiler, S. W., et al. (2000). Vasoregulation by the beta1 subunit of the calcium-activated potassium channel. *Nature, 407*, 870–876.
Broughton, G., 2nd. (1994). Chenodeoxycholate: The bile acid. The drug. A review. *American Journal of Medical Sciences, 307*, 54–63.
Brown, D. A., & London, E. (1998). Functions of lipid rafts in biological membranes. *Annual Review of Cell and Developmental Biology, 14*, 111–136.
Bukiya, A. N., Belani, J. D., Rychnovsky, S., & Dopico, A. M. (2011). Specificity of cholesterol and analogs to modulate BK channels points to direct sterol-channel protein interactions. *Journal of General Physiology, 137*, 93–110.
Bukiya, A., Dopico, A. M., Leffler, C. W., & Fedinec, A. (2014). Dietary cholesterol protects against alcohol-induced cerebral artery constriction. *Alcoholism: Clinical and Experimental Research, 38*, 1216–1226.
Bukiya, A. N., Kuntamallappanavar, G., Edwards, J., Singh, A. K., Shivakumar, B., & Dopico, A. M. (2014). An alcohol-sensing site in the calcium- and voltage-gated, large conductance potassium (BK) channel. *Proceedings of the National Academy of Sciences of the United States of America, 111*, 9313–9318.
Bukiya, A. N., Liu, J., Toro, L., & Dopico, A. M. (2007). Beta1 (*KCNMB1*) subunits mediate lithocholate activation of large-conductance Ca^{2+}-activated K^{+} channels and dilation in small, resistance-size arteries. *Molecular Pharmacology, 72*, 359–369.
Bukiya, A. N., McMillan, J. E., Fedinec, A. L., Patil, S. A., Miller, D. D., Leffler, C. W., et al. (2013). Cerebrovascular dilation via selective targeting of the cholane steroid-recognition site in the BK channel β1-subunit by a novel nonsteroidal agent. *Molecular Pharmacology, 83*, 1030–1044.
Bukiya, A. N., McMillan, J., Liu, J., Shivakumar, B., Parrill, A. L., & Dopico, A. M. (2014). Activation of calcium- and voltage-gated potassium channels of large conductance by leukotriene B4. *Journal of Biological Chemistry, 289*, 35314–35325.
Bukiya, A. N., McMillan, J., Parrill, A. L., & Dopico, A. M. (2008). Structural determinants of monohydroxylated bile acids to activate beta 1 subunit-containing BK channels. *Journal of Lipid Research, 49*, 2441–2451.
Bukiya, A. N., Patil, S. A., Li, W., Miller, D. D., & Dopico, A. M. (2012). Calcium- and voltage-gated potassium (BK) channel activators in the 5β-cholanic acid-3α-ol analogue series with modifications in the lateral chain. *ChemMedChem*, 7, 1784–1792.
Bukiya, A., & Rosnehouse-Dantsker, A. (2015). Hypercholesterolemia effect on potassium channels. In Dr. Sekar Ashok Kumar (Ed.), *Hypercholesterolemia*. Rijeka, Croatia: InTech. http://dx.doi.org/10.5772/59761.
Bukiya, A. N., Singh, A. K., Parrill, A. L., & Dopico, A. M. (2011). The steroid interaction site in transmembrane domain 2 of the large conductance, voltage- and calcium-gated potassium (BK) channel accessory β1 subunit. *Proceedings of the National Academy of Sciences of the United States of America, 108*, 20207–20212.
Bukiya, A. N., Vaithianathan, T., Kuntamallappanavar, G., Asuncion-Chin, M., & Dopico, A. M. (2011). Smooth muscle cholesterol enables BK β1 subunit-mediated channel inhibition and subsequent vasoconstriction evoked by alcohol. *Arteriosclerosis, Thrombosis, and Vascular Biology, 31*, 2410–2423.
Bukiya, A. N., Vaithianathan, T., Toro, L., & Dopico, A. M. (2008). The second transmembrane domain of the large conductance, voltage- and calcium-gated potassium channel beta(1) subunit is a lithocholate sensor. *FEBS Letters, 582*, 673–678.

Bukiya, A. N., Vaithianathan, T., Toro, L., & Dopico, A. M. (2009). Channel beta2–4 subunits fail to substitute for beta1 in sensitizing BK channels to lithocholate. *Biochemical and Biophysical Research Communications, 390*, 995–1000.

Cabral, D. J., Hamilton, J. A., & Small, D. M. (1986). The ionization behavior of bile acids in different aqueous environments. *Journal of Lipid Research, 27*, 334–343.

Cantor, R. S. (1999). Lipid composition and the lateral pressure profile in bilayers. *Biophysical Journal, 76*, 2625–2639.

Carey, M. C. (1985). Physical-chemical properties of bile acids and their salts. In H. Danielsson & J. Sjovall (Eds.), *Sterols and bile acids* (pp. 345–403). Amsterdam: Elsevier.

Carey, M. C., & Small, D. M. (1972). Micelle formation by bile salts. Physical-chemical and thermodynamic considerations. *Archives of Internal Medicine, 130*, 506–527.

Ceryak, S., Bouscarel, B., & Fromm, H. (1993). Comparative binding of bile acids to serum lipoproteins and albumin. *Journal of Lipid Research, 34*, 1661–1674.

Chang, H. M., Reitstetter, R., Mason, R. P., & Gruener, R. (1995). Attenuation of channel kinetics and conductance by cholesterol: An interpretation using structural stress as a unifying concept. *Journal of Membrane Biology, 143*, 51–63.

Choudhury, S., Garg, S. K., Singh, T. U., & Mishra, S. K. (2011). Functional and molecular characterization of maxi K^+-channels ($BK_{(Ca)}$) in buffalo myometrium. *Animal Reproduction Science, 126*, 173–178.

Clarke, J. A., Heron, A. J., Seddon, J. M., & Law, R. V. (2006). The diversity of the liquid ordered (Lo) phase of phosphatidylcholine/cholesterol membranes: A variable temperature multinuclear solid-state NMR and X-ray diffraction study. *Biophysical Journal, 90*, 2383–2393.

Clay, M. A., & Barter, P. J. (1996). Formation of new HDL particles from lipid-free apolipoprotein A-I. *Journal of Lipid Research, 37*, 1722–1732.

Copple, B. L., & Li, T. (2016). Pharmacology of bile acid receptors: Evolution of bile acids from simple detergents to complex signaling molecules. *Pharmacological Research, 104*, 9–21.

Crowley, J. J., Treistman, S. N., & Dopico, A. M. (2003). Cholesterol antagonizes ethanol potentiation of human brain BK_{Ca} channels reconstituted into phospholipid bilayers. *Molecular Pharmacology, 64*, 365–372.

Dai, S., Hall, D. D., & Hell, J. W. (2009). Supramolecular assemblies and localized regulation of voltage-gated ion channels. *Physiological Reviews, 89*, 411–452.

Davies, J. P., & Ioannou, Y. A. (2000). Topological analysis of Niemann–Pick C1 protein reveals that the membrane orientation of the putative sterol-sensing domain is identical to those of 3-hydroxy-3-methylglutaryl–CoA reductase and sterol regulatory element binding protein cleavage-activating protein. *Journal of Biological Chemistry, 275*, 24367–24374.

Dawson, P. A., & Karpen, S. J. (2015). Intestinal transport and metabolism of bile acids. *Journal of Lipid Research, 56*, 1085–1099.

de Aguiar Vallim, T. Q., Tarling, E. J., & Edwards, P. A. (2013). Pleiotropic roles of bile acids in metabolism. *Cell Metabolism, 17*, 657–669.

de Meyer, F., & Smit, B. (2009). Effect of cholesterol on the structure of a phospholipid bilayer. *Proceedings of the National Academy of Sciences of the United States of America, 106*, 3654–3658.

Degirolamo, C., Modica, S., Palasciano, G., & Moschetta, A. (2011). Bile acids and colon cancer: Solving the puzzle with nuclear receptors. *Trends in Molecular Medicine, 17*, 564–572.

Dopico, A. M., & Bukiya, A. N. (2014). Lipid regulation of BK channel function. *Frontiers in Physiology, 5*, 312.

Dopico, A., Bukiya, A., & Singh, A. (2012a). Steroid regulation of BK channels: Of channel protein and surrounding lipids. In I. Levitan & F. Barrantes (Eds.), *Cholesterol regulation of ion channels and receptors* (pp. 109–134). Hoboken, NJ: J. Wiley and Sons.

Dopico, A. M., Bukiya, A. N., & Singh, A. K. (2012b). Large conductance, calcium- and voltage-gated potassium (BK) channels: Regulation by cholesterol. *Pharmacology & Therapeutics*, *135*, 133–150.

Dopico, A. M., Kirber, M. T., Singer, J. J., & Walsh, J. V., Jr. (1994). Membrane stretch directly activates large conductance $Ca^{(2+)}$-activated K^+ channels in mesenteric artery smooth muscle cells. *American Journal of Hypertension*, 7, 82–89.

Dopico, A. M., & Tigyi, G. J. (2007). A glance at the structural and functional diversity of membrane lipids. *Methods in Molecular Biology*, *400*, 1–13.

Dopico, A. M., Walsh, J. V., Jr., & Singer, J. J. (2002). Natural bile acids and synthetic analogues modulate large conductance Ca^{2+}-activated K^+ (BK_{Ca}) channel activity in smooth muscle cells. *Journal of General Physiology*, *119*, 251–273.

Dunham, C. M., Fealk, M. H., & Sever, W. E., 3rd. (2003). Following severe injury, hypocholesterolemia improves with convalescence but persists with organ failure or onset of infection. *Critical Care*, 7, R145–R153.

El-Sayyad, H. I., Elmansi, A. A., & Bakr, E. H. (2015). Hypercholesterolemia-induced ocular disorder: Ameliorating role of phytotherapy. *Nutrition*, *31*, 1307–1316.

Epand, R. M. (2006). Cholesterol and the interaction of proteins with membrane domains. *Progress in Lipid Research*, *45*, 279–294.

Fakler, B., & Adelman, J. P. (2008). Control of $K_{(Ca)}$ channels by calcium nano/microdomains. *Neuron*, *59*, 873–881.

Falany, C. N., Johnson, M. R., Barnes, S., & Diasio, R. B. (1994). Glycine and taurine conjugation of bile acids by a single enzyme. Molecular cloning and expression of human liver bile acid CoA: Amino acid N-acyltransferase. *Journal of Biological Chemistry*, *269*, 19375–19379.

Fantini, J., & Barrantes, F. J. (2013). How cholesterol interacts with membrane proteins: An exploration of cholesterol-binding sites including CRAC, CARC, and tilted domains. *Frontiers in Physiology*, *4*, 31.

Fettiplace, R., & Fuchs, P. A. (1999). Mechanisms of hair cell tuning. *Annual Review of Physiology*, *61*, 809–834.

Fodor, A. A., & Aldrich, R. W. (2009). Convergent evolution of alternative splices at domain boundaries of the BK channel. *Annual Review of Physiology*, *71*, 19–36.

Forman, B. M., Goode, E., Chen, J., Oro, A. E., Bradley, D. J., Perlmann, T., et al. (1995). Identification of a nuclear receptor that is activated by farnesol metabolites. *Cell*, *81*, 687–693.

France, M., Bhattarai, Y., Galligan, J. J., & Xu, H. (2012). Impaired propulsive motility in the distal but not proximal colon of BK channel β1-subunit knockout mice. *Neurogastroenterology and Motility*, *24*, e450–e459.

Geng, Y., & Magleby, K. L. (2015). Single-channel kinetics of BK (Slo1) channels. *Frontiers in Physiology*, *5*, 532.

Gimpl, G., Burger, K., & Fahrenholz, F. (1997). Cholesterol as modulator of receptor function. *Biochemistry*, *36*, 10959–10974.

Golding, N. L., Jung, H. Y., Mickus, T., & Spruston, N. (1999). Dendritic calcium spike initiation and repolarization are controlled by distinct potassium channel subtypes in CA1 pyramidal neurons. *Journal of Neuroscience*, *19*, 8789–8798.

Goluszko, P., & Nowicki, B. (2005). Membrane cholesterol: A crucial molecule affecting interactions of microbial pathogens with mammalian cells. *Infection and Immunity*, *73*, 7791–7796.

Greco, A. V., & Mingrone, G. (1993). Serum bile acid concentrations in mild liver cirrhosis. *Clinica Chimica Acta*, *221*, 183–189.

Grimm, P. R., Foutz, R. M., Brenner, R., & Sansom, S. C. (2007). Identification and localization of BK-beta subunits in the distal nephron of the mouse kidney. *American Journal of Physiology. Renal Physiology*, *293*, F350–F359.

Guimarães, S. M., Lima, E. Q., Cipullo, J. P., Lobo, S. M., & Burdmann, E. A. (2008). Low insulin-like growth factor-1 and hypocholesterolemia as mortality predictors in acute kidney injury in the intensive care unit. *Critical Care Medicine*, *36*, 3165–3170.

Güldütuna, S., You, T., Kurts, W., & Leuschner, U. (1993). High performance liquid chromatographic determination of free and conjugated bile acids in serum, liver biopsies, bile, gastric juice and feces by fluorescence labeling. *Clinica Chimica Acta*, *214*, 195–207.

Hamdan, H., & Stacey, N. H. (1993). Mechanism of trichloroethylene-induced elevation of individual serum bile acids. I. Correlation of trichloroethylene concentrations to bile acids in rat serum. *Toxicology and Applied Pharmacology*, *121*, 291–295.

He, F., Li, J., Mu, Y., Kuruba, R., Ma, Z., Wilson, A., et al. (2006). Downregulation of endothelin-1 by farnesoid X receptor in vascular endothelial cells. *Circulation Research*, *98*, 192–199.

Hénin, J., Salari, R., Murlidaran, S., & Brannigan, G. (2014). A predicted binding site for cholesterol on the $GABA_A$ receptor. *Biophysical Journal*, *106*, 1938–1949.

Heppner, T. J., Herrera, G. M., Bonev, A. D., Hill-Eubanks, D., & Nelson, M. T. (2003). Ca^{2+} sparks and $K_{(Ca)}$ channels: Novel mechanisms to relax urinary bladder smooth muscle. *Advances in Experimental Medicine and Biology*, *539*, 347–357.

Hermann, A., Sitdikova, G. F., & Weiger, T. M. (2015). Oxidative stress and maxi calcium-activated potassium (BK) channels. *Biomolecules*, *5*, 1870–1911.

Hofmann, A. F., & Borgstroem, B. (1964). The intraluminal phase of fat digestion in man: The lipid content of micellar and oil phases of intestinal content obtained during fat digestion and absorbtion. *Journal of Clinical Investigation*, *43*, 247–257.

Horrigan, F. T., & Aldrich, R. W. (2002). Coupling between voltage sensor activation, Ca^{2+} binding and channel opening in large conductance (BK) potassium channels. *Journal of General Physiology*, *120*, 267–305.

Hoshi, T., & Heinemann, S. H. (2016). Modulation of BK channels by small endogenous molecules and pharmaceutical channel openers. *International Review of Neurobiology*, *128*, 193–237.

Hou, S., Heinemann, S. H., & Hoshi, T. (2009). Modulation of BK_{Ca} channel gating by endogenous signaling molecules. *Physiology (Bethesda)*, *24*, 26–35.

Hou, S., Xu, R., Heinemann, S. H., & Hoshi, T. (2008). The RCK1 high-affinity Ca^{2+} sensor confers carbon monoxide sensitivity to Slo1 BK channels. *Proceedings of the National Academy of Sciences of the United States of America*, *105*, 4039–4043.

Hu, T., Chouinard, M., Cox, A. L., Sipes, P., Marcelo, M., Ficorilli, J., et al. (2006). Farnesoid X receptor agonist reduces serum asymmetric dimethylarginine levels through hepatic dimethylarginine dimethylaminohydrolase-1 gene regulation. *Journal of Biological Chemistry*, *281*, 39831–39838.

Inagaki, T., Moschetta, A., Lee, Y. K., Peng, L., Zhao, G., Downes, M., et al. (2006). Regulation of antibacterial defense in the small intestine by the nuclear bile acid receptor. *Proceedings of the National Academy of Sciences of the United States of America*, *103*, 3920–3925.

Isaacson, J. S., & Murphy, G. J. (2001). Glutamate-mediated extrasynaptic inhibition: Direct coupling of NMDA receptors to $Ca^{(2+)}$ activated K^+ channels. *Neuron*, *31*, 1027–1034.

Jacobs, D., et al. (1992). Report of the conference on low blood cholesterol: Mortality associations. *Circulation*, *86*, 1046–1060.

Jaggar, J. H., Li, A., Parfenova, H., Liu, J., Umstot, E. S., Dopico, A. M., et al. (2005). Heme is a carbon monoxide receptor for large-conductance Ca^{2+}-activated K^+ channels. *Circulation Research*, *97*, 805–812.

Jaggar, J. H., Porter, V. A., Lederer, W. J., & Nelson, M. T. (2000). Calcium sparks in smooth muscle. *American Journal of Physiology. Cell Physiology*, *278*, C235–C256.

Jeremy, R. W., & McCarron, H. (2000). Effect of hypercholesterolemia on $Ca^{(2+)}$-dependent $K^{(+)}$ channel-mediated vasodilatation in vivo. *American Journal of Physiology Heart and Circulatory Physiology*, *279*, H1600–H1608.

Jönsson, G., Hedenborg, G., Wisén, O., & Norman, A. (1992). Serum concentrations and excretion of bile acids in cirrhosis. *Scandinavian Journal of Clinical & Laboratory Investigation, 52*, 599–605.

Kaiser, H. J., Lingwood, D., Levental, I., Sampaio, J. L., Kalvodova, L., Rajendran, L., et al. (2009). Order of lipid phases in model and plasma membranes. *Proceedings of the National Academy of Sciences of the United States of America, 106*, 16645–16650.

Kallen, J., Schlaeppi, J. M., Bitsch, F., Delhon, I., & Fournier, B. (2004). Crystal structure of the human RORalpha ligand binding domain in complex with cholesterol sulfate at 2.2 A. *Journal of Biological Chemistry, 279*, 14033–14038.

Kato, M., Tanaka, N., Ishii, H., Yin, C., & Sakuma, Y. (2009). Ca^{2+} channels and Ca^{2+}-activated K^{+} channels in adult rat gonadotrophin-releasing hormone neurones. *Journal of Neuroendocrinology, 21*, 312–315.

Kawamata, Y., Fujii, R., Hosoya, M., Harada, M., Yoshida, H., Miwa, M., et al. (2003). A G protein-coupled receptor responsive to bile acids. *Journal of Biological Chemistry, 278*, 9435–9440.

Keitel, V., Cupisti, K., Ullmer, C., Knoefel, W. T., Kubitz, R., & Häussinger, D. (2009). The membrane-bound bile acid receptor TGR5 is localized in the epithelium of human gallbladders. *Hepatology, 50*, 861–870.

Khaitan, D., Sankpal, U. T., Weksler, B., Meister, E. A., Romero, I. A., Couraud, P. O., et al. (2009). Role of KCNMA1 gene in breast cancer invasion and metastasis to brain. *BMC Cancer, 9*, 258.

Kida, T., Murata, T., Hori, M., & Ozaki, H. (2009). Chronic stimulation of farnesoid X receptor impairs nitric oxide sensitivity of vascular smooth muscle. *American Journal of Physiology Heart and Circulatory Physiology, 296*, H195–H201.

King, J. T., Lovell, P. V., Rishniw, M., Kotlikoff, M. I., Zeeman, M. L., & McCobb, D. P. (2006). Beta2 and beta4 subunits of BK channels confer differential sensitivity to acute modulation by steroid hormones. *Journal of Neurophysiology, 95*, 2878–2888.

Ko, J., Hamilton, J. A., Ton-Nu, H. T., Schteingart, C. D., Hofmann, A. F., & Small, D. M. (1994). Effects of side chain length on ionization behavior and transbilayer transport of unconjugated dihydroxy bile acids: A comparison of nor-chenodeoxycholic acid and chenodeoxycholic acid. *Journal of Lipid Research, 35*, 883–892.

Korinek, M., Vyklicky, V., Borovska, J., Lichnerova, K., Kaniakova, M., Krausova, B., et al. (2015). Cholesterol modulates open probability and desensitization of NMDA receptors. *Journal of Physiology, 593*, 2279–2293.

Korman, M. G., Hofmann, A. F., & Summerskill, W. H. (1974). Assessment of activity in chronic active liver disease. Serum bile acids compared with conventional tests and histology. *New England Journal of Medicine, 290*, 1399–1402.

Kotlikoff, M. I. (1993). Potassium channels in airway smooth muscle: A tale of two channels. *Pharmacology & Therapeutics, 58*, 1–12.

Kwakye, J. B., Barnes, S., & Diasio, R. B. (1993). Identification of bile acid coenzyme A synthetase in rat kidney. *Journal of Lipid Research, 34*, 95–99.

Kwon, H. J., Abi-Mosleh, L., Wang, M. L., Deisenhofer, J., Goldstein, J. L., Brown, M. S., et al. (2009). Structure of N-terminal domain of NPC1 reveals distinct subdomains for binding and transfer of cholesterol. *Cell, 137*, 1213–1224.

Kyle, B. D., & Braun, A. P. (2014). The regulation of BK channel activity by pre- and post-translational modifications. *Frontiers in Physiology, 5*, 316.

Lam, R. S., Shaw, A. R., & Duszyk, M. (2004). Membrane cholesterol content modulates activation of BK channels in colonic epithelia. *Biochimica et Biophysica Acta, 1667*, 241–248.

Langhi, C., Le May, C., Kourimate, S., Caron, S., Staels, B., Krempf, M., et al. (2008). Activation of the farnesoid X receptor represses PCSK9 expression in human hepatocytes. *FEBS Letters, 582*, 949–955.

Lascombe, M. B., Ponchet, M., Venard, P., Milat, M. L., Blein, J. P., & Prangé, T. (2002). The 1.45 Å resolution structure of the cryptogein-cholesterol complex: A close-up view of a sterol carrier protein (SCP) active site. *Acta Crystallographica. Section D, Biological Crystallography, 58*, 1442–1447.

Latorre, R., & Brauchi, S. (2006). Large conductance Ca^{2+}-activated K^{+} (BK) channel: Activation by Ca^{2+} and voltage. *Biological Research, 39*, 385–401.

Latorre, R., Morera, F. J., & Zaelzer, C. (2010). Allosteric interactions and the modular nature of the voltage- and Ca^{2+}-activated (BK) channel. *Journal of Physiology, 588*, 3141–3148.

Lee, U. S., & Cui, J. (2010). BK channel activation: Structural and functional insights. *Trends in Neurosciences, 33*, 415–423.

Lefebvre, P., Cariou, B., Lien, F., Kuipers, F., & Staels, B. (2009). Role of bile acids and bile acid receptors in metabolic regulation. *Physiological Reviews, 89*, 147–191.

Lefèvre, C. M., Diakov, A., Haerteis, S., Korbmacher, C., Gründer, S., & Wiemuth, D. (2014). Pharmacological and electrophysiological characterization of the human bile acid-sensitive ion channel (hBASIC). *Pflügers Archiv, 466*, 253–263.

Levitan, I., Fang, Y., Rosenhouse-Dantsker, A., & Romanenko, V. (2010). Cholesterol and ion channels. *Subcellular Biochemistry, 51*, 509–549.

Levitan, I., Singh, D. K., & Rosenhouse-Dantsker, A. (2014). Cholesterol binding to ion channels. *Frontiers in Physiology, 5*, 65.

Lewis, B., Tabaqchali, S., Panveliwalla, D., & Wootton, I. D. (1969). Serum-bile-acids in the stagnant-loop syndrome. *Lancet, 1*, 219–220.

Li, T., & Chiang, J. Y. (2015). Bile acids as metabolic regulators. *Current Opinion in Gastroenterology, 31*, 159–165.

Li, T., Holmstrom, S. R., Kir, S., Umetani, M., Schmidt, D. R., Kliewer, S. A., et al. (2011). The G protein-coupled bile acid receptor, TGR5, stimulates gallbladder filling. *Molecular Endocrinology, 25*, 1066–1071.

Li, F., Liu, J., Valls, L., Hiser, C., & Ferguson-Miller, S. (2015). Identification of a key cholesterol binding enhancement motif in translocator protein 18 kDa. *Biochemistry, 54*, 1441–1443.

Li, Y. T., Swales, K. E., Thomas, G. J., Warner, T. D., & Bishop-Bailey, D. (2007). Farnesoid x receptor ligands inhibit vascular smooth muscle cell inflammation and migration. *Arteriosclerosis, Thrombosis, and Vascular Biology, 27*, 2606–2611.

Li, J., Wilson, A., Kuruba, R., Zhang, Q., Gao, X., He, F., et al. (2008). FXR-mediated regulation of eNOS expression in vascular endothelial cells. *Cardiovascular Research, 77*, 169–177.

Li, H., Yao, Z., Degenhardt, B., Teper, G., & Papadopoulos, V. (2001). Cholesterol binding at the cholesterol recognition/interaction amino acid consensus (CRAC) of the peripheral-type benzodiazepine receptor and inhibition of steroidogenesis by an HIV TAT-CRAC peptide. *Proceedings of the National Academy of Sciences of the United States of America, 98*, 1267–1272.

Lindblad, L., Lundholm, K., & Scherstén, T. (1977). Influence of cholic and chenodeoxycholic acid on biliary cholesterol secretion in man. *European Journal of Clinical Investigation*, 7, 383–388.

Liu, J., Liu, P., Asuncion-Chin, M. T., & Dopico, A. M. (2005). In *Cloning and functional characterization of BK channel-forming subunits from rat cerebral artery myocytes. Society for neuroscience 35th annual meeting, Washington, DC, November 12–16, 2005. Abstract online.*

Lo, Y. H., Bellis, S. L., Cheng, L. J., Pang, J., Bradley, T. M., & Rhoads, D. E. (1994). Signal transduction for taurocholic acid in the olfactory system of Atlantic salmon. *Chemical Senses, 19*, 371–380.

Löhn, M., Fürstenau, M., Sagach, V., Elger, M., Schulze, W., Luft, F. C., et al. (2000). Ignition of calcium sparks in arterial and cardiac muscle through caveolae. *Circulation Research*, *87*, 1034–1039.

Lorenz, S., Heils, A., Kasper, J. M., & Sander, T. (2007). Allelic association of a truncation mutation of the KCNMB3 gene with idiopathic generalized epilepsy. *American Journal of Medical Genetics Part B: Neuropsychiatric Genetics*, *144B*, 10–13.

Lu, R., Alioua, A., Kumar, Y., Eghbali, M., Stefani, E., & Toro, L. (2006). MaxiK channel partners: Physiological impact. *Journal of Physiology*, *570*, 65–72.

Lundbaek, J. A., Andersen, O. S., Werge, T., & Nielsen, C. (2003). Cholesterol-induced protein sorting: An analysis of energetic feasibility. *Biophysical Journal*, *84*, 2080–2089.

Makishima, M., Lu, T. T., Xie, W., Whitfield, G. K., Domoto, H., Evans, R. M., et al. (2002). Vitamin D receptor as an intestinal bile acid sensor. *Science*, *296*, 1313–1316.

Makishima, M., Okamoto, A. Y., Repa, J. J., Tu, H., Learned, R. M., Luk, A., et al. (1999). Identification of a nuclear receptor for bile acids. *Science*, *284*, 1362–1365.

Marrion, N. V., & Tavalin, S. J. (1998). Selective activation of Ca^{2+}-activated K^+ channels by co-localized Ca^{2+} channels in hippocampal neurons. *Nature*, *395*, 900–905.

Maruyama, T., Miyamoto, Y., Nakamura, T., Tamai, Y., Okada, H., Sugiyama, E., et al. (2002). Identification of membrane-type receptor for bile acids (M-BAR). *Biochemical and Biophysical Research Communications*, *298*, 714–719.

Mast, N., White, M. A., Bjorkhem, I., Johnson, E. F., Stout, C. D., & Pikuleva, I. A. (2008). Crystal structures of substrate-bound and substrate-free cytochrome P450 46A1, the principal cholesterol hydroxylase in the brain. *Proceedings of the National Academy of Sciences of the United States of America*, *105*, 9546–9551.

Maxfield, F. R. (2014). Role of endosomes and lysosomes in human disease. *Cold Spring Harbor Perspectives in Biology*, *6*, a016931.

McMillan, J. E., Bukiya, A. N., Terrell, C. L., Patil, S. A., Miller, D. D., Dopico, A. M., et al. (2014). Multi-generational pharmacophore modeling for ligands to the cholane steroid-recognition site in the β_1 modulatory subunit of the BK_{Ca} channel. *Journal of Molecular Graphics and Modeling*, *54*, 174–183.

McMillin, M., & DeMorrow, S. (2016). Effects of bile acids on neurological function and disease. *FASEB Journal*, *30*, 3658–3668. in press.

Meng, X. F., Yu, J. T., Wang, H. F., Tan, M. S., Wang, C., Tan, C. C., et al. (2014). Midlife vascular risk factors and the risk of Alzheimer's disease: A systematic review and meta-analysis. *Journal of Alzheimers Disease*, *42*, 1295–1310.

Najibi, S., & Cohen, R. A. (1995). Enhanced role of K^+ channels in relaxations of hypercholesterolemic rabbit carotid artery to NO. *American Journal of Physiology*, *269*, H805–H811.

Nakajima, M., Hutchinson, H. G., Fujinaga, M., Hayashida, W., Morishita, R., Zhang, L., et al. (1995). The angiotensin II type 2 (AT2) receptor antagonizes the growth effects of the AT1 receptor: Gain-of-function study using gene transfer. *Proceedings of the National Academy of Sciences of the United States of America*, *92*, 10663–10667.

Navi, B. B., & Segal, A. Z. (2009). The role of cholesterol and statins in stroke. *Current Cardiology Reports*, *11*, 4–11.

Nes, W. D. (2011). Biosynthesis of cholesterol and other sterols. *Chemical Reviews*, *111*, 6423–6451.

Nowaczyk, M. J., & Irons, M. B. (2012). Smith–Lemli–Opitz syndrome: Phenotype, natural history, and epidemiology. *American Journal of Medical Genetics Part C: Seminars in Medical Genetics*, *160C*, 250–262.

Oddi, S., Dainese, E., Fezza, F., Lanuti, M., Barcaroli, D., De Laurenzi, V., et al. (2011). Functional characterization of putative cholesterol binding sequence (CRAC) in human type-1 cannabinoid receptor. *Journal of Neurochemistry*, *116*, 858–865.

Ohvo-Rekilä, H., Ramstedt, B., Leppimäki, P., & Slotte, J. P. (2002). Cholesterol interactions with phospholipids in membranes. *Progress in Lipid Research, 41*, 66–97.

Orio, P., Rojas, P., Ferreira, G., & Latorre, R. (2002). New disguises for an old channel: MaxiK channel beta-subunits. *News in Physiological Sciences, 17*, 156–161.

Ostrow, J. D. (1993). Metabolism of bile salts in cholestasis in humans. In N. Tavoloni & P. D. Berk (Eds.), *Hepatic transport and bile secretion* (pp. 673–712). New York: Raven Press.

Pak, J. M., & Lee, S. S. (1993). Vasoactive effects of bile salts in cirrhotic rats: In vivo and in vitro studies. *Hepatology, 18*, 1175–1181.

Palmer, M. (2004). Cholesterol and the activity of bacterial toxins. *FEMS Microbiology Letters, 238*, 281–289.

Palmer, C. P., Mahen, R., Schnell, E., Djamgoz, M. B., & Aydar, E. (2007). Sigma-1 receptors bind cholesterol and remodel lipid rafts in breast cancer cell lines. *Cancer Research, 67*, 11166–11175.

Pang, L., Graziano, M., & Wang, S. (1999). Membrane cholesterol modulates galanin–GalR2 interaction. *Biochemistry, 38*, 12003–12011.

Pape, H. C., Munsch, T., & Budde, T. (2004). Novel vistas of calcium-mediated signalling in the thalamus. *Pflügers Archiv, 448*, 131–138.

Parks, D. J., Blanchard, S. G., Bledsoe, R. K., Chandra, G., Consler, T. G., Kliewer, S. A., et al. (1999). Bile acids: Natural ligands for an orphan nuclear receptor. *Science, 284*, 1365–1368.

Patil, S., Bukiya, A. N., Li, W., Dopico, A. M., & Miller, D. (2008). Design and synthesis of hydroxy-alkynoic acids and their methyl esters as novel activators of BK channels. *Bioorganic and Medicinal Chemistry Letters, 18*, 3427–3430.

Pfrieger, F. W. (2003). Role of cholesterol in synapse formation and function. *Biochimica et Biophysica Acta, 1610*, 271–280.

Pikuleva, I. A. (2008). Cholesterol-metabolizing cytochromes P450: Implications for cholesterol lowering. *Expert Opinion on Drug Metabolism & Toxicology, 4*, 1403–1414.

Pikuleva, I. A., Mackman, R. L., Kagawa, N., Waterman, M. R., & Ortiz de Montellano, P. R. (1995). Active-site topology of bovine cholesterol side-chain cleavage cytochrome P450 (P450scc) and evidence for interaction of tyrosine 94 with the side chain of cholesterol. *Archives of Biochemistry and Biophysics, 322*, 189–197.

Pine, S. (1987). Lipids. In *Organic chemistry* (5th ed., pp. 858–908). New York: McGraw-Hill Company.

Pols, T. W., Noriega, L. G., Nomura, M., Auwerx, J., & Schoonjans, K. (2011). The bile acid membrane receptor TGR5: A valuable metabolic target. *Digestive Diseases, 29*, 37–44.

Porez, G., Prawitt, J., Gross, B., & Staels, B. (2012). Bile acid receptors as targets for the treatment of dyslipidemia and cardiovascular disease. *Journal of Lipid Research, 53*, 1723–1737.

Purcell, E. K., Liu, L., Thomas, P. V., & Duncan, R. K. (2011). Cholesterol influences voltage-gated calcium channels and BK-type potassium channels in auditory hair cells. *PLoS One, 6*, e26289.

Quinn, P. J., & Wolf, C. (2009). The liquid-ordered phase in membranes. *Biochimica et Biophysica Acta, 1788*, 33–46.

Ramanathan, K., & Fuchs, P. A. (2002). Modeling hair cell tuning by expression gradients of potassium channel beta subunits. *Biophysical Journal, 82*, 64–75.

Ramprasad, O. G., Srinivas, G., Rao, K. S., Joshi, P., Thiery, J. P., Dufour, S., et al. (2007). Changes in cholesterol levels in the plasma membrane modulate cell signaling and regulate cell adhesion and migration on fibronectin. *Cell Motility and the Cytoskeleton, 64*, 199–216.

Rask-Madsen, C., & King, G. L. (2007). Mechanisms of disease: Endothelial dysfunction in insulin resistance and diabetes. *Nature Clinical Practice. Endocrinology & Metabolism, 3*, 46–56.
Ridlon, J. M., Kang, D. J., & Hylemon, P. B. (2006). Bile salt biotransformations by human intestinal bacteria. *Journal of Lipid Research, 47*, 241–259.
Rosenson, R. S., Brewer, H. B., Jr., Ansell, B. J., Barter, P., Chapman, M. J., Heinecke, J. W., et al. (2016). Dysfunctional HDL and atherosclerotic cardiovascular disease. *Nature Reviews Cardiology, 13*, 48–60.
Russell, D. W. (2003). The enzymes, regulation, and genetics of bile acid synthesis. *Annual Review of Biochemistry, 72*, 137–174.
Russell, D. W., & Setchell, K. D. (1992). Bile acid biosynthesis. *Biochemistry, 31*, 4737–4749.
Saher, G., Brügger, B., Lappe-Siefke, C., Möbius, W., Tozawa, R., Wehr, M. C., et al. (2005). High cholesterol level is essential for myelin membrane growth. *Nature Neuroscience, 8*, 468–475.
Saini, H. K., Arneja, A. S., & Dhalla, N. S. (2004). Role of cholesterol in cardiovascular dysfunction. *Canadian Journal of Cardiology, 20*, 333–346.
Salkoff, L., Butler, A., Ferreira, G., Santi, C., & Wei, A. (2006). High-conductance potassium channels of the SLO family. *Nature Reviews Neuroscience, 7*, 921–931.
Sausbier, M., Matos, J. E., Sausbier, U., Beranek, G., Arntz, C., Neuhuber, W., et al. (2006). Distal colonic $K^{(+)}$ secretion occurs via BK channels. *Journal of the American Society of Nephrology, 17*, 1275–1282.
Schaap, F. G., Trauner, M., & Jansen, P. L. (2014). Bile acid receptors as targets for drug development. *Nature Reviews. Gastroenterology & Hepatology, 11*, 55–67.
Scharschmidt, B. F., Blanckaert, N., Farina, F. A., Kabra, P. M., Stafford, B. E., & Weisiger, R. A. (1982). Measurement of serum bilirubin and its mono- and diconjugates: Application to patients with hepatobiliary disease. *Gut, 23*, 643–649.
Schroeder, C. (2010). Cholesterol-binding viral proteins in virus entry and morphogenesis. *Subcellular Biochemistry, 51*, 77–108.
Schubert, R., & Nelson, M. T. (2001). Protein kinases: Tuners of the BK_{Ca} channel in smooth muscle. *Trends in Pharmacological Sciences, 22*, 505–512.
Semenov, I., Wang, B., Herlihy, J. T., & Brenner, R. (2006). BK channel beta1-subunit regulation of calcium handling and constriction in tracheal smooth muscle. *American Journal of Physiology. Lung Cellular and Molecular Physiology, 291*, L802–L810.
Setchell, K. D., Lawson, A. M., Tanida, N., & Sjövall, J. (1983). General methods for the analysis of metabolic profiles of bile acids and related compounds in feces. *Journal of Lipid Research, 24*, 1085–1100.
Shao, L. R., Halvorsrud, R., Borg-Graham, L., & Storm, J. F. (1999). The role of BK-type Ca^{2+}-dependent K^{+} channels in spike broadening during repetitive firing in rat hippocampal pyramidal cells. *Journal of Physiology, 521*, 135–146.
Shmygol, A., Noble, K., & Wray, S. (2007). Depletion of membrane cholesterol eliminates the Ca2+-activated component of outward potassium current and decreases membrane capacitance in rat uterine myocytes. *Journal of Physiology, 581*, 445–456.
Simons, K., & Sampaio, J. L. (2011). Membrane organization and lipid rafts. *Cold Spring Harbor Perspectives in Biology, 3*, a004697.
Sinal, C. J., Tohkin, M., Miyata, M., Ward, J. M., Lambert, G., & Gonzalez, F. J. (2000). Targeted disruption of the nuclear receptor FXR/BAR impairs bile acid and lipid homeostasis. *Cell, 102*, 731–744.
Singh, A. K., McMillan, J., Bukiya, A. N., Burton, B., Parrill, A. L., & Dopico, A. M. (2012). Multiple cholesterol recognition/interaction amino acid consensus (CRAC) motifs in cytosolic C tail of Slo1 subunit determine cholesterol sensitivity of Ca^{2+}- and voltage-gated K^{+} (BK) channels. *Journal of Biological Chemistry, 287*, 20509–20521.

Sjövall, J. (2004). Fifty years with bile acids and steroids in health and disease. *Lipids*, *39*, 703–722.

Sonnino, S., & Prinetti, A. (2013). Membrane domains and the "lipid raft" concept. *Current Medicinal Chemistry*, *20*, 4–21.

Sontheimer, H. (2008). An unexpected role for ion channels in brain tumor metastasis. *Experimental Biology and Medicine (Maywood, NJ)*, *233*, 779–791.

Stachon, A., Böning, A., Weisser, H., Laczkovics, A., Skipka, G., & Krieg, M. (2000). Prognostic significance of low serum cholesterol after cardiothoracic surgery. *Clinical Chemistry*, *46*, 1114–1120.

Staels, B., & Fonseca, V. A. (2009). Bile acids and metabolic regulation: Mechanisms and clinical responses to bile acid sequestration. *Diabetes Care*, *32*, S237–S245.

Staels, B., & Kuipers, F. (2007). Bile acid sequestrants and the treatment of type 2 diabetes mellitus. *Drugs*, *67*, 1383–1392.

Staudinger, J., Liu, Y., Madan, A., Habeebu, S., & Klaassen, C. D. (2001). Coordinate regulation of xenobiotic and bile acid homeostasis by pregnane X receptor. *Drug Metabolism and Disposition*, *29*, 1467–1472.

Tajima, N., Itokazu, Y., Korpi, E. R., Somerharju, P., & Käkelä, R. (2011). Activity of $BK_{(Ca)}$ channel is modulated by membrane cholesterol content and association with Na+/K+-ATPase in human melanoma IGR39 cells. *Journal of Biological Chemistry*, *286*, 5624–5638.

Taoka, H., Yokoyama, Y., Morimoto, K., Kitamura, N., Tanigaki, T., Takashina, Y., et al. (2016). Role of bile acids in the regulation of the metabolic pathways. *World Journal of Diabetes*, *7*, 260–270.

Tarantino, G., Cambri, S., Ferrara, A., Marzano, M., Liberti, A., Vellone, G., et al. (1989). Serum concentration of bile acids and portal hypertension in cirrhotic patients. Possible correlations. *Riv European Science Medical Pharmacology*, *11*, 195–205.

Teodoro, J. S., Rolo, A. P., & Palmeira, C. M. (2011). Hepatic FXR: Key regulator of whole-body energy metabolism. *Trends in Endocrinology and Metabolism*, *22*, 458–466.

Thomas, C., Pellicciari, R., Pruzanski, M., Auwerx, J., & Schoonjans, K. (2008). Targeting bile-acid signalling for metabolic diseases. *Nature Reviews Drug Discovery*, *7*, 678–693.

Tian, L., Coghill, L. S., MacDonald, S. H., Armstrong, D. L., & Shipston, M. J. (2003). Leucine zipper domain targets cAMP-dependent protein kinase to mammalian BK channels. *Journal of Biological Chemistry*, *278*, 8669–8677.

Tian, L., Duncan, R. R., Hammond, M. S., Coghill, L. S., Wen, H., Rusinova, R., et al. (2001). Alternative splicing switches potassium channel sensitivity to protein phosphorylation. *Journal of Biological Chemistry*, *276*, 7717–7720.

Tillman, T. S., & Cascio, M. (2003). Effects of membrane lipids on ion channel structure and function. *Cell Biochemistry and Biophysics*, *38*, 161–190.

Toro, L., Li, M., Zhang, Z., Singh, H., Wu, Y., & Stefani, E. (2014). MaxiK channel and cell signalling. *Pflügers Archiv*, *466*, 875–886.

Torres, Y. P., Granados, S. T., & Latorre, R. (2014). Pharmacological consequences of the coexpression of BK channel α and auxiliary β subunits. *Frontiers in Physiology*, *5*, 383.

Torres, Y. P., Morera, F. J., Carvacho, I., & Latorre, R. (2007). A marriage of convenience: Beta-subunits and voltage-dependent K^+ channels. *Journal of Biological Chemistry*, *282*, 24485–24489.

Ujčíková, H., Brejchová, J., Vošahlíková, M., Kagan, D., Dlouhá, K., Sýkora, J., et al. (2014). Opioid-receptor (OR) signaling cascades in rat cerebral cortex and model cell lines: The role of plasma membrane structure. *Physiological Research*, *63*, S165–S176.

Vaithianathan, T., Bukiya, A., Liu, J., Liu, P., Asuncion-Chin, M., Fan, Z., et al. (2008). Direct regulation of BK channels by phosphatidylinositol 4,5-bisphosphate as a novel signaling pathway. *Journal of General Physiology*, *132*, 13–28.

van der Kant, R., Zondervan, I., Janssen, L., & Neefjes, J. (2013). Cholesterol-binding molecules MLN64 and ORP1L mark distinct late endosomes with transporters ABCA3 and NPC1. *Journal of Lipid Research, 54*, 2153–2165.

Vyroubal, P., Chiarla, C., Giovannini, I., Hyspler, R., Ticha, A., Hrnciarikova, D., et al. (2008). Hypocholesterolemia in clinically serious conditions—Review. *Biomedical papers of the Medical Faculty of the University Palacky, Olomouc, Czechoslovakia, 152*, 181–189.

Wahlström, A., Sayin, S. I., Marschall, H. U., & Bäckhed, F. (2016). Intestinal crosstalk between bile acids and microbiota and its impact on host metabolism. *Cell Metabolism, 24*, 41–50.

Wallner, M., Meera, P., & Toro, L. (1996). Determinant for beta-subunit regulation in high-conductance voltage-activated and $Ca^{(2+)}$-sensitive K^+ channels: An additional transmembrane region at the N terminus. *Proceedings of the National Academy of Sciences of the United States of America, 93*, 14922–14927.

Wang, B., & Brenner, R. (2006). An S6 mutation in BK channels reveals beta1 subunit effects on intrinsic and voltage-dependent gating. *Journal of General Physiology, 128*, 731–744.

Wang, Y. W., Ding, J. P., Xia, X. M., & Lingle, C. J. (2002). Consequences of the stoichiometry of Slo1 alpha and auxiliary beta subunits on functional properties of large-conductance Ca^{2+}-activated K^+ channels. *Journal of Neuroscience, 22*, 1550–1561.

Wang, L., & Sigworth, F. J. (2009). Structure of the BK potassium channel in a lipid membrane from electron cryomicroscopy. *Nature, 461*, 292–295.

Wang, X. L., Ye, D., Peterson, T. E., Cao, S., Shah, V. H., Katusic, Z. S., et al. (2005). Caveolae targeting and regulation of large conductance $Ca^{(2+)}$-activated K^+ channels in vascular endothelial cells. *Journal of Biological Chemistry, 280*, 11656–11664.

Weaver, A. K., Olsen, M. L., McFerrin, M. B., & Sontheimer, H. (2007). BK channels are linked to inositol 1,4,5-triphosphate receptors via lipid rafts: A novel mechanism for coupling $[Ca^{(2+)}]_{(i)}$ to ion channel activation. *Journal of Biological Chemistry, 282*, 31558–31568.

Wei, A., Solaro, C., Lingle, C., & Salkoff, L. (1994). Calcium sensitivity of BK-type K_{Ca} channels determined by a separable domain. *Neuron, 13*, 671–681.

Westover, E. J., Covey, D. F., Brockman, H. L., Brown, R. E., & Pike, L. J. (2003). Cholesterol depletion results in site-specific increases in epidermal growth factor receptor phosphorylation due to membrane level effects. Studies with cholesterol enantiomers. *Journal of Biological Chemistry, 278*, 51125–51133.

Wiecha, J., Schläger, B., Voisard, R., Hannekum, A., Mattfeldt, T., & Hombach, V. (1997). $Ca^{(2+)}$-activated K^+ channels in human smooth muscle cells of coronary atherosclerotic plaques and coronary media segments. *Basic Research in Cardiology, 92*, 233–239.

Wu, W., Wang, Y., Deng, X. L., Sun, H. Y., & Li, G. R. (2013). Cholesterol down-regulates BK channels stably expressed in HEK 293 cells. *PLoS One, 8*, e79952.

Xia, X. M., Ding, J. P., & Lingle, C. J. (1999). Molecular basis for the inactivation of Ca^{2+}- and voltage-dependent BK channels in adrenal chromaffin cells and rat insulinoma tumor cells. *Journal of Neuroscience, 19*, 5255–5264.

Xu, S., Benoff, B., Liou, H.-L., Lobel, P., & Stock, A. M. (2007). Structural basis of sterol binding by NPC2, a lysosomal protein deficient in Niemann-Pick type C2 disease. *Journal of Biological Chemistry, 282*, 23525–23531.

Yan, J., & Aldrich, R. W. (2010). LRRC26 auxiliary protein allows BK channel activation at resting voltage without calcium. *Nature, 466*, 513–516.

Yan, J., Olsen, J. V., Park, K. S., Li, W., Bildl, W., Schulte, U., et al. (2008). Profiling the phospho-status of the BK_{Ca} channel alpha subunit in rat brain reveals unexpected patterns and complexity. *Molecular & Cellular Proteomics, 7*, 2188–2198.

Yang, H., Zhang, G., & Cui, J. (2015). BK channels: Multiple sensors, one activation gate. *Frontiers in Physiology, 6*, 29.

Yazejian, B., Sun, X. P., & Grinnell, A. D. (2000). Tracking presynaptic Ca^{2+} dynamics during neurotransmitter release with Ca^{2+}-activated K^+ channels. *Nature Neuroscience, 3*, 566–571.

Yu, Q., Zhang, Y., & Xu, C. B. (2015). Apolipoprotein B, the villain in the drama? *European Journal of Pharmacology, 748*, 166–169.

Yuan, C., Chen, M., Covey, D. F., Johnston, L. J., & Treistman, S. N. (2011). Cholesterol tuning of BK ethanol response is enantioselective, and is a function of accompanying lipids. *PLoS One, 6*, e27572.

Yuan, P., Leonetti, M. D., Pico, A. R., Hsiung, Y., & MacKinnon, R. (2010). Structure of the human BK channel Ca^{2+}-activation apparatus at 3.0 A resolution. *Science, 329*, 182–186.

Zhang, Q., He, F., Kuruba, R., Gao, X., Wilson, A., Li, J., et al. (2008). FXR-mediated regulation of angiotensin type 2 receptor expression in vascular smooth muscle cells. *Cardiovascular Research*, 77, 560–569.

Zhang, J., & Yan, J. (2014). Regulation of BK channels by auxiliary γ subunits. *Frontiers in Physiology, 5*, 401.

Zhang, Y., Yin, L., Anderson, J., Ma, H., Gonzalez, F. J., Willson, T. M., et al. (2010). Identification of novel pathways that control farnesoid X receptor-mediated hypocholesterolemia. *Journal of Biological Chemistry, 285*, 3035–3043.

Zhou, X. B., Wulfsen, I., Utku, E., Sausbier, U., Sausbier, M., Wieland, T., et al. (2010). Dual role of protein kinase C on BK channel regulation. *Proceedings of the National Academy of Sciences of the United States of America, 107*, 8005–8010.

Zidovetzki, R., & Levitan, I. (2007). Use of cyclodextrins to manipulate plasma membrane cholesterol content: Evidence, misconceptions and control strategies. *Biochimica et Biophysica Acta, 1768*, 1311–1324.

FURTHER READING

Du, P., Cui, G. B., Wang, Y. R., Zhang, X. Y., Ma, K. J., & Wei, J. G. (2006). Down regulated expression of the beta1 subunit of the big-conductance Ca^{2+} sensitive K^+ channel in sphincter of Oddi cells from rabbits fed with a high cholesterol diet. *Acta Biochimica et Biophysica Sinica Shanghai, 38*, 893–899.

Hanson, M. A., Cherezov, V., Griffith, M. T., Roth, C. B., Jaakola, V. P., Chien, E. Y., et al. (2008). A specific cholesterol binding site is established by the 2.8 A structure of the human beta2-adrenergic receptor. *Structure, 16*, 897–905.

Li, W., Scott, A. P., Siefkes, M. J., Yan, H., Liu, Q., Yun, S. S., et al. (2002). Bile acid secreted by male sea lamprey that acts as a sex pheromone. *Science, 296*, 138–141.

Lin, M. W., Wu, A. Z., Ting, W. H., Li, C. L., Cheng, K. S., & Wu, S. N. (2006). Changes in membrane cholesterol of pituitary tumor (GH3) cells regulate the activity of large-conductance Ca^{2+}-activated K^+ channels. *Chinese Journal of Physiology, 49*, 1–13.

Prendergast, C., Quayle, J., Burdyga, T., & Wray, S. (2010). Cholesterol depletion alters coronary artery myocyte $Ca^{(2+)}$ signalling in a stimulus-specific manner. *Cell Calcium, 47*, 84–91.

Wiemuth, D., Sahin, H., Lefèvre, C. M., Wasmuth, H. E., & Gründer, S. (2013). Strong activation of bile acid-sensitive ion channel (BASIC) by ursodeoxycholic acid. *Channels (Austin, TX)*, 7, 38–42.

CHAPTER FOUR

The Role of Cholesterol in the Activation of Nicotinic Acetylcholine Receptors

John E. Baenziger[1], Jaimee A. Domville, J.P. Daniel Therien
University of Ottawa, Ottawa, ON, Canada
[1]Corresponding author: e-mail address: john.baenziger@uottawa.ca

Contents

Abstract

Cholesterol is a potent modulator of the nicotinic acetylcholine receptor (nAChR) from *Torpedo*. Here, we review current understanding of the mechanisms underlying cholesterol–nAChR interactions in the context of increasingly available high-resolution structural and functional data. Cholesterol and other lipids influence function by conformational selection and kinetic mechanisms, stabilizing varying proportions of activatable vs nonactivatable conformations, as well as influencing the rates of

Current Topics in Membranes, Volume 80
ISSN 1063-5823
http://dx.doi.org/10.1016/bs.ctm.2017.05.002

transitions between conformational states. In the absence of cholesterol and anionic lipids, the nAChR adopts an uncoupled conformation that binds agonist but does not undergo agonist-induced conformational transitions—unless the nAChR is located in a relatively thick lipid bilayer, such as that found in cholesterol-rich lipid rafts. We highlight different sites of cholesterol action, including the lipid-exposed M4 transmembrane α-helix. Cholesterol and other lipids likely alter function by modulating interactions between M4 and the adjacent transmembrane α-helices, M1 and M3. These same interactions have been implicated in both the folding and trafficking of nAChRs to the cell surface. We evaluate the nature of cholesterol–nAChR interactions, considering the evidence supporting the roles of both direct binding to allosteric sites and cholesterol-induced changes in bulk membrane physical properties.

1. INTRODUCTION

The adult human brain is estimated to contain between 100 and 500 trillion chemical connections or synapses (Drachman, 2005), which allow neurons to form the complex circuits that are essential to higher brain functions, such as perception, thought, memory, and learning. Signaling across chemical synapses is initiated when an action potential in the presynaptic neuron triggers the fusion of neurotransmitter-filled vesicles with the presynaptic membrane, thus releasing neurotransmitter into the synaptic cleft. Neurotransmitter then diffuses across the synaptic cleft to activate receptors located on the target cell membrane and thus initiate a postsynaptic response. The target neurotransmitter receptors are classified as either "fast" or "slow," with fast ionotropic receptors transiently opening either cation- or anion-selective ion channels across the membrane and slow metabotropic neurotransmitter receptors initiating cytoplasmic second messenger signaling cascades. Ionotropic receptors include a super family of pentameric ligand-gated ion channels (pLGICs), whose members respond to either excitatory neurotransmitters, such as acetylcholine or serotonin, or inhibitory neurotransmitters, such as γ-aminobutyric acid or glycine. These pLGICs perform a central role in information processing, have been implicated in a number of neurological disorders, and are the targets of numerous pharmaceutical products (Baenziger & Corringer, 2011; Dani & Bertrand, 2007; Kalamida et al., 2007; Mineur & Picciotto, 2008; Mufson, Counts, Perez, & Ginsberg, 2008; Taly, Corringer, Guedin, Lestage, & Changeux, 2009).

Nicotinic acetylcholine receptors (nAChRs) are excitatory cation-selective pLGICs that are found in pre-, post-, and nonsynaptic membranes throughout the central and peripheral nervous system where they respond to the neurotransmitter, acetylcholine. There are 17 genes in vertebrates encoding 17 different subunits (α1–α10, β1–β4, γ, ε, and δ) that combine to form a variety of either homo- or heteropentameric structures (Jones & Sattelle, 2010). Due to its high natural abundance in electric fish, the *Torpedo* nAChR is the best characterized of the nAChRs in terms of its structure and function. The *Torpedo* nAChR is most similar to the muscle-type nAChR found at the neuromuscular junction, being formed from four distinct subunit types organized in an $(\alpha1)_2\beta1\gamma\delta$ pentamer—although in adult muscle the fetal γ-subunit is replaced by the ε-subunit. Both heteromeric α4β2 and homomeric α7 nAChRs are abundant throughout the human brain. Less abundant combinations, such as α3β4, α3β2, and α6β2β3 nAChRs, etc., are targeted to specific brain regions (Dani & Bertrand, 2007; Taly et al., 2009).

Pioneering studies in the late 1970s and early 1980s aimed at reconstituting acetylcholine-induced channel function into model membranes revealed a functional sensitivity of the *Torpedo* nAChR to lipids, including cholesterol (Briley & Changeux, 1978; Criado, Eibl, & Barrantes, 1982; Dalziel, Rollins, & McNamee, 1980; Epstein & Racker, 1978; Fong & McNamee, 1986; Heidmann, Sobel, Popot, & Changeux, 1980; Schiebler & Hucho, 1978). Since then, the lipid sensitivity of the nAChR has been thoroughly documented, as summarized in several reviews (Baenziger, Henault, Therien, & Sun, 2015; Barrantes, 2002, 2004, 2015; Fantini & Barrantes, 2009; Henault, Juranka, Baenziger, et al., 2015). In the past decade or so, however, there have been three broad advances that have altered our mechanistic understanding of cholesterol–nAChR interactions. First, recent studies have highlighted the importance of interpreting lipid–nAChR interactions in the context of a thermodynamic framework, with an additional conformation, the lipid-dependent uncoupled conformation, required to explain lipid sensing (daCosta & Baenziger, 2009; daCosta, Dey, Therien, & Baenziger, 2013; daCosta et al., 2009). The existence of a lipid-dependent uncoupled conformation requires a reevaluation of previously obtained data that were interpreted exclusively in terms of the canonical resting, open, and desensitized conformations. Second, a 4-Å resolution cryoelectron microscopy model of the *Torpedo* nAChR published in 2005 (Unwin, 2005), as well as X-ray crystal structures of a number of homologs pLGICs (Althoff, Hibbs, Banerjee, & Gouaux, 2014; Bocquet et al., 2009; Hassaine et al., 2014; Hibbs & Gouaux, 2011; Hilf & Dutzler, 2008, 2009; Miller & Aricescu, 2014; Nury et al., 2011; Sauguet et al., 2013) including the $\alpha4\beta2$ neuronal nAChR (Morales-Perez, Noviello, & Hibbs, 2016), has provided an increasingly detailed picture of both nAChR structure and the nature of ligand-induced conformational change. Atomic models now allow data on lipid–nAChR interactions to be placed in a structural context. Finally, increasing research highlights complex interactions between the nAChR and its lipid environment and suggests that lipids not only modulate nAChR function though allosteric mechanisms, but by altering cell-surface trafficking and localization (Baier, Gallegos, Levi, & Barrantes, 2010; Barrantes, 2010). In a biological setting, there are complex mechanisms by which the lipid environment influences nAChR function.

Insight into lipid–nAChR interactions provides a basis for understanding lipid sensing in the broader pLGIC superfamily. In this review, we focus on the role of cholesterol as an allosteric modulator of nAChR function. The goal is not to present a comprehensive survey of the literature surrounding cholesterol–nAChR interactions, but instead to critically assess the available literature in order to provide a clear sense of current mechanistic understanding. Note that lipid–nAChR interactions are of broad interest because even subtle changes in nAChR activity have profound effects on human biology (Schaaf, 2014; Shen, Deymeer, Sine, & Engel, 2006). Dynamic changes in lipid composition occur both as the nAChR traffics from intracellular compartments to the plasma membrane and as a consequence of diet, aging, and neurodegenerative disease (Ledesma, Martin, & Dotti, 2012; Martin, Dotti, & Ledesma, 2010; Sebastiao, Colino-Oliveira, Assaife-Lopes, Dias, & Ribeiro, 2013). The effects of cholesterol on nAChR function are particularly prescient given the abundance of cholesterol in neuronal membranes, and its importance to human health and disease. It is likely that lipid-induced changes in nAChR activity occur during both normal and abnormal brain function to alter synaptic communication.

2. CHOLESTEROL–nAChR INTERACTIONS: THE LIPID PERSPECTIVE

2.1 An Essential Role in nAChR Function?

Lipids, including cholesterol, are potent modulators of nAChR function (Fig. 1). The importance of cholesterol in nAChR function was first highlighted in 1980 when it was shown that the inclusion of 25 or 50 mol% cholesterol in a reconstituted membrane composed of 3:1 (mol: mol) phosphatidylethanolamine (PE):phosphatidylserine (PS) promotes both agonist-induced cation flux across the membrane and transitions from the

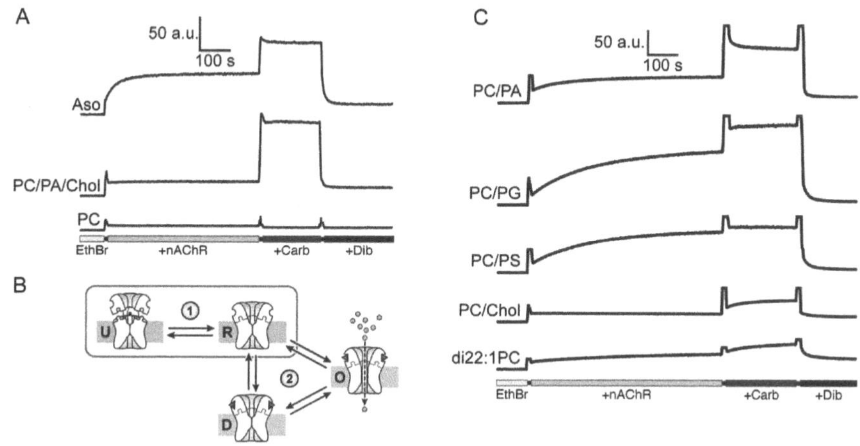

Fig. 1 Functional sensitivity of the *Torpedo* nAChR to lipids. (A) An ethidium-binding assay illustrates the functional sensitivity of the nAChR to cholesterol and anionic lipids, such as phosphatidic acid (PA). Ethidium bromide (EthBr; 0.3 μM) is first mixed with the nAChR reconstituted into soybean asolectin (Aso, *top trace*), PC/PA/cholesterol 3:1:1 (mol:mol:mol) (*middle trace*), or PC (*lower trace*). At this concentration, no binding of ethidium is observed to the resting state ($K_d \cong 1$ mM) —ethidium binding prior to agonist addition is mainly to desensitized nAChRs ($K_d = 0.3$ μM) and leads to a 15-fold increase in ethidium fluorescence (Herz, Johnson, & Taylor, 1987). The addition of 500 μM agonist, carbamylcholine (Carb), leads to rapid ethidium binding to nAChRs that transition from resting to open and then desensitized conformations. Traces are normalized to that of PC/PA/Chol–nAChR, where the dibucaine (Dib) displaceable fluorescence is 100 arbitrary units. PC/PA/Chol stabilizes predominantly activatable resting nAChRs, while asolectin stabilizes both resting and desensitized nAChRs. In PC membranes (16:0/18:1 acyl chains), the nAChR is locked in an uncoupled conformation that does not bind ethidium. (B) nAChR function is usually interpreted in terms of a conformational scheme involving resting (R), open (O), and desensitized (D) conformations (scheme 2). The membrane-reconstituted *Torpedo* nAChR can also adopt a non-responsive uncoupled (U) conformation (scheme 1). (C) Ethidium-binding assays recorded from the nAChR in 3:2 PC/PA, 3:2 PC/phosphatidyl glycerol, 3:2 PC/PS, 3:2 PC/Chol, and di22:1PC membranes show that anionic lipids stabilize varying proportions of activatable resting vs desensitized conformations (see daCosta et al., 2009). Biophysical studies suggest that di22:1PC-nAChR is uncoupled but undergoes slow transitions to the desensitized state (daCosta et al., 2013).

low-affinity agonist-binding resting to the high-affinity agonist-binding desensitized state (Dalziel et al., 1980). The ability of cholesterol to enhance nAChR function was subsequently documented in numerous studies, with some of these identifying synergistic roles for both cholesterol and anionic lipids (Criado, Eibl, & Barrantes, 1984; Criado et al., 1982; Fong & McNamee, 1986; Ochoa, Dalziel, & McNamee, 1983). In fact, these early studies led to the pervasive view in the literature that cholesterol is "essential" for the nAChR to adopt a structure that undergoes agonist-induced conformational transitions, with some studies also identifying an essential role for anionic lipids.

While there is no doubt that cholesterol is an effective potentiator of function, there is currently disagreement in the literature as to whether cholesterol is "essential" for the nAChR to undergo conformational transitions, with this disagreement depending on whether one views the role of cholesterol from a biological or a purely mechanistic perspective. Like most membranes, the *Torpedo* postsynaptic membrane is composed of cholesterol, glycerophospholipids, and sphingolipids (Gonzalez-Ros, Llanillo, Paraschos, & Martinez-Carrion, 1982; Popot, Demel, Sobel, Van Deenen, & Changeux, 1978; Rotstein, Arias, Barrantes, & Aveldano, 1987). Cholesterol is the most abundant representing 40–50 mol% of the total lipid composition. Glycerophospholipids make up the remaining 50–60 mol%, with sugar containing lipids making up less than 2 mol% of the total lipid composition. The zwitterionic phosphatidylcholine (~40%) and ethanolamine phosphoglycerides (30%–40%) are the major glycerophospholipids, with sphingomyelins representing less than 5% of the glycerophospholipid content. Anionic lipids make up the remaining 15–20 mol% of the phospholipids, with PS being the most abundant anionic lipid (~10 mol%) and phosphatidylinositols (PI), phosphatidic acid (PA), and cardiolipin found in lesser amounts. Significantly, when the nAChR is reconstituted into membranes with a broad phospholipid content that mimics that of its natural *Torpedo* membrane, close to 40 mol% cholesterol is required to promote agonist-induced conformational transitions (Rankin, Addona, Kloczewiak, Bugge, & Miller, 1997). In the natural *Torpedo* membrane environment where a diverse lipid composition is required to facilitate a broad range of postsynaptic membrane functions, 40 mol% cholesterol is *essential* to stabilize an activatable nAChR.

Many of the original assays that defined the essential role for cholesterol and anionic lipids in nAChR function, however, only probed whether the inclusion of a particular lipid at specific level in a PC membrane stabilized a measurable pool of functional nAChRs. As these assays could not accurately quantify the relative proportions of activatable vs nonactivatable conformations, membranes were typically characterized as either supporting channel function or not. For example, PC membranes containing cholesterol and an anionic lipid at a 3:1:1 M ratio were generally found to stabilize a large pool of agonist-responsive nAChRs, while a weak or no response was observed with 3:1 M ratios of PC/cholesterol, PC/PS, PC/PA, or PC/cardiolipin (Fong & McNamee, 1986; Sunshine & McNamee, 1992). The binary mixtures of PC and either cholesterol or anionic lipids were thus deemed nonfunctional, leading to the suggestion that cholesterol and anionic lipids in a PC membrane are both essential.

Subsequent studies using more quantifiable assays of channel function showed that increasing levels of either cholesterol or PA in a PC membrane increase the proportion of agonist-activatable nAChRs, with even low dosages of either lipid having measurable effects (Baenziger, Morris, Darsaut, & Ryan, 2000; Ryan, Demers, Chew, & Baenziger, 1996). In other words, a measurable agonist-induced response can be observed in PC membranes lacking cholesterol or anionic lipids, as long as sufficient quantities of either cholesterol or PA are present. In fact, high levels of PA in a PC membrane are almost as effective as more complex mixtures of PC, PA, and cholesterol at stabilizing a pool of functional receptors. Even the nAChR reconstituted into PA alone bilayers gives a measureable agonist-induced response (Sturgeon & Baenziger, 2010). Although the levels of PA required to optimally promote agonist-induced conformational change are not biologically relevant, these newer studies suggested that the nAChR can undergo conformational transitions in the absence of cholesterol or anionic lipids, but not both.

Finally, it was recently shown that the nAChR reconstituted into PC membranes with relatively thick hydrophobic cores undergoes agonist-induced conformational transitions even in the absence of both cholesterol and anionic lipids, suggesting that an appropriate membrane physical environment can substitute, to some extent, for specific lipid chemistry in promoting nAChR function (daCosta et al., 2013).

Note that the lack of an absolute functional requirement for cholesterol is supported further by the observation that several cholesterol analogs and other neutral lipids, such as diacylglycerol, effectively replace cholesterol in PC/anionic lipid bilayers to support channel function (Addona, Sandermann, Kloczewiak, & Miller, 2003; Kilian et al., 1980; Ryan et al., 1996; Sunshine & McNamee, 1992). On the other hand, the mechanistic requirement for anionic lipids remains less clear. Although anionic lipids are broadly effective at promoting agonist-induced flux in the presence of cholesterol (Sunshine & McNamee, 1992), distinct efficacies emerge in its absence. Membranes composed of mixtures of PC and PA stabilize a large pool of agonist-responsive nAChRs, while membranes composed of PC and PS exhibit little, if any response (daCosta, Ogrel, McCardy, Blanton, &

Baenziger, 2002; daCosta, Wagg, McKay, & Baenziger, 2004). There is clearly more to the effects of anionic lipids on nAChR function than the net anionic lipid charge.

Note also that the lipid requirements of the nAChR in a simple reconstituted membrane may not accurately reflect the lipid requirements of the nAChR in a natural membrane environment. Biological membranes are asymmetric with an uneven distribution of lipids in the two leaflets of the lipid bilayer (Barrantes, 2004; Perillo, Penalva, Vitale, Barrantes, & Antollini, 2016). The possibility that lipids have different effects on nAChR function when included in different leaflets has not been rigorously tested. Natural membranes also exhibit lateral separations of lipids into microdomains/lipid rafts, with the lipid composition in the microenvironment surrounding the nAChR not matching the bulk membrane lipid composition. The inclusion of only 1.6 mol% PA in a complex synthetic membrane composed of PC, PS, phosphatidylinositol, and cholesterol has a measurable effect on the proportion of agonist-responsive nAChRs, while the inclusion of 1.6% PA in a minimal PC membrane does not (Hamouda, Sanghvi, Sauls, Machu, & Blanton, 2006). These studies collectively show that the levels of a particular lipid required to measurably influence nAChR function in a simple homogeneous reconstituted membrane do not necessarily match the levels of the same lipid that are required in a biological membrane, where lipid distributions are heterogeneous and where synergistic effects on function likely occur. Even trace lipids may have an important role in nAChR function in biological membranes. The possibility that lipids involved in signaling influence nAChR function is obviously of considerable importance, particularly if signaling lipids are enriched in the nAChR's lipid microenvironment.

2.2 Cholesterol Effects on Transmembrane Domain Secondary Structure

Using established membrane reconstitution protocols, the nAChR incorporates into almost any synthetic lipid membrane and maintains agonist-binding capabilities (Criado et al., 1982, 1984; Fong & McNamee, 1986). As binding sites occur in the extracellular domain (ECD) at the interfaces between subunits, the functional consequences of cholesterol cannot be attributed to the stabilization of nAChR quaternary structure. Electron micrographs also show that the nonfunctional nAChR in PC membranes (PC-nAChR) exhibits the quaternary structure expected for a pLGIC, even in the absence of cholesterol (Bhushan & McNamee, 1990).

Early studies probed whether cholesterol influences the secondary structure of the nAChR. Infrared spectra recorded from the nAChR in membranes with or without cholesterol revealed spectral changes that were initially interpreted to suggest a slight cholesterol-induced increase in α-helix content (Butler & McNamee, 1993; Fong & McNamee, 1987). Cholesterol was proposed to intercalate between grooves to stabilize transmembrane α-helical structures. Another study reported that cholesterol dramatically increases the α-helical content leading to enhanced agonist-induced flux (Fernandez-Ballester et al., 1994). It was suggested that cholesterol exerts its effects on nAChR function by stabilizing elements of nAChR transmembrane secondary structure that are essential for conformational transitions.

Subsequent studies, however, suggested that the detected changes in secondary structure were due to experimental artifacts (Methot, Demers, & Baenziger, 1995). Infrared spectra are susceptible to the presence of water vapor, which can influence the analysis of protein secondary structure—especially if the levels of water vapor in the spectra vary from one sample to another (Reid, Moffat, & Baenziger, 1996). Infrared spectra are also typically recorded in ^2H_2O, with the resulting peptide backbone N–^{1}H to N–^{2}H exchange leading to subtle alterations in the shape of the secondary structure-sensitive amide I band (Methot, Ritchie, Blanton, & Baenziger, 2001). If samples are not exposed to ^2H_2O for similar lengths of time, these changes in amide I band shape can be easily misinterpreted in terms of altered secondary structure. Significantly, spectra of the nAChR recorded in the presence and absence of cholesterol at equivalent levels of peptide N–^{1}H to N–^{2}H exchange reveal essentially identical amide I band shapes, showing that cholesterol has no measurable effect on nAChR secondary structure (Methot et al., 1995). The thermal denaturation temperature of the nAChR is also similar in the presence and absence of cholesterol, although the cooperativity of unfolding is slightly reduced (daCosta & Baenziger, 2009; daCosta, Kaiser, & Baenziger, 2005). Cholesterol thus does not enhance channel function by stabilizing specific secondary structural features.

Note that a serendipitous discovery from these infrared measurements was that both cholesterol and anionic lipids in a PC membrane alter the kinetics of nAChR peptide backbone N–^{1}H to N–^{2}H exchange (Baenziger, Darsaut, & Morris, 1999; daCosta & Baenziger, 2009; Methot et al., 1995). The rates at which peptide hydrogens exchange with those in the bulk solvent depend on both hydrogen bonding and solvent exposure. Extensive data suggest that in PC membranes lacking cholesterol and anionic lipids, a region of the polypeptide backbone that is normally buried from solvent becomes exposed to the aqueous environment. As discussed later, the increased peptide backbone solvent exposure has been rationalized in terms of nAChR structure to provide insight into the mechanisms of both cholesterol– and anionic lipid–nAChR interactions.

2.3 A Thermodynamic Framework for Cholesterol–nAChR Interactions

In native *Torpedo* membranes, the nAChR exists in equilibrium between activatable resting and nonactivatable desensitized states, with the equilibrium strongly favoring the desensitized conformation (roughly 80:20 resting:desensitized) (Boyd & Cohen, 1980; Heidmann & Changeux, 1979). Brief exposure to agonist leads to rapid transient transitions from resting to channel open states, with prolonged exposure leading to the formation of one or more poorly defined channel inactive desensitized conformations, the latter characterized by a relatively high affinity for agonist (Changeux & Edelstein, 2005) (Fig. 1B). In the absence of clear effects on nAChR structure, lipid–nAChR functional data were increasingly interpreted in terms of the canonical resting, open, and desensitized conformations (Baenziger et al., 2000; Fernandez Nievas, Barrantes, & Antollini, 2008; Hamouda, Sanghvi, et al., 2006; Rankin et al., 1997).

Chemical labeling using the photoactivatable hydrophobic probe, 3-trifluoromethyl-3-(*m*-[^{125}I]iodophenyl) diazirine (TID) suggested that in the absence of cholesterol and anionic lipids, PC-nAChR adopts a desensitized conformation (McCarthy & Moore, 1992). Subsequent studies built on this assertion to show that increasing levels of cholesterol and/or anionic lipids in a PC membrane stabilize increasing proportions of activatable resting vs nonactivatable desensitized nAChRs (Baenziger et al., 2000; Hamouda, Sanghvi, et al., 2006; Ryan et al., 1996). The latter led to a "conformational selection" model, which proposed that lipids, including cholesterol, modulate the natural equilibrium between resting and desensitized states to influence channel function. As noted, the 80:20 equilibrium between resting and desensitized nAChRs in native membranes shows that the energy difference between the two conformations is relatively small. Subtle differences in the interactions between lipids/bulk membrane properties and either the resting or desensitized conformations could thus easily shift this equilibrium in favor of one conformation over the other.

One problem with a conformational selection model involving only resting and desensitized nAChRs, however, emerged from the above-described infrared structural studies of the nAChR in different membranes. As noted, these studies showed that the presumed "desensitized" PC-nAChR exhibits enhanced peptide N–^{1}H to N–^{2}H exchange kinetics compared to the agonist-desensitized nAChR in PC/PA/cholesterol or soybean asolectin membranes. The different peptide N–^{1}H to N–^{2}H exchange kinetics suggest that PC-nAChR is stabilized in a conformation that is distinct from the agonist-desensitized nAChR. Although many studies rationalized this observation by describing PC-nAChR as "desensitized-like," it was eventually shown that PC-nAChR does not exhibit the high agonist-binding affinity characteristic of the desensitized state (Baenziger et al., 2008; daCosta & Baenziger, 2009). In fact, additional pharmacological characterization has further defined the uncoupled PC-nAChR as a distinct conformation from both the resting and desensitized conformations (Sun, Comeau, & Baenziger, 2017). In light of these and other biochemical data, the canonical conformational scheme was expanded to include a lipid-dependent uncoupled conformation. The uncoupled PC-nAChR is characterized by a resting state-like agonist affinity, yet an inability to undergo agonist-induced conformational transitions (daCosta & Baenziger, 2009). It was thus suggested that lipids modulate the relative proportions of resting vs uncoupled conformations, with effects on this equilibrium essentially defining the functional response of the nAChR in different membrane environments.

Two subsequent studies suggested an even greater complexity to the mechanisms underlying lipid–nAChR interactions. First, different anionic lipids in a reconstituted PC membrane were each found to stabilize different proportions of uncoupled, resting, and desensitized conformations (daCosta et al., 2009) (Fig. 1C). For example, PC/PA/cholesterol and PC/PA membranes stabilize a large population of activatable resting state nAChRs, while PC/PS membranes stabilize a large proportion of desensitized nAChRs. The capacity of the nAChR to flux cations in a particular reconstituted membrane thus depends on the ability of that membrane to preferentially stabilize resting over both desensitized and uncoupled states. This distinction is important because it demonstrates that a nonfunctional nAChR in a particular membrane can result from different mechanisms. For example, both PC-nAChR and PC/PS-nAChR are inactive, but the former is inactive because it is predominantly uncoupled, while the latter is inactive because it is predominantly desensitized (daCosta et al., 2009, 2004).

Another study showed that lipids can alter the rates of transitions between conformations. The nAChR reconstituted into PC membranes lacking cholesterol and anionic lipids is stabilized almost exclusively in the uncoupled conformation, regardless of the thickness of the hydrophobic core of the membrane (dictated by the lengths of the fatty acyl chains). In thin PC membranes (PC with 14:1 or 18:1 diacyl chains), the uncoupled PC-nAChR does not transition upon agonist binding to the desensitized state over the time course of a typical biochemical experiment, despite the fact that the desensitized conformation binds agonist with much higher affinity. This implies that the activation energy barrier between uncoupled and coupled (resting, open, and desensitized) conformations is insurmountably high. In thick PC membranes (PC with 22:1 diacyl chains) the uncoupled PC-nAChR, however, undergoes slow agonist-induced transitions to ultimately the desensitized state (Fig. 1C). In these thick PC bilayers, the activation energy barrier between uncoupled and coupled conformations is lowered sufficiently to allow agonist-induced conformational transitions on an experimentally accessible time scale (daCosta et al., 2013). In this case, lipids influence nAChR function via a kinetic, as opposed to a conformational selection, mechanism.

The above discussion illustrates the evolution of our mechanistic understanding of lipid–nAChR interactions. While early studies sought to elucidate the "mechanism" by which lipids, including cholesterol, influence nAChR function, the above studies show that lipid–nAChR interactions must be interpreted in terms of a comprehensive thermodynamic framework (Fig. 2). Lipids/membranes can influence function by interacting preferentially with and thus lowering the energy of any conformational state to alter the relative proportions of activatable resting vs nonactivatable desensitized or uncoupled conformations. Lipids can also influence function by interacting preferentially with and thus lowering the energies of conformational transition states to promote conformational transitions. In this thermodynamic framework, there are a plethora of mechanisms by which lipids can influence nAChR function. In fact, a full understanding of the mechanisms underlying lipid–nAChR interactions will require intimate knowledge of the structure of each nAChR conformation and transition state, as well as an understanding of how lipids/membranes interact with each of these structures to influence function (Fig. 2B).

While understanding lipid–nAChR interactions in a thermodynamic framework is a complex undertaking, it is important to note that the situation is further complicated by the possibility that even a single lipid could influence function via both direct and indirect mechanisms. For example, a lipid could bind preferentially to a site on the resting conformation to stabilize an increased proportion of activatable nAChRs and thus promote function, while at the same time could alter the bulk membrane physical

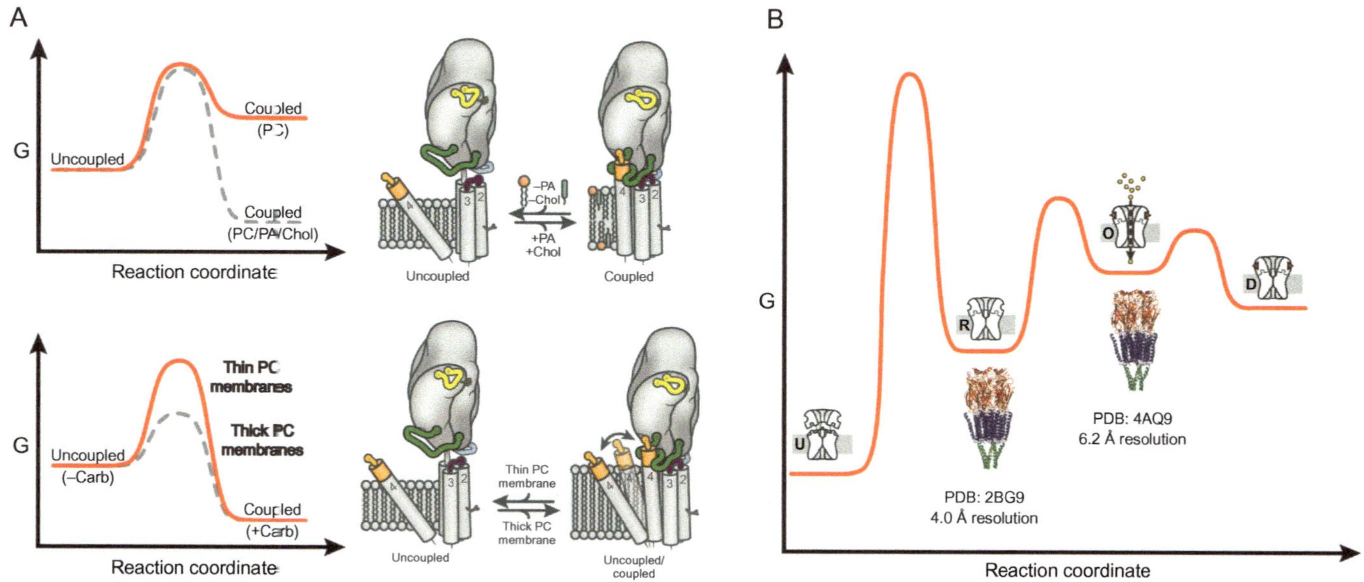

Fig. 2 A thermodynamic framework for interpreting lipid–nAChR interactions. (A) *Left panel:* Reaction coordinate diagrams illustrating qualitatively the effects of lipids on the structure and function of the nAChR. Membranes and/or specific lipids can influence the relative proportion of uncoupled vs coupled conformations (i.e., conformations where agonist binding and gating are allosterically coupled), a *conformational selection* mechanism. Membranes and/or specific lipids can also influence the activation energy barriers between conformations to alter the rates of conformational transitions, a *kinetic mechanism*. Note that the reaction coordinate diagrams are qualitative representations. *Right panel:* Schematic diagram of proposed models of uncoupling in the context of conformational selection vs kinetic mechanisms of lipid–nAChR interactions. Only one subunit is shown. Thicker membranes may reduce the activation energy between uncoupled (M4 tilted away to interact with lipids) and coupled (M4 bound to M1/M3) conformations. (B) A thermodynamic framework for interpreting lipid–nAChR interactions. Lipids can influence function by altering the relative proportions of resting, open, desensitized, and uncoupled conformations and/or the transitions between any of these states. Structures of the nAChR at 4 and 6.2 Å are available for the resting and open states (Unwin, 2005; Unwin & Fujiyoshi, 2012).

properties of the bilayer in a manner that favors a nonactivatable desensitized or uncoupled conformation. Dissecting the molecular mechanisms underlying lipid–nAChR interactions thus remains a daunting task.

2.4 Cholesterol Effects in a Conformational and Kinetic Framework

A few reports are notable in the insight they provide into cholesterol–nAChR interactions within the context of the thermodynamic framework discussed earlier. Rankin et al. were the first to probe in detail the kinetics of nAChR conformational transitions in different membrane environments, which they did by monitoring the rapid (ms to s) binding of the conformationally selective probe, ethidium, using a stopped flow fluorimeter (Rankin et al., 1997). When increasing levels of cholesterol were added to a reconstituted PC/PA membrane, an increasing component of rapid ethidium binding to the open state was observed. As agonist-binding studies showed that the proportion of high- (desensitized) vs low- (presumed to be resting) affinity sites did not markedly change in the presence of cholesterol, it was suggested that cholesterol enhances function by promoting conformational transitions from resting to open states—i.e., by a kinetic mechanism (see Fig. 2A, lower panel). The above data, however, were interpreted without knowledge of the low-affinity agonist-binding uncoupled conformation. So while the study suggested that cholesterol enhances the rates of conformational transitions between resting and open states, it does not exclude the possibility that cholesterol increases the proportion of activatable resting vs nonactivatable uncoupled conformations, which both bind agonist with relatively low affinity. Some of the observed effects regarding nAChR activation could thus reflect a conformational selection for resting vs uncoupled nAChRs, as opposed to a lowering of the kinetic barrier between resting and open/desensitized nAChRs (see Fig. 2A, top panel).

Hamouda et al. also examined the effects of increasing cholesterol on function in a reconstituted PC/PA membrane, but using the photoactivatable probe, TID (Hamouda, Sanghvi, et al., 2006). Resting state nAChRs exhibit relatively intense TID labeling, particularly of the α and γ subunits. In contrast, the desensitized nAChR exhibits weak stoichiometric labeling of the four subunits, α, β, γ, and δ. Although this TID-labeling study was interpreted without the knowledge of the uncoupled state, the work clearly shows that increasing levels of cholesterol in a PC/PA membrane lead to an increased proportion of resting state nAChRs that undergo agonist-induced conformational transitions.

Note that the ability of cholesterol to stabilize the resting conformation appears to be enhanced in the presence of anionic lipids (daCosta et al., 2002; Fong & McNamee, 1986; Sunshine & McNamee, 1992). In the absence of anionic lipids, infrared difference measurements suggest that increasing levels of cholesterol in a PC membrane still stabilize an increasing proportion of activatable nAChRs, although cholesterol is less efficacious than PA in this regard (Baenziger et al., 2000). Rankin et al. detected agonist-induced conformational transitions in PC/cholesterol membranes, but the proportion of nAChRs that become active in the presence of cholesterol is not clear from the presented data (Rankin et al., 1997). On the other hand, TID labeling does not detect significant stabilization of resting state nAChRs in the absence of anionic lipids, even with 40 mol% cholesterol in a PC membrane (daCosta et al., 2002; McCarthy & Moore, 1992). So although the addition of cholesterol to a PC membranes leads to the stabilization of activatable nAChRs, the activating effects of cholesterol in the absence of anionic lipids appears to be limited.

A recent ethidium-binding study suggested additional features regarding the mechanisms by which cholesterol influences nAChR function in the absence of anionic lipids. No ethidium binding to the nAChR in PC/cholesterol 3:2 (mol:mol) membranes is observed in the absence of agonist, suggesting that membranes composed of PC/cholesterol 3:2 do not stabilize an appreciable proportion of desensitized nAChRs (Fig. 1C). Subsequent agonist addition leads to only a small proportion of nAChRs transitioning ultimately to the high ethidium-binding affinity desensitized state. Of these, there is a small fast component of ethidium binding (i.e., within the mixing time of the manual ethidium addition) likely reflecting resting state nAChRs that rapidly transition to the open and then desensitized state. Intriguingly, there is also a slower component of ethidium binding (minutes time scale) that mimics the binding observed when agonist is added to the uncoupled nAChR in hydrophobically thick PC membranes. As noted earlier, the slow binding of ethidium to the nAChR in hydrophobically thick PC membranes suggests a slow conformational transition from the uncoupled to the desensitized state. One interpretation of the slow cholesterol binding is that cholesterol increases bilayer hydrophobic thickness to promote slow transitions from uncoupled to coupled conformations. Although the effects of cholesterol are modest in the absence of anionic lipids, cholesterol may enhance function by stabilizing activatable resting state nAChRs and by lowering the activation energy barrier between uncoupled and resting states.

2.5 Mechanism of Cholesterol Action: An Allosteric Site and/or Bulk Membrane Effects?

The above data suggest that cholesterol influences nAChR function by stabilizing a small proportion of resting nAChRs. Cholesterol may also promote conformational transitions leading to the open state. These effects on function could result from binding to a specific site or sites on the nAChR or from indirect effects on bulk membrane physical properties. Note that the above discussion shows that while cholesterol in *Torpedo* membranes is essential for stabilizing an activatable nAChR, other lipids can substitute for cholesterol to promote agonist-induced conformational transitions. This distinction is important from a mechanistic perspective, as an "essential" requirement implies that there is an allosteric site that must be occupied specifically with cholesterol to promote nAChR function. Although a nonessential modulatory role could still result from binding to a site with loose lipid specificity, a more likely interpretation is that cholesterol influences function by modulating the bulk physical properties of the membrane environment in a manner that can be mimicked by other neutral or anionic lipids. In this context, we discuss the evidence supporting the allosteric site and/or bulk membrane models of cholesterol action.

2.5.1 Cholesterol-Binding Sites on the nAChR

It has been suggested that cholesterol exerts its effects on function by interacting with annular and/or nonannular sites on the nAChR (Fig. 3). The nAChR incorporates preferentially into cholesterol-rich monolayers suggesting a strong affinity between cholesterol and the nAChR surface (Popot et al., 1978). A photoactivatable cholesterol analog labels sites on the lipid-exposed surface of the nAChR, including predominantly residues on M4 of each subunit (Blanton, Xie, Dangott, & Cohen, 1999; Corbin, Wang, & Blanton, 1998; Hamouda, Chiara, Sauls, Cohen, & Blanton, 2006). Annular lipid-binding sites have been observed in crystal structures of the prokaryotic homolog, GLIC (Bocquet et al., 2009) (see below).

Spin-labeling studies confirm that cholesterol exhibits a strong preference over other lipids for the annulus of lipid molecules that surround the outer surface of the nAChR (Ellena, Blazing, & McNamee, 1983; Mantipragada et al., 2003; Marsh & Barrantes, 1978; Marsh, Watts, & Barrantes, 1981). Although these cholesterol molecules are in fast exchange with cholesterol in the bulk lipid environment, at any given moment in time they are "bound" to the nAChR surface. The entire lipid-exposed surface of

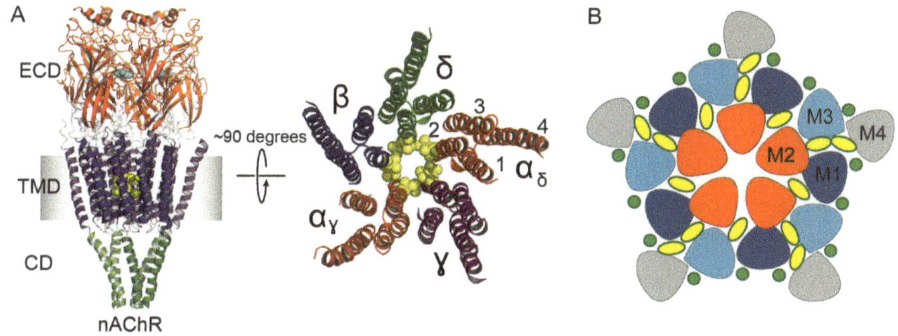

Fig. 3 Structure of the nAChR highlighting potential cholesterol-binding sites. (A) *Left panel:* A side view of the *Torpedo* nAChR structure (PDB ID: 2BG9) is shown on the left with the agonist-binding extracellular domain (ECD) in *red*, the transmembrane domain (TMD) in *blue*, and the cytoplasmic domain (CD) in *green*. Residues contributing to the proposed channel gate (αLeu251, αVal255, and homologs residues in the β, γ, and δ subunits) are shown as *yellow* spheres. The agonist-binding site αTrp149 is shown as *cyan* spheres. *Right panel:* Top-down view of the TMD highlighting individual subunits, as well as the four TMD α-helices in each subunit. (B) schematic top-down view of the TMD showing putative cholesterol-binding sites both between the TMD helices (*ovals;* Brannigan, Henin, Law, Eckenhoff, & Klein, 2008) and at the nAChR–lipid interface (*circles;* Narayanaswami & McNamee, 1993). *Figure modified from Brannigan, G., Henin, J., Law, R., Eckenhoff, R., Klein, M.L. Embedded cholesterol in the nicotinic acetylcholine receptor.* Proceedings of the National Academy of Sciences of the United States of America, 105*(2008), 14418–14423.*

the nAChR may serve as an "allosteric site" that is sensitive to bulk membrane physical properties, such as membrane hydrophobic thickness, etc. We do not yet understand if and how these fast exchanging lipids influence nAChR function.

Nonannular cholesterol-binding sites imbedded within the transmembrane domain (TMD) between α-helices were originally inferred based on fluorescence quenching studies with brominated lipids (Jones & McNamee, 1988). Dibromocholesterol quenches intrinsic nAChR fluorescence beyond that observed with dibromo-PC, with the latter presumably binding to annular sites on the protein surface. The enhanced quenching with dibromocholesterol was interpreted in terms of nonannular cholesterol-binding sites located within the core of the TMD, with additional evidence for the existence of distinct cholesterol-binding sites coming subsequently from similar fluorescence-based approaches using different fluorescent probes (Antollini & Barrantes, 1998; Narayanaswami & McNamee, 1993; see also Fernandez Nievas et al., 2008; Nievas, Barrantes, & Antollini, 2007). Two pools of cholesterol have also been identified in native nAChR-rich membranes, an easily extracted pool associated with the bulk membrane lipid and a nonextractable fraction thought to be tightly associated with the nAChR (Leibel, Firestone, Legler, Braswell, & Miller, 1987).

A recent study noted the existence of gaps in the electron density map of the nAChR, which can accommodate up to 15 cholesterol molecules per nAChR pentamer (Brannigan et al., 2008; see also Henin, Salari, Murlidaran, & Brannigan, 2014). Molecular dynamics simulations suggest that cholesterol binding to these nonannular cavities located between α-helices within the TMD α-helical bundles stabilizes the TMD structure. Significantly, the simulations suggest that cholesterol binding to these sites facilitates interactions between the transmembrane and ECDs (Brannigan et al., 2008), likely to promote the coupling of agonist binding and channel gating. A bioinformatics study also highlighted the existence of CRAC motifs (cholesterol recognition/interaction amino acid consensus) within the nAChR. A CRAC motif has the rather loose pattern of "-L/V-(X)(1–5)-Y-(X)(1–5)-R/K-," where "(X)(1–5)" represents between one and five residues of any amino acid. This motif has been suggested to facilitate the propensity of membrane proteins to incorporate into cholesterol-rich domains. Although a CRAC motif is found just prior to the transmembrane helix, M1, in the muscle-type nAChR, an inverted CRAC motif, termed CARC, is located on the M4-facing surface of M1 adjacent to one of the proposed cholesterol-binding cavities (Baier, Fantini, & Barrantes, 2011; Fantini, Di Scala, Baier, & Barrantes, 2016).

While the above all point to an important role for specific cholesterol-binding sites in nAChR function, other studies are less supportive. A number of neutral lipids, including diacylglycerol, α-tocopherol, coenzyme Q, and vitamins D_3 and K_1, as well as a host of cholesterol analogs including those with a negative charge are capable of replacing cholesterol to support function (Addona et al., 2003; Kilian et al., 1980; Labriola et al., 2010; Ryan et al., 1996; Sunshine & McNamee, 1992). The nAChR retains the capacity to undergo agonist-induced conformational transitions in membranes composed of PC/PA, PC/PG, and PA (Baenziger & Chew, 1997; Baenziger et al., 2000; daCosta et al., 2009, 2002, 2004; Labriola et al., 2010). Even the nAChR reconstituted into PC membranes lacking both activating cholesterol and anionic lipids still retains some capacity to undergo agonist induced conformational transitions, as long as the thickness of the bilayer is sufficient to lower the activation energy barrier between uncoupled and coupled (resting, open, and desensitized) conformations. In addition, cholesterol hemisuccinate covalently tethered to the glycerol backbone of PC is equally as effective as free cholesterol in supporting nAChR function in reconstituted PC/PA membranes (Addona, Sandermann, Kloczewiak, Husain, & Miller, 1998). The latter would presumably restrict cholesterol binding to only nonannular sites within 6 Å of the protein surface.

Finally, the structure of the human α4β2 nAChR was recently solved at 3.9 Å resolution. Although the best crystals were formed in the presence of both nicotine and a cholesterol analog—thus highlighting the importance of cholesterol in nAChR structure and function—electron density for specifically bound cholesterol was not observed in the structure (Morales-Perez et al., 2016). This contrasts the crystal structure of the glutamate-activated chloride channel, GluCl, where a PC molecule is wedged between α-helices at the interface between subunits to promote channel gating (Althoff et al., 2014).

Thus, while there is evidence both for and against the existence of specific cholesterol-binding sites, the occupation of such sites with cholesterol does not appear to be essential for *Torpedo* nAChR activity. If such modulatory sites exist, they exhibit rather lax structural requirements and can be filled by a variety of lipids.

2.5.2 Cholesterol and Bulk Membrane Physical Properties

Cholesterol could also exert its effects on nAChR function by altering bulk membrane physical properties. Cholesterol acts as a bidirectional regulator of the membrane physical environment in that it prevents membranes from becoming too fluid at high temperatures, while intercalating between phospholipids to prevent lipid associations and thus formation of the highly ordered gel state (Barrantes, 2004; Lundbaek, Andersen, Werge, & Nielsen, 2003). Above the gel-to-liquid crystal phase transition of a bilayer, cholesterol increases the order of the reconstituted bilayers, to ultimately alter other bulk membrane properties, such as bilayer thickness and elasticity, curvature stress, etc. (Andersen & Koeppe, 2007; Lee, 2003, 2004).

The role of bulk membrane physical properties in nAChR function has been debated in numerous publications, with some studies concluding that bulk membrane effects are the key modulators of channel function, and others concluding that lipid chemistry, and thus presumably the binding of lipids to specific sites on the nAChR, is more important (daCosta et al., 2004; Fong & McNamee, 1986; Sunshine & McNamee, 1994). Biophysical studies of the ordering of fluorescent probes have ruled out the effects of cholesterol on bilayer fluidity/membrane order as a modulatory mechanism for cholesterol action on nAChR function (Fernandez-Ballester et al., 1994; Sunshine & McNamee, 1994). In fact, the nAChR retains the ability to undergo conformational transitions both above and below the gel-to-liquid crystal phase transition of a reconstituted membrane (daCosta et al., 2013). On the other hand, single measurements of bulk membrane order may NOT be sufficiently robust to capture the complex physical environment of a reconstituted membrane thus making it difficult to correlate changes in membrane order with altered function.

It is also difficult to probe the influence of bulk membrane properties, such as membrane hydrophobic thickness, curvature stress, bilayer ordering, etc. without changing the bilayer lipid chemistry, complicating the interpretation of the experimental findings. In this context it is important to reiterate that changing the membrane hydrophobic thickness alters the ability of the nAChR to transition from uncoupled to coupled conformations, even in the absence of cholesterol or anionic lipids (daCosta et al., 2013). The latter provides definitive evidence that bulk membrane properties can influence agonist-induced state transitions. In addition, the slow conformational transitions, possibly from uncoupled to coupled conformations, observed upon agonist binding to the nAChR in PC/cholesterol membranes mimic the slow conformational transitions from uncoupled to coupled conformations observed in thick PC bilayers (daCosta et al., 2013). The latter supports the assertion that cholesterol influences conformational transitions via a bulk membrane effect—although not simply by altering membrane fluidity/order.

A role for bulk membrane physical properties in nAChR function could explain the enhanced ability of cholesterol to modulate nAChR function in the presence of anionic lipids. The anionic lipid charge may be required to promote effective electrostatic interactions between the agonist-binding ECD and the channel gating TMD (Xiu, Hanek, Wang, Lester, & Dougherty, 2005) (see below), but such interactions may be particularly effective when the nAChR is immersed in an appropriate bulk membrane physical environment—such as that created ideally in the presence of roughly 40 mol% cholesterol.

Bulk membrane effects could also explain the distinct efficacies of the anionic lipids, PA, and PS, in supporting nAChR function. While anionic lipids are broadly effective at promoting agonist-induced flux in PC/cholesterol membranes (Sunshine & McNamee, 1992), distinct efficacies emerge in the absence of cholesterol. As noted, membranes composed of PC and PA stabilize a large pool of agonist-responsive nAChRs, while those composed of PC and PS do not (daCosta et al., 2009, 2002, 2004). High levels of the anionic lipid PA in a PC membrane may be uniquely effective at stabilizing a functional nAChR because PA has both a net anionic charge and a small head group. Given the small head group, incorporation of PA into a PC bilayer leads to an ordering of the membrane and an increase in its gel-to-liquid phase transitions (daCosta et al., 2002), which may mimic to some extent the ordering effects of cholesterol. In contrast, high levels of PS, which has a larger head group, in a PC membrane, have little effect on reconstituted membrane physical properties (daCosta et al., 2004).

One study tested whether a rapid flipping of cholesterol between leaflets of the bilayer could relieve a putative mechanical bending strain in the bilayer induced by an asymmetric change in nAChR cross-sectional area upon activation, but found that cholesterol analogs that do not flip from one leaflet of the bilayer to the other are just as effective as cholesterol in supporting nAChR function (Addona et al., 1998). As noted, the same study showed that a PC-tethered cholesterol, which presumably resides within the bulk membrane environment, is as effective as free cholesterol in supporting channel function. While this study does not rule out the existence of cholesterol-binding sites close to the lipid–nAChR interface, it is consistent with cholesterol exerting its effects via the bulk membrane environment (Addona et al., 1998). Note that this tethered cholesterol analog did not replace free cholesterol in its ability to support the activity of the (Ca^{2+}–Mg^{2+})-ATPase—the latter study supporting the existence of a functionally important nonannular cholesterol-binding site(s) (Ding, Starling, East, & Lee, 1994).

Based on the ability of cholesterol analogs and other neutral lipids to substitute for cholesterol in promoting channel function, as well as the arguments presented earlier, it seems likely that bulk membrane physical properties have an important, yet poorly understood role in cholesterol action at the nAChR.

2.6 A Biological Role for Lipid-Dependent Uncoupling

A dominant feature of cholesterol–nAChR interactions is the existence of a lipid-dependent uncoupled conformation. The uncoupled nAChR is formed in PC membranes lacking cholesterol and anionic lipids. It exhibits low resting state-like agonist-binding affinity, but in most membranes does not transition to open or desensitized conformations (daCosta & Baenziger, 2009). Understanding lipid-dependent uncoupling is of particular interest because of anecdotal evidence suggesting that uncoupled nAChRs are relevant to human cholinergic biology. A large proportion of α4β2 nAChRs expressed in both human embryonic kidney cells and frog oocytes bind agonist, but do not undergo agonist-induced channel gating, consistent with the existence of a large pool of uncoupled receptors (Fenster, Whitworth, Sheffield, Quick, & Lester, 1999; Li & Steinbach, 2010). Some nAChR subtypes exist in oligomeric structures on cell surfaces but do not flux cations in response to acetylcholine binding (Broadbent et al., 2006; Drenan et al., 2008). The normalized current (current per acetylcholine-binding site on the surface of the cell) increases with time after injection of muscle-type cRNA into oocytes, suggesting a time-dependent maturation in "folding" to activatable conformations (Li, Schuchard, Palma, Pradier, & McNamee, 1990). Chronic nicotine exposure upregulates nAChR trafficking to the cell surface, but there appears to be a role for inactive nAChRs in this process (Nashmi et al., 2007), with chronic nicotine exposure potentially leading to the activation of previously assembled but inactive nAChRs (Govind, Walsh, & Green, 2012; Vallejo, Buisson, Bertrand, & Green, 2005).

Uncoupled nAChRs may also be relevant to nAChR folding and trafficking. As noted, thicker membranes, such as those found in lipid rafts (Lundbaek et al., 2003; Vuong, Baenziger, & Johnston, 2010), promote slow conformational transitions from uncoupled to coupled conformations (daCosta et al., 2013). Neuronal nAChRs require lipid rafts for trafficking to the cell surface (Bruses, Chauvet, & Rutishauser, 2001; Marchand, Devillers-Thiery, Pons, Changeux, & Cartaud, 2002). A lipid raft-dependent transition from uncoupled to coupled conformations could represent a final step in the folding and trafficking of nAChRs to the cell surface.

3. CHOLESTEROL–nAChR INTERACTIONS: THE PROTEIN PERSPECTIVE

3.1 nAChR Structure

The solution of a 4-Å resolution structure of the *Torpedo* nAChR in 2005 and subsequent structures of other homologs provides a structural framework for understanding lipid–nAChR interactions, including the lipid-dependent uncoupled conformation (Fig. 3). The *Torpedo* nAChR is formed from four distinct subunits organized in a pentamer, with the five subunits arranged pseudosymmetrically around a central axis that functions as an ion channel (Fig. 1A) (Althoff et al., 2014; Bocquet et al., 2009;

Hassaine et al., 2014; Hibbs & Gouaux, 2011; Hilf & Dutzler, 2008, 2009; Miller & Aricescu, 2014; Nury et al., 2011; Sauguet et al., 2013; Unwin, 2005). Each subunit contributes three distinct domains: an N-terminal ECD responsible for agonist binding, a TMD responsible for ion channel conductance, and a cytoplasmic domain of variable length that links to the cytoskeleton. The ECD of each subunit consists of 10 β-strands (β1–β10) forming 2 β-sheets that fold into a classic β-sandwich. The agonist-binding sites are located in the ECD, with the α-subunit contributing the principle face of each site (Kao & Karlin, 1986) and the adjacent γ- and δ-subunits contributing the complementary face (Corringer et al., 1995; Czajkowski, Kaufmann, & Karlin, 1993). The TMD of each subunit contributes four transmembrane α-helices (M1–M4) organized in a four-helix bundle. The five M2 α-helices line the channel pore, while M1 and M3 from each subunit form a ring of α-helices that shield M2 from the membrane (Corringer, Le Novere, & Changeux, 2000; Miyazawa, Fujiyoshi, & Unwin, 2003; Unwin, 2005). M4 is on the periphery of each subunit exposed to the lipid environment. As discussed later, considerable evidence suggests that M4 is involved in lipid sensing.

Lipids influence the capacity of the nAChR to couple agonist binding into channel gating. Coupling between the agonist-binding ECD and channel-gating TMD is mediated primarily by the covalent link between the C-terminus of β10 in the ECD and the N-terminus of M1 in the TMD, as well as by noncovalent connections between the β1/β2 and β6/β7 loops (the latter is referred to as the Cys-loop in eukaryotic pLGICs) of the ECD and the M2–M3 linker of the TMD (Fig. 2) (Jha, Cadugan, Purohit, & Auerbach, 2007; Lee & Sine, 2005). Although the detailed structural changes that occur when agonist-binding couples to channel gating remain to be defined, it is thought that concerted movements of the two β-sheets in the ECD lead to changes in structure of both the β1–β2 and β6–β7 loops, which are then proposed to translate into movement of the pore-lining M2 α-helices via direct interactions with the M2–M3 linker (Bocquet et al., 2009; Hibbs & Gouaux, 2011; Hilf & Dutzler, 2009; Lee & Sine, 2005; Lummis et al., 2005; Sauguet et al., 2014; Unwin & Fujiyoshi, 2012). In the closed resting conformation, the ion channel is occluded by hydrophobic residues located near the center of the bilayer that provide a barrier to the flow of a solvated ion (Fig. 1) (Beckstein & Sansom, 2006). A twisting and tilting of the M2 α-helices is thought to widen the pore by ~3 Å, allowing the flow of solvated ions down their electrochemical gradient.

3.2 M4 as a Lipid Sensor

Chemical labeling studies performed in the early 1990s originally proposed that the M4 transmembrane α-helix forms the predominant interface between the TMD of each subunit and the lipid environment (Blanton & Cohen, 1992, 1994). The location of M4 at the periphery of the TMD led to the suggestion that M4 plays a central role in lipid sensing (Antollini, Xu, Jiang, & Barrantes, 2005; Barrantes, 2003; Xu et al., 2005). In support of this assertion, numerous mutations of M4 lipid-facing residues on both *Torpedo* and muscle-type nAChRs, which alter M4–lipid interactions, also alter channel function (Bouzat, Roccamo, Garbus, & Barrantes, 1998; Caballero-Rivera et al., 2012; Diaz-De Leon, Otero-Cruz, Torres-Nunez, Casiano, & Lasalde-Dominicci, 2008; Lasalde et al., 1996; Lee et al., 1994; Li, Lee, Pappone, Palma, & McNamee, 1992; Ortiz-Acevedo et al., 2004; Shen et al., 2006; Tamamizu et al., 2000), typically affecting channel opening and closing rates (Bouzat, Barrantes, & Sine, 2000; Bouzat, Gumilar, del Carmen Esandi, & Sine, 2002; Bouzat et al., 1998). M4-swapped chimeric constructs of the homologous glycine receptor α1 and α3 subunits demonstrate that subunit-specific agonist efficacy is driven in large part by M4, with agonist sensitivity mediated by differences in M4–lipid interactions (Chen, Webb, & Lynch, 2009). Two of the three lipid-binding sites observed in the crystal structure of prokaryotic homolog, GLIC, are found at the interfaces between M4 and the adjacent TMD α-helices M1 and M3 (see Section 3.6).

More specifically, considerable evidence suggests that M4 is involved in cholesterol–nAChR interactions. As noted, a photoactivatable cholesterol analog labels sites on the lipid-exposed surface of the nAChR, including predominantly residues on M4 (Corbin et al., 1998; Hamouda, Chiara, et al., 2006). Fluorescence energy transfer experiments suggest interactions between residues on M4 and cholesterol (Narayanaswami & McNamee, 1993). Varying levels of cholesterol also have different effects on the activities of wild-type *Torpedo* nAChR vs *Torpedo* nAChR with a congenital myasthenic syndrome-causing Cys418Trp mutation (Santiago et al., 2001). The Cys418Trp mutation is located on the lipid-exposed surface of the M4 transmembrane α-helix of the α subunit and likely influences nAChR function by altering nAChR–lipid interactions (Shen et al., 2006).

Single-channel measurements of nAChR gating show that the structure of the lipid-exposed α-helix, M4, changes in four of five subunits when the nAChR transitions from resting to open conformations. When gating, the two αM4 α-helices in the $(\alpha1)_2\beta1\varepsilon\delta$ pentamer move synchronously, each moving as a unit roughly halfway along the reaction coordinate between agonist binding and the open transition state (Mitra, Bailey, & Auerbach, 2004). The motions of the εM4 and βM4 follow that of αM4, while δM4 has no apparent motion during gating. The temporal differences in various M4 positions should permit the pentamer to present a unique-binding surface to the lipid bilayer in each distinct conformation and/or while transitioning between them. Movement of M4 has also been suggested upon desensitization (Velisetty, Chalamalasetti, & Chakrapani, 2014). Such structural changes in M4 could expose or mask lipid-binding sites, allowing for stronger or weaker interactions with the lipid bilayer that preferentially stabilize different conformations or promote/inhibit transitions between these states. In addition, biophysical measurements and molecular dynamics simulations suggest that the orientation of M4 relative to the membrane normal depends on the presence of cholesterol and bilayer thickness (Antollini et al., 2005; Xu et al., 2005). Lipids may preferentially stabilize different orientations of M4 relative to the remainder of the TMD, altering gating kinetics as M4 moves along the reaction coordinate leading from agonist binding to the open state.

3.3 The M4 Lipid-Sensor Model of Uncoupling

The one defining biophysical property of the uncoupled PC-nAChR is that it undergoes more extensive peptide $N–^1H/N–^2H$ exchange when exposed to ^2H_2O buffer than either the resting or desensitized nAChRs. In fact, hydrogen exchange studies show that a population of peptide hydrogens (a roughly estimate points to 20%) that are buried from solvent and thus $N–^1H/N–^2H$ exchange-resistant in resting and desensitized states become exposed to solvent and thus exchange rapidly in the uncoupled state (daCosta & Baenziger, 2009). Although the regions of the nAChR that become solvent-exposed have not been defined, the critical importance of the ECD/TMD interface in channel function (Jha et al., 2007; Lee, Free, & Sine, 2009) led to the hypothesis that uncoupling results from weakened interactions leading to an increased physical separation between the ECD and TMD. Weakened interactions leading to a physical separation between the two domains would account for both the loss of function and the increased solvent accessibility, as well as both the lack of change in secondary/quaternary structure and the slight loss of cooperativity in the thermal unfolding of the uncoupled state (daCosta & Baenziger, 2009). Several crystallographic studies support the proposed structural independence of the ECD (Brejc et al., 2001; Dellisanti, Yao, Stroud, Wang, & Chen, 2007; Nury et al., 2010).

How could lipids influence interactions between the ECD and TMD via the proposed M4 lipid sensor? The cryo-EM structure shows that the C-terminus of M4 extends beyond the lipid bilayer to interact with a conserved phenylalanine in the Cys-loop (Fig. 2). Given that the Cys-loop plays a central role at the interface between the ECD and TMD, the simplest interpretation is that interactions between M4 and the Cys-loop are important for coupling agonist binding to channel gating. The M4 lipid-sensor model proposed that lipids influence M4 binding to M1/M3 to modulate M4–Cys-loop interactions and thus channel function. Ineffective M4–Cys-loop interactions could lead to partial dissociation of the ECD from the TMD to form the uncoupled state (daCosta & Baenziger, 2009).

3.4 Insight From Crystal Structures of Prokaryotic pLGICs

Crystal structures of the homopentameric prokaryotic pLGICs, GLIC and ELIC, reveal two features that impact on our understanding of lipid–nAChR interactions (Bocquet et al., 2009, 2007; Hilf & Dutzler, 2008, 2009; Sauguet et al., 2013; Tasneem, Iyer, Jakobsson, & Aravind, 2005; Zimmermann & Dutzler, 2011) (Fig. 4):

First, crystal structures of GLIC exhibit electron density located at the periphery of the TMD, suggesting the presence of partially ordered lipid molecules (Bocquet et al., 2009). Two lipid-binding sites on each subunit are found at the interfaces between M4 and the adjacent α-helices, M1 and M3, with one of these bridging interactions between the M4 C-terminus and the β6–β7 loop. Intriguingly, the lipid bridging M4 to the β6–β7 loop in the GLIC structure is displaced slightly by the inhibitory anesthetic propofol, suggesting that this lipid occupies an allosteric site formed by M4 and the β6–β7 loop (Nury et al., 2011).

Second, crystal structures of ELIC solved in the presence of agonist exhibit electron density in the agonist site but show no agonist-induced structural rearrangements in the TMD pore (Gonzalez-Gutierrez & Grosman, 2010; Gonzalez-Gutierrez et al., 2012; Zimmermann & Dutzler, 2011). In fact, no movement of the pore-lining M2 α-helices was detected with mutants that exhibit no propensity to desensitize. The ELIC structure thus reflects a conformation that is refractory to agonist binding (Gonzalez-Gutierrez & Grosman, 2010)—i.e., a conformation that is

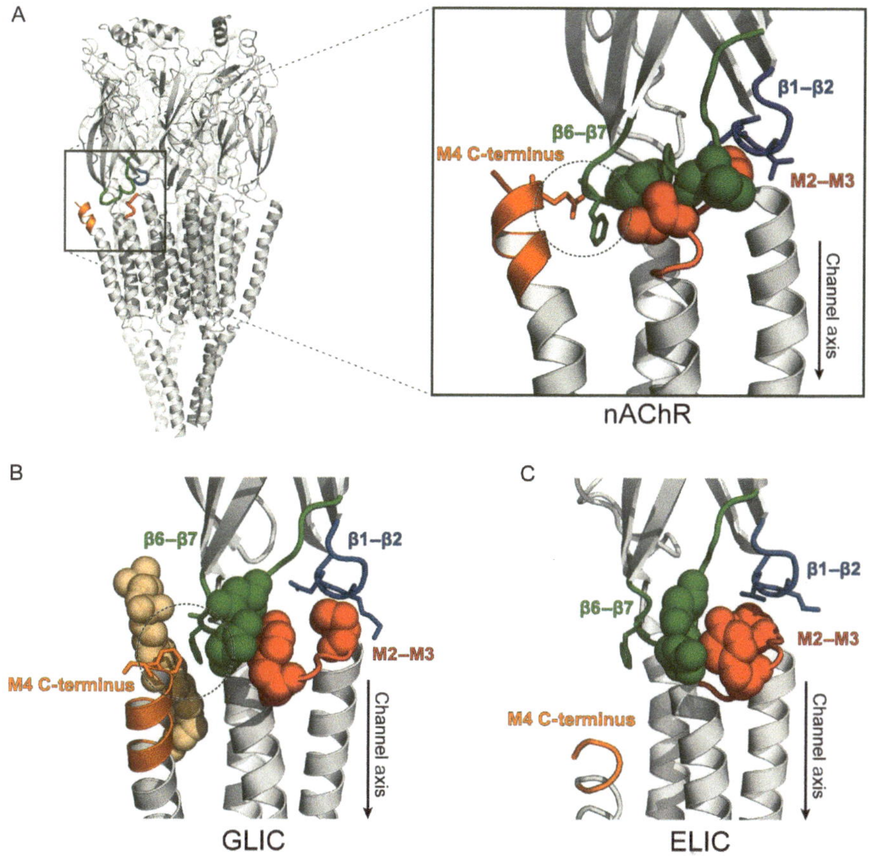

Fig. 4 Conformations of M4 at the ECD/TMD interface of the nAChR, GLIC, and ELIC. Structures are for the (A) nAChR (PDB ID: 2BG9), (B) GLIC (PDB ID: 3EAM), and (C) ELIC (PDB ID: 2VL0). In each case, M4, the β1–β2 loop, the β6–β7 loop (Cys-loop), and the M2–M3 linker are highlighted in *orange*, *light blue*, *green*, and *red*, respectively. For the nAChR, direct interactions between M4 (Gln435) and the Cys-loop (Phe137) are *circled*. For GLIC, a lipid molecule (beige space filling model) bridges interactions between M4 (Phe315) and the β6–β7 loop (Phe121).

functionally equivalent to the lipid-dependent uncoupled state of the *Torpedo* nAChR (daCosta & Baenziger, 2013). Significantly, the ELIC structures exhibit several of the features that have been proposed for the uncoupled nAChR (Fig. 4): (1) The M4 α-helix is partially unwound and tilted away from the remaining TMD, with several C-terminal M4 residues unresolved. (2) ELIC shows no direct contact between M4 C-terminus and either the adjacent TMD α-helix, M3, or the β6–β7 loop, and (3) The tilting of M4 away from the remainder of the TMD is accompanied by reduced contact, relative to the *Torpedo* nAChR, between both the β1–β2/β6–β7 loops of the ECD and the M2–M3 linker of the TMD. In effect, reduced interactions between M4 and M1/M3 are accompanied by weakened interactions between the ECD and TMD—as predicted by the M4 lipid-sensor model.

3.5 Do M4–M1/M3 Interactions Influence Channel Function?

The noted difference in the conformation of M4 in the crystal structures of ELIC and GLIC could result from different levels of aromatic interactions at the M4–M1/M3 interface. Aromatic residues at this interface play an essential role in the binding of M4 to M1/M3 during folding of the homologs glycine receptor (Haeger et al., 2010). GLIC has nine aromatic residues at the M4–M1/M3 interfaces that form an extensive network of π–π and CH–π interactions. Four of these aromatic residues are conserved in ELIC, with the remaining five aromatics replaced in ELIC by aliphatic residues (Fig. 5B). In particular, ELIC lacks a cluster of three aromatic residues located near the M4 C-terminus. The lower number of aromatic interactions at the M4–M1/M3 interface may render M4 binding to M1/M3 weaker in ELIC, thus making ELIC more susceptible to the perturbing effects of detergent solubilization prior to crystallization (daCosta & Baenziger, 2013).

Aromatic substitutions were used in GLIC and ELIC to test the effects of altered M4–M1/M3 interactions on pLGIC activity (Carswell, Henault, et al., 2015). Ala substitutions of aromatic residues were introduced into the M4–M1/M3 interface of GLIC to weaken M4 binding (Haeger et al., 2010), with every aromatic substitution leading to a reduction in channel function. Conversely, aliphatic to aromatic substitutions were introduced at the same interface in ELIC to promote M4 binding, with every aromatic addition leading to an enhancement in channel function. The introduction of a single pair of interacting aromatic residues on the C-terminus of M4 and the N-terminus of M3 led to a roughly 10-fold potentiation of channel function. In GLIC and ELIC, enhanced interactions between M4 and M1/M3 promote coupling between the agonist-binding site and channel gate.

Additional studies suggest that modulatable M4 binding to M1/M3 plays a role in pLGIC lipid sensing. Unlike the nAChR, both GLIC and ELIC can be purified in the absence of added lipid prior to reconstitution into lipid bilayers (Carswell, Sun, & Baenziger, 2015; Dellisanti et al., 2013; Labriola et al., 2013; Velisetty & Chakrapani, 2012; Velisetty,

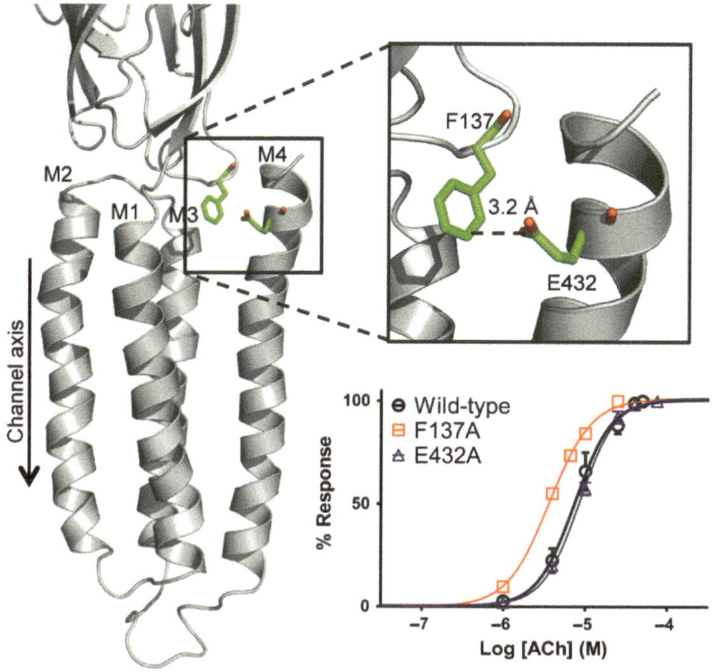

Fig. 5 Functional effects residues thought to mediate M4–Cys-loop interactions. Homology model of the human α-subunit based upon the *Torpedo* nAChR α-subunit (PDB ID: 2BG9). Residues F137 and E432 are shown as *sticks*, with carbon atoms in *green* and oxygen atoms in *red*. Averaged dose–response curves are presented for wild-type ($EC_{50}=7.61\pm1.25\,\mu M$) and both the F137A ($EC_{50}=3.47\pm0.26\,\mu M$) and E432A ($EC_{50}=8.83\pm0.62\,\mu M$) mutants, with $n\geq8$.

Chalamalasetti, & Chakrapani, 2012). GLIC retains robust agonist-induced gating in the minimal PC membranes that stabilize the uncoupled nAChR (Labriola et al., 2013). In fact, electrophysiological measurements suggest that GLIC function is relatively insensitive to its lipid environment (Dellisanti et al., 2013; Velisetty & Chakrapani, 2012). The abundance of aromatic interactions at the M4–M1/M3 interface in GLIC may be sufficient to maintain effective M4–M1/M3 interactions and retain channel function, even in PC membranes lacking cholesterol and anionic lipids. In contrast, ELIC does not gate open in response to agonist in these same minimal PC membranes. Significantly, engineering GLIC's M4 C-terminal aromatic cluster into ELIC restores the ability of ELIC to gate open in PC bilayers (Carswell, Sun, et al., 2015). Modulatable M4–M1/M3 interactions are thus relevant to ELIC lipid sensing.

3.6 How Do M4 Interactions With M1/M3 Influence Channel Function?

The M4 lipid-sensor model proposes that lipids promote M4–M1/M3 interactions to ultimately enhance interactions between M4 and the β6–β7 loop to promote channel function. In GLIC, a cluster of four interacting aromatic residues located on the C-terminus of M4, the N-terminus of M3 and the β6–β7 loop play an essential role in trafficking and function. Ala mutation of any one of these residues leads to a loss of function, while simultaneous mutation of two of these residues leads to a complete loss of cell surface expression and/or function, as does deletion of the two M4 C-terminal aromatics (Henault, Juranka, & Baenziger, 2015). An analogous set of interacting residues linking the M4 C-terminus, the M3 N-terminus, and the β6–β7 loop is not present in ELIC. In fact, sequential deletion of up to seven residues from the M4 C-terminus has little effect or even leads to a slight gain in channel function. So while interactions between the M4 C-terminus and the β6–β7 loop are essential in GLIC, similar interactions are not found in ELIC. Paradoxically, enhancing interactions between the M4 C-terminus and M3 in ELIC both promotes channel function and reduces the sensitivity of the pLGIC to its surrounding lipid environment (Carswell, Henault, et al., 2015; Carswell, Sun, et al., 2015).

To directly test the role of M4–β6–β7 loop interactions in the function of muscle-type nAChRs, we mutated two of the putative interacting residues, Phe137 and Gln435, to Ala (Fig. 5). Surprisingly, neither mutation led to a loss-of-function phenotype, suggesting that these residues are not essential for channel function. Although additional mutagenesis studies are required to fully elucidate the role of M4–β6–β7 loop interactions in nAChR lipid sensing, the data suggest that enhanced M4–M1/M3 interactions may promote channel function via mechanisms other than direct interactions between M4 and the β6–β7 loop.

3.7 Role of Cholesterol and M4–M1/M3 Interactions in the Trafficking and Clustering of nAChRs

The proposed role of M4 binding to M1/M3 may also be relevant to the folding, assembly, and trafficking of nAChRs to the cell surface. Although many factors are known to influence mammalian nAChR trafficking (Amici et al., 2012; Beeson et al., 2003; Henault, Juranka, et al., 2015; Henderson & Lester, 2015; Tsetlin, Kuzmin, & Kasheverov, 2011), a role for M4 binding to M1/M3 has been demonstrated in the homologous glycine receptor. When expressed in oocytes, the glycine receptor is cleaved within the intracellular loop but still expresses on the cell surface and undergoes agonist-induced channel gating. A truncated glycine receptor containing both the ectodomain and the first three TMD α-helices fails to traffic to the cell surface (Haeger et al., 2010). Coexpression with the M4 α-helix rescues both cell surface expression and channel function. Aromatic interactions at the interface between M4 and M1/M3 were found to be critical for M4 binding and thus in the folding and trafficking of these pLGICs.

The M4 C-terminus is critical for the surface expression of *Torpedo* nAChRs in frog oocytes (Tobimatsu et al., 1987). A similar role for the M4 C-terminus has been elucidated for the homologous $GABA_{\rho 1}$ (Reyes-Ruiz, Ochoa-de la Paz, Martinez-Torres, & Miledi, 2010) and 5-HT_{3A} (Butler et al., 2009) receptors, with the elimination of two or more residues preventing trafficking of 5-HT_{3A} receptors to the cell surface. It has also been suggested that M4 binding to M1/M3 leads to direct interactions between M4 and the Cys-loop to lock a homomeric $\alpha 7/5\text{HT}_{3A}$ chimera in a functional conformation (Pons et al., 2004).

As noted earlier, thick membranes support slow agonist-induced conformational transitions from uncoupled to coupled conformations. Lipid rafts/microdomains are typically enriched in cholesterol, which orders and increases the hydrophobic thickness of the bilayer (Lundbaek et al., 2003; Simons & Sampaio, 2011). Both muscle-type and neuronal nAChRs require lipid rafts for trafficking to the plasma membrane, with the raft-forming lipids cholesterol, sphingolipids, and ceramides being important (Baier & Barrantes, 2007; Baier et al., 2010; Bruses et al., 2001; Gallegos, Pediconi, & Barrantes, 2008; Marchand et al., 2002). Disruption of these rafts leads to both altered cell surface exposure and altered nAChR function (Baier et al., 2010; Borroni & Barrantes, 2011). Integral membrane proteins with shorter transmembrane α-helices tend to remain in intracellular membrane compartments, possibly because they hydrophobically match the thinner intracellular membranes more favorably (Bretscher & Munro, 1993; Lundbaek et al., 2003). Partitioning of the nAChR into thicker lipid rafts could favor transitions from uncoupled to coupled conformations, as we observe in the thicker di22:1PC model membranes. If this transition leads to a preferential alignment of M4 perpendicular to the membrane surface, the increased hydrophobic length of M4 should promote trafficking to the cell surface. Such a mechanism could explain why nicotine acts as a chaperone to promote cell-surface expression of the high-affinity $(\alpha 4)_2(\beta 2)_3$

nAChR vs the lower-affinity $(\alpha4)_3(\beta2)_2$ nAChR (Xiao et al., 2009). Preferential nicotine binding could lead to maturation of the high-affinity $(\alpha4)_2(\beta2)_3$ nAChRs toward coupled (resting or desensitized) conformations, where M4 is aligned closer to the bilayer normal to promote cell-surface expression. In this scenario, limited binding would thus have minimal effect on the trafficking of the low-affinity $(\alpha4)_3(\beta2)_2$ nAChR.

The clustering of nAChRs on the plasma membrane is also lipid raft-dependent (Zhu, Xiong, & Mei, 2006). Lowering cholesterol levels to disrupt lipid rafts leads to altered nAChR clustering and mobility on cell surfaces, rapid internalization, and ultimately enhances function of the surface-retained nAChRs (Almarza, Sanchez, & Barrantes, 2014; Borroni & Barrantes, 2011; Borroni et al., 2007). Although the molecular mechanisms underlying nAChR lipid-raft associations remain obscure, it has been shown that incorporation of the nAChR into model membranes containing anionic lipids, particularly PA, leads to a change in the packing properties of the surrounding bilayer (daCosta et al., 2002; Wenz & Barrantes, 2005). The nAChR concentrates cations, including protons, at the bilayer surface, facilitating the ordering of nearby anionic lipids (daCosta et al., 2004; Sturgeon & Baenziger, 2010). There appear to be favorable interactions between the nAChR and ordered lipids, such as those found in lipid rafts.

4. CONCLUSIONS

Three decades of research have revealed both an exquisite sensitivity of the *Torpedo* nAChR to its surrounding lipid environment and a tremendous complexity to lipid–nAChR interactions. Cholesterol plays a vital role stabilizing the nAChR in a functional conformation in native membranes. Although cholesterol may bind to sites on the nAChR, it likely exerts effects on nAChR function by altering bulk membrane physical properties. The mechanisms by which the membrane environment couples with the nAChR to influence function remain poorly understood.

Lipids, including cholesterol, influence nAChR function by stabilizing different proportions of resting, desensitized, and uncoupled conformations, as well as by promoting the transitions between these conformational states. Although the mechanisms by which cholesterol interacts preferentially with different conformations/transition states remain obscure, increasing evidence points to a role for the outermost transmembrane α-helix in lipid sensing. Both lipid binding and bulk membrane effects likely influence M4 interactions with M1/M3 to alter function. One hypothesis is that lipid-sensitive interactions between M4 and M1/M3 ultimately influence interactions between M4 and the Cys-loop to regulate nAChR activity, although other mechanisms are possible.

A lipid-dependent uncoupled conformation has been identified in membranes lacking cholesterol and anionic lipids. While this conformation has only been conclusively demonstrated for *Torpedo* nAChRs in reconstituted membranes, anecdotal evidence suggests that neuronal nAChRs also adopt an uncoupled conformation. An important goal will be to test for existence of uncoupled nAChRs in biological membranes, and whether cholesterol or other allosteric modulators enhance transitions from uncoupled to coupled conformations to promote cell-surface trafficking and thus ultimately enhance synaptic communication.

ACKNOWLEDGMENTS

This work was supported with funding provided to J.E.B. from the Canadian Institutes of Health Research (CIHR; Grant #111243) and the Natural Sciences and Engineering Research Council (Grant #113312).

REFERENCES

Addona, G. H., Sandermann, H., Jr., Kloczewiak, M. A., Husain, S. S., & Miller, K. W. (1998). Where does cholesterol act during activation of the nicotinic acetylcholine receptor? *Biochimica et Biophysica Acta, 1370*, 299–309.

Addona, G. H., Sandermann, H., Jr., Kloczewiak, M. A., & Miller, K. W. (2003). Low chemical specificity of the nicotinic acetylcholine receptor sterol activation site. *Biochimica et Biophysica Acta, 1609*, 177–182.

Almarza, G., Sanchez, F., & Barrantes, F. J. (2014). Transient cholesterol effects on nicotinic acetylcholine receptor cell-surface mobility. *PloS One, 9*, e100346.

Althoff, T., Hibbs, R. E., Banerjee, S., & Gouaux, E. (2014). X-ray structures of GluCl in apo states reveal a gating mechanism of Cys-loop receptors. *Nature, 512*, 333–337.

Amici, S. A., McKay, S. B., Wells, G. B., Robson, J. I., Nasir, M., Ponath, G., et al. (2012). A highly conserved cytoplasmic cysteine residue in the alpha4 nicotinic acetylcholine receptor is palmitoylated and regulates protein expression. *The Journal of Biological Chemistry, 287*, 23119–23127.

Andersen, O. S., & Koeppe, R. E., 2nd. (2007). Bilayer thickness and membrane protein function: An energetic perspective. *Annual Review of Biophysics and Biomolecular Structure, 36*, 107–130.

Antollini, S. S., & Barrantes, F. J. (1998). Disclosure of discrete sites for phospholipid and sterols at the protein-lipid interface in native acetylcholine receptor-rich membrane. *Biochemistry, 37*, 16653–16662.

Antollini, S. S., Xu, Y., Jiang, H., & Barrantes, F. J. (2005). Fluorescence and molecular dynamics studies of the acetylcholine receptor γM4 transmembrane peptide in reconstituted systems. *Molecular Membrane Biology, 22*, 471–483.

Baenziger, J. E., & Chew, J. P. (1997). Desensitization of the nicotinic acetylcholine receptor mainly involves a structural change in solvent-accessible regions of the polypeptide backbone. *Biochemistry, 36*, 3617–3624.

Baenziger, J. E., & Corringer, P. J. (2011). 3D structure and allosteric modulation of the transmembrane domain of pentameric ligand-gated ion channels. *Neuropharmacology, 60,* 116–125.

Baenziger, J. E., Darsaut, T. E., & Morris, M. L. (1999). Internal dynamics of the nicotinic acetylcholine receptor in reconstituted membranes. *Biochemistry, 38,* 4905–4911.

Baenziger, J. E., Henault, C. M., Therien, J. P., & Sun, J. (2015). Nicotinic acetylcholine receptor-lipid interactions: Mechanistic insight and biological function. *Biochimica et Biophysica Acta, 1848,* 1806–1817.

Baenziger, J. E., Morris, M. L., Darsaut, T. E., & Ryan, S. E. (2000). Effect of membrane lipid composition on the conformational equilibria of the nicotinic acetylcholine receptor. *The Journal of Biological Chemistry, 275,* 777–784.

Baenziger, J. E., Ryan, S. E., Goodreid, M. M., Vuong, N. Q., Sturgeon, R. M., & daCosta, C. J. B. (2008). Lipid composition alters drug action at the nicotinic acetylcholine receptor. *Molecular Pharmacology, 73,* 880–890.

Baier, C. J., & Barrantes, F. J. (2007). Sphingolipids are necessary for nicotinic acetylcholine receptor export in the early secretory pathway. *Journal of Neurochemistry, 101,* 1072–1084.

Baier, C. J., Fantini, J., & Barrantes, F. J. (2011). Disclosure of cholesterol recognition motifs in transmembrane domains of the human nicotinic acetylcholine receptor. *Scientific Reports, 1,* 69.

Baier, C. J., Gallegos, C. E., Levi, V., & Barrantes, F. J. (2010). Cholesterol modulation of nicotinic acetylcholine receptor surface mobility. *European Biophysics Journal, 39,* 213–227.

Barrantes, F. J. (2002). Lipid matters: Nicotinic acetylcholine receptor-lipid interactions. *Molecular Membrane Biology, 19,* 277–284.

Barrantes, F. J. (2003). Modulation of nicotinic acetylcholine receptor function through the outer and middle rings of transmembrane domains. *Current Opinion in Drug Discovery & Development, 6,* 620–632.

Barrantes, F. J. (2004). Structural basis for lipid modulation of nicotinic acetylcholine receptor function. *Brain Research. Brain Research Reviews, 47,* 71–95.

Barrantes, F. J. (2010). Cholesterol effects on nicotinic acetylcholine receptor: Cellular aspects. *Sub-Cellular Biochemistry, 51,* 467–487.

Barrantes, F. J. (2015). Phylogenetic conservation of protein-lipid motifs in pentameric ligand-gated ion channels. *Biochimica et Biophysica Acta, 1848,* 1796–1805.

Beckstein, O., & Sansom, M. S. (2006). A hydrophobic gate in an ion channel: The closed state of the nicotinic acetylcholine receptor. *Physical Biology, 3,* 147–159.

Beeson, D., Webster, R., Ealing, J., Croxen, R., Brownlow, S., Brydson, M., et al. (2003). Structural abnormalities of the AChR caused by mutations underlying congenital myasthenic syndromes. *Annals of the New York Academy of Sciences, 998,* 114–124.

Bhushan, A., & McNamee, M. G. (1990). Differential scanning calorimetry and Fourier transform infrared analysis of lipid-protein interactions involving the nicotinic acetylcholine receptor. *Biochimica et Biophysica Acta, 1027,* 93–101.

Blanton, M. P., & Cohen, J. B. (1992). Mapping the lipid-exposed regions in the *Torpedo californica* nicotinic acetylcholine receptor. *Biochemistry, 31,* 3738–3750.

Blanton, M. P., & Cohen, J. B. (1994). Identifying the lipid-protein interface of the Torpedo nicotinic acetylcholine receptor: Secondary structure implications. *Biochemistry, 33,* 2859–2872.

Blanton, M. P., Xie, Y., Dangott, L. J., & Cohen, J. B. (1999). The steroid promegestone is a noncompetitive antagonist of the Torpedo nicotinic acetylcholine receptor that interacts with the lipid-protein interface. *Molecular Pharmacology, 55,* 269–278.

Bocquet, N., Nury, H., Baaden, M., Le Poupon, C., Changeux, J. P., Delarue, M., et al. (2009). X-ray structure of a pentameric ligand-gated ion channel in an apparently open conformation. *Nature, 457,* 111–114.

Bocquet, N., Prado de Carvalho, L., Cartaud, J., Neyton, J., Le Poupon, C., Taly, A., et al. (2007). A prokaryotic proton-gated ion channel from the nicotinic acetylcholine receptor family. *Nature, 445*, 116–119.

Borroni, V., Baier, C. J., Lang, T., Bonini, I., White, M. M., Garbus, I., et al. (2007). Cholesterol depletion activates rapid internalization of submicron-sized acetylcholine receptor domains at the cell membrane. *Molecular Membrane Biology, 24*, 1–15.

Borroni, V., & Barrantes, F. J. (2011). Cholesterol modulates the rate and mechanism of acetylcholine receptor internalization. *The Journal of Biological Chemistry, 286*, 17122–17132.

Bouzat, C., Barrantes, F., & Sine, S. (2000). Nicotinic receptor fourth transmembrane domain: Hydrogen bonding by conserved threonine contributes to channel gating kinetics. *The Journal of General Physiology, 115*, 663–672.

Bouzat, C., Gumilar, F., del Carmen Esandi, M., & Sine, S. M. (2002). Subunit-selective contribution to channel gating of the M4 domain of the nicotinic receptor. *Biophysical Journal, 82*, 1920–1929.

Bouzat, C., Roccamo, A. M., Garbus, I., & Barrantes, F. J. (1998). Mutations at lipid-exposed residues of the acetylcholine receptor affect its gating kinetics. *Molecular Pharmacology, 54*, 146–153.

Boyd, N. D., & Cohen, J. B. (1980). Kinetics of binding of [3H]acetylcholine to Torpedo postsynaptic membranes: Association and dissociation rate constants by rapid mixing and ultrafiltration. *Biochemistry, 19*, 5353–5358.

Brannigan, G., Henin, J., Law, R., Eckenhoff, R., & Klein, M. L. (2008). Embedded cholesterol in the nicotinic acetylcholine receptor. *Proceedings of the National Academy of Sciences of the United States of America, 105*, 14418–14423.

Brejc, K., van Dijk, W. J., Klaassen, R. V., Schuurmans, M., van Der Oost, J., Smit, A. B., et al. (2001). Crystal structure of an ACh-binding protein reveals the ligand-binding domain of nicotinic receptors. *Nature, 411*, 269–276.

Bretscher, M. S., & Munro, S. (1993). Cholesterol and the Golgi apparatus. *Science, 261*, 1280–1281.

Briley, M. S., & Changeux, J. P. (1978). Recovery of some functional properties of the detergent-extracted cholinergic receptor protein from Torpedo marmorata after reintegration into a membrane environment. *European Journal of Biochemistry, 84*, 429–439.

Broadbent, S., Groot-Kormelink, P. J., Krashia, P. A., Harkness, P. C., Millar, N. S., Beato, M., et al. (2006). Incorporation of the beta3 subunit has a dominant-negative effect on the function of recombinant central-type neuronal nicotinic receptors. *Molecular Pharmacology, 70*, 1350–1357.

Bruses, J. L., Chauvet, N., & Rutishauser, U. (2001). Membrane lipid rafts are necessary for the maintenance of the (alpha)7 nicotinic acetylcholine receptor in somatic spines of ciliary neurons. *The Journal of Neuroscience, 21*, 504–512.

Butler, A. S., Lindesay, S. A., Dover, T. J., Kennedy, M. D., Patchell, V. B., Levine, B. A., et al. (2009). Importance of the C-terminus of the human 5-HT3A receptor subunit. *Neuropharmacology, 56*, 292–302.

Butler, D. H., & McNamee, M. G. (1993). FTIR analysis of nicotinic acetylcholine receptor secondary structure in reconstituted membranes. *Biochimica et Biophysica Acta, 1150*, 17–24.

Caballero-Rivera, D., Cruz-Nieves, O. A., Oyola-Cintron, J., Torres-Nunez, D. A., Otero-Cruz, J. D., & Lasalde-Dominicci, J. A. (2012). Tryptophan scanning mutagenesis reveals distortions in the helical structure of the deltaM4 transmembrane domain of the Torpedo californica nicotinic acetylcholine receptor. *Channels (Austin, Tex.), 6*, 111–123.

Carswell, C. L., Henault, C. M., Murlidaran, S., Therien, J. P., Juranka, P. F., Surujballi, J. A., et al. (2015a). Role of the fourth transmembrane α helix in the allosteric modulation of pentameric ligand-gated ion channels. *Structure, 23*, 1655–1664.

Carswell, C. L., Sun, J., & Baenziger, J. E. (2015b). Intramembrane aromatic interactions influence the lipid sensitivities of pentameric ligand-gated ion channels. *The Journal of Biological Chemistry, 290*, 2496–2507.

Changeux, J. P., & Edelstein, S. J. (2005). Allosteric mechanisms of signal transduction. *Science, 308*, 1424–1428.

Chen, X., Webb, T. I., & Lynch, J. W. (2009). The M4 transmembrane segment contributes to agonist efficacy differences between alpha1 and alpha3 glycine receptors. *Molecular Membrane Biology, 26*, 321–332.

Corbin, J., Wang, H. H., & Blanton, M. P. (1998). Identifying the cholesterol binding domain in the nicotinic acetylcholine receptor with [125I]azido-cholesterol. *Biochimica et Biophysica Acta, 1414*, 65–74.

Corringer, P. J., Galzi, J. L., Eisele, J. L., Bertrand, S., Changeux, J. P., & Bertrand, D. (1995). Identification of a new component of the agonist binding site of the nicotinic alpha 7 homooligomeric receptor. *The Journal of Biological Chemistry, 270*, 11749–11752.

Corringer, P. J., Le Novere, N., & Changeux, J. P. (2000). Nicotinic receptors at the amino acid level. *Annual Review of Pharmacology and Toxicology, 40*, 431–458.

Criado, M., Eibl, H., & Barrantes, F. J. (1982). Effects of lipids on acetylcholine receptor. Essential need of cholesterol for maintenance of agonist-induced state transitions in lipid vesicles. *Biochemistry, 21*, 3622–3629.

Criado, M., Eibl, H., & Barrantes, F. J. (1984). Functional properties of the acetylcholine receptor incorporated in model lipid membranes. Differential effects of chain length and head group of phospholipids on receptor affinity states and receptor-mediated ion translocation. *The Journal of Biological Chemistry, 259*, 9188–9198.

Czajkowski, C., Kaufmann, C., & Karlin, A. (1993). Negatively charged amino acid residues in the nicotinic receptor delta subunit that contribute to the binding of acetylcholine. *Proceedings of the National Academy of Sciences of the United States of America, 90*, 6285–6289.

daCosta, C. J. B., & Baenziger, J. E. (2009). A lipid-dependent uncoupled conformation of the acetylcholine receptor. *The Journal of Biological Chemistry, 284*, 17819–17825.

daCosta, C. J., & Baenziger, J. E. (2013). Gating of pentameric ligand-gated ion channels: Structural insights and ambiguities. *Structure, 21*, 1271–1283.

daCosta, C. J. B., Dey, L., Therien, J. P., & Baenziger, J. E. (2013). A distinct mechanism for activating uncoupled nicotinic acetylcholine receptors. *Nature Chemical Biology, 9*, 701–707.

daCosta, C. J. B., Kaiser, D. E., & Baenziger, J. E. (2005). Role of glycosylation and membrane environment in nicotinic acetylcholine receptor stability. *Biophysical Journal, 88*, 1755–1764.

daCosta, C. J. B., Medaglia, S. A., Lavigne, N., Wang, S., Carswell, C. L., & Baenziger, J. E. (2009). Anionic lipids allosterically modulate multiple nicotinic acetylcholine receptor conformational equilibria. *The Journal of Biological Chemistry, 284*, 33841–33849.

daCosta, C. J. B., Ogrel, A. A., McCardy, E. A., Blanton, M. P., & Baenziger, J. E. (2002). Lipid-protein interactions at the nicotinic acetylcholine receptor. A functional coupling between nicotinic receptors and phosphatidic acid-containing lipid bilayers. *The Journal of Biological Chemistry, 277*, 201–208.

daCosta, C. J. B., Wagg, I. D., McKay, M. E., & Baenziger, J. E. (2004). Phosphatidic acid and phosphatidylserine have distinct structural and functional interactions with the nicotinic acetylcholine receptor. *The Journal of Biological Chemistry, 279*, 14967–14974.

Dalziel, A. W., Rollins, E. S., & McNamee, M. G. (1980). The effect of cholesterol on agonist-induced flux in reconstituted acetylcholine receptor vesicles. *FEBS Letters, 122*, 193–196.

Dani, J. A., & Bertrand, D. (2007). Nicotinic acetylcholine receptors and nicotinic cholinergic mechanisms of the central nervous system. *Annual Review of Pharmacology and Toxicology, 47*, 699–729.

Dellisanti, C. D., Ghosh, B., Hanson, S. M., Raspanti, J. M., Grant, V. A., Diarra, G. M., et al. (2013). Site-directed spin labeling reveals pentameric ligand-gated ion channel gating motions. *PLoS Biology*, *11*, e1001714.

Dellisanti, C. D., Yao, Y., Stroud, J. C., Wang, Z. Z., & Chen, L. (2007). Crystal structure of the extracellular domain of nAChR alpha1 bound to alpha-bungarotoxin at 1.94 A resolution. *Nature Neuroscience*, *10*, 953–962.

Diaz-De Leon, R., Otero-Cruz, J. D., Torres-Nunez, D. A., Casiano, A., & Lasalde-Dominicci, J. A. (2008). Tryptophan scanning of the acetylcholine receptor's betaM4 transmembrane domain: Decoding allosteric linkage at the lipid-protein interface with ion-channel gating. *Channels (Austin, Tex.)*, *2*, 439–448.

Ding, J., Starling, A. P., East, J. M., & Lee, A. G. (1994). Binding sites for cholesterol on ca(2+)-ATPase studied by using a cholesterol-containing phospholipid. *Biochemistry*, *33*, 4974–4979.

Drachman, D. A. (2005). Do we have brain to spare? *Neurology*, *64*, 2004–2005.

Drenan, R. M., Nashmi, R., Imoukhuede, P., Just, H., McKinney, S., & Lester, H. A. (2008). Subcellular trafficking, pentameric assembly, and subunit stoichiometry of neuronal nicotinic acetylcholine receptors containing fluorescently labeled alpha6 and beta3 subunits. *Molecular Pharmacology*, *73*, 27–41.

Ellena, J. F., Blazing, M. A., & McNamee, M. G. (1983). Lipid-protein interactions in reconstituted membranes containing acetylcholine receptor. *Biochemistry*, *22*, 5523–5535.

Epstein, M., & Racker, E. (1978). Reconstitution of carbamylcholine-dependent sodium ion flux and desensitization of the acetylcholine receptor from *Torpedo californica*. *The Journal of Biological Chemistry*, *253*, 6660–6662.

Fantini, J., & Barrantes, F. J. (2009). Sphingolipid/cholesterol regulation of neurotransmitter receptor conformation and function. *Biochimica et Biophysica Acta*, *1788*, 2345–2361.

Fantini, J., Di Scala, C., Baier, C. J., & Barrantes, F. J. (2016). Molecular mechanisms of protein-cholesterol interactions in plasma membranes: Functional distinction between topological (tilted) and consensus (CARC/CRAC) domains. *Chemistry and Physics of Lipids*, *199*, 52–60.

Fenster, C. P., Whitworth, T. L., Sheffield, E. B., Quick, M. W., & Lester, R. A. (1999). Upregulation of surface alpha4beta2 nicotinic receptors is initiated by receptor desensitization after chronic exposure to nicotine. *The Journal of Neuroscience*, *19*, 4804–4814.

Fernandez Nievas, G. A., Barrantes, F. J., & Antollini, S. S. (2008). Modulation of nicotinic acetylcholine receptor conformational state by free fatty acids and steroids. *The Journal of Biological Chemistry*, *283*, 21478–21486.

Fernandez-Ballester, G., Castresana, J., Fernandez, A. M., Arrondo, J. L., Ferragut, J. A., & Gonzalez-Ros, J. M. (1994). A role for cholesterol as a structural effector of the nicotinic acetylcholine receptor. *Biochemistry*, *33*, 4065–4071.

Fong, T. M., & McNamee, M. G. (1986). Correlation between acetylcholine receptor function and structural properties of membranes. *Biochemistry*, *25*, 830–840.

Fong, T. M., & McNamee, M. G. (1987). Stabilization of acetylcholine receptor secondary structure by cholesterol and negatively charged phospholipids in membranes. *Biochemistry*, *26*, 3871–3880.

Gallegos, C. E., Pediconi, M. F., & Barrantes, F. J. (2008). Ceramides modulate cell-surface acetylcholine receptor levels. *Biochimica et Biophysica Acta*, *1778*, 917–930.

Gonzalez-Gutierrez, G., & Grosman, C. (2010). Bridging the gap between structural models of nicotinic receptor superfamily ion channels and their corresponding functional states. *Journal of Molecular Biology*, *403*, 693–705.

Gonzalez-Gutierrez, G., Lukk, T., Agarwal, V., Papke, D., Nair, S. K., & Grosman, C. (2012). Mutations that stabilize the open state of the Erwinia chrisanthemi ligand-gated ion channel fail to change the conformation of the pore domain in crystals. *Proceedings of the National Academy of Sciences of the United States of America*, *109*, 6331–6336.

Gonzalez-Ros, J. M., Llanillo, M., Paraschos, A., & Martinez-Carrion, M. (1982). Lipid environment of acetylcholine receptor from Torpedo californica. *Biochemistry*, *21*, 3467–3474.

Govind, A. P., Walsh, H., & Green, W. N. (2012). Nicotine-induced upregulation of native neuronal nicotinic receptors is caused by multiple mechanisms. *The Journal of Neuroscience*, *32*, 2227–2238.

Haeger, S., Kuzmin, D., Detro-Dassen, S., Lang, N., Kilb, M., Tsetlin, V., et al. (2010). An intramembrane aromatic network determines pentameric assembly of Cys-loop receptors. *Nature Structural & Molecular Biology*, *17*, 90–98.

Hamouda, A. K., Chiara, D. C., Sauls, D., Cohen, J. B., & Blanton, M. P. (2006b). Cholesterol interacts with transmembrane alpha-helices M1, M3, and M4 of the Torpedo nicotinic acetylcholine receptor: Photolabeling studies using [(3)H]Azicholesterol. *Biochemistry*, *45*, 976–986.

Hamouda, A. K., Sanghvi, M., Sauls, D., Machu, T. K., & Blanton, M. P. (2006a). Assessing the lipid requirements of the *Torpedo californica* nicotinic acetylcholine receptor. *Biochemistry*, *45*, 4327–4337.

Hassaine, G., Deluz, C., Grasso, L., Wyss, R., Tol, M. B., Hovius, R., et al. (2014). X-ray structure of the mouse serotonin 5-HT3 receptor. *Nature*, *512*, 276–281.

Heidmann, T., & Changeux, J. P. (1979). Fast kinetic studies on the interaction of a fluorescent agonist with the membrane-bound acetylcholine receptor from Torpedo marmorata. *European Journal of Biochemistry*, *94*, 255–279.

Heidmann, T., Sobel, A., Popot, J. L., & Changeux, J. P. (1980). Reconstitution of a functional acetylcholine receptor. Conservation of the conformational and allosteric transitions and recovery of the permeability response; role of lipids. *European Journal of Biochemistry*, *110*, 35–55.

Henault, C. M., Juranka, P. F., & Baenziger, J. E. (2015b). The M4 transmembrane α-helix contributes differently to both the maturation and function of two prokaryotic pentameric ligand-gated ion channels. *The Journal of Biological Chemistry*, *290*, 25118–25128.

Henault, C. M., Sun, J., Therien, J. P., daCosta, C. J. B., Carswell, C. L., Labriola, J. M., et al. (2015a). The role of the M4 lipid-sensor in the folding, trafficking, and allosteric modulation of nicotinic acetylcholine receptors. *Neuropharmacology*, *96*, 157–168.

Henderson, B. J., & Lester, H. A. (2015). Inside-out neuropharmacology of nicotinic drugs. *Neuropharmacology*, *96*, 178–193.

Henin, J., Salari, R., Murlidaran, S., & Brannigan, G. (2014). A predicted binding site for cholesterol on the GABAA receptor. *Biophysical Journal*, *106*, 1938–1949.

Herz, J. M., Johnson, D. A., & Taylor, P. (1987). Interaction of noncompetitive inhibitors with the acetylcholine receptor. The site specificity and spectroscopic properties of ethidium binding. *The Journal of Biological Chemistry*, *262*, 7238–7247.

Hibbs, R. E., & Gouaux, E. (2011). Principles of activation and permeation in an anion-selective Cys-loop receptor. *Nature*, *474*, 54–60.

Hilf, R. J., & Dutzler, R. (2008). X-ray structure of a prokaryotic pentameric ligand-gated ion channel. *Nature*, *452*, 375–379.

Hilf, R. J., & Dutzler, R. (2009). Structure of a potentially open state of a proton-activated pentameric ligand-gated ion channel. *Nature*, *457*, 115–118.

Jha, A., Cadugan, D. J., Purohit, P., & Auerbach, A. (2007). Acetylcholine receptor gating at extracellular transmembrane domain interface: The cys-loop and M2-M3 linker. *The Journal of General Physiology*, *130*, 547–558.

Jones, O. T., & McNamee, M. G. (1988). Annular and nonannular binding sites for cholesterol associated with the nicotinic acetylcholine receptor. *Biochemistry*, *27*, 2364–2374.

Jones, A. K., & Sattelle, D. B. (2010). Diversity of insect nicotinic acetylcholine receptor subunits. *Advances in Experimental Medicine and Biology*, *683*, 25–43.

Kalamida, D., Poulas, K., Avramopoulou, V., Fostieri, E., Lagoumintzis, G., Lazaridis, K., et al. (2007). Muscle and neuronal nicotinic acetylcholine receptors. Structure, function and pathogenicity. *The FEBS Journal*, *274*, 3799–3845.

Kao, P. N., & Karlin, A. (1986). Acetylcholine receptor binding site contains a disulfide cross-link between adjacent half-cystinyl residues. *The Journal of Biological Chemistry*, *261*, 8085–8088.

Kilian, P. L., Dunlap, C. R., Mueller, P., Schell, M. A., Huganir, R. L., & Racker, E. (1980). Reconstitution of acetylcholine receptor from Torpedo californica with highly purified phospholipids: Effect of alpha-tocopherol, phylloquinone, and other terpenoid quinones. *Biochemical and Biophysical Research Communications*, *93*, 409–414.

Labriola, J. M., daCosta, C. J. B., Wang, S., Figeys, D., Smith, J. C., Sturgeon, R. M., et al. (2010). Phospholipase C activity affinity purifies with the *Torpedo* nicotinic acetylcholine receptor. *The Journal of Biological Chemistry*, *285*, 10337–10343.

Labriola, J. M., Pandhare, A., Jansen, M., Blanton, M. P., Corringer, P. J., & Baenziger, J. E. (2013). Structural sensitivity of a prokaryotic pentameric ligand-gated ion channel to its membrane environment. *The Journal of Biological Chemistry*, *288*, 11294–11303.

Lasalde, J. A., Tamamizu, S., Butler, D. H., Vibat, C. R., Hung, B., & McNamee, M. G. (1996). Tryptophan substitutions at the lipid-exposed transmembrane segment M4 of Torpedo californica acetylcholine receptor govern channel gating. *Biochemistry*, *35*, 14139–14148.

Ledesma, M. D., Martin, M. G., & Dotti, C. G. (2012). Lipid changes in the aged brain: Effect on synaptic function and neuronal survival. *Progress in Lipid Research*, *51*, 23–35.

Lee, A. G. (2003). Lipid-protein interactions in biological membranes: A structural perspective. *Biochimica et Biophysica Acta*, *1612*, 1–40.

Lee, A. G. (2004). How lipids affect the activities of integral membrane proteins. *Biochimica et Biophysica Acta*, *1666*, 62–87.

Lee, W. Y., Free, C. R., & Sine, S. M. (2009). Binding to gating transduction in nicotinic receptors: Cys-loop energetically couples to pre-M1 and M2-M3 regions. *The Journal of Neuroscience*, *29*, 3189–3199.

Lee, Y. H., Li, L., Lasalde, J., Rojas, L., McNamee, M., Ortiz-Miranda, S. I., et al. (1994). Mutations in the M4 domain of Torpedo californica acetylcholine receptor dramatically alter ion channel function. *Biophysical Journal*, *66*, 646–653.

Lee, W. Y., & Sine, S. M. (2005). Principal pathway coupling agonist binding to channel gating in nicotinic receptors. *Nature*, *438*, 243–247.

Leibel, W. S., Firestone, L. L., Legler, D. C., Braswell, L. M., & Miller, K. W. (1987). Two pools of cholesterol in acetylcholine receptor-rich membranes from Torpedo. *Biochimica et Biophysica Acta*, *897*, 249–260.

Li, L., Lee, Y. H., Pappone, P., Palma, A., & McNamee, M. G. (1992). Site-specific mutations of nicotinic acetylcholine receptor at the lipid-protein interface dramatically alter ion channel gating. *Biophysical Journal*, *62*, 61–63.

Li, L., Schuchard, M., Palma, A., Pradier, L., & McNamee, M. G. (1990). Functional role of the cysteine 451 thiol group in the M4 helix of the gamma subunit of Torpedo californica acetylcholine receptor. *Biochemistry*, *29*, 5428–5436.

Li, P., & Steinbach, J. H. (2010). The neuronal nicotinic alpha4beta2 receptor has a high maximal probability of being open. *British Journal of Pharmacology*, *160*, 1906–1915.

Lummis, S. C., Beene, D. L., Lee, L. W., Lester, H. A., Broadhurst, R. W., & Dougherty, D. A. (2005). Cis-trans isomerization at a proline opens the pore of a neurotransmitter-gated ion channel. *Nature*, *438*, 248–252.

Lundbaek, J. A., Andersen, O. S., Werge, T., & Nielsen, C. (2003). Cholesterol-induced protein sorting: An analysis of energetic feasibility. *Biophysical Journal*, *84*, 2080–2089.

Mantipragada, S. B., Horvath, L. I., Arias, H. R., Schwarzmann, G., Sandhoff, K., Barrantes, F. J., et al. (2003). Lipid-protein interactions and effect of local anesthetics in acetylcholine receptor-rich membranes from Torpedo marmorata electric organ. *Biochemistry, 42*, 9167–9175.

Marchand, S., Devillers-Thiery, A., Pons, S., Changeux, J. P., & Cartaud, J. (2002). Rapsyn escorts the nicotinic acetylcholine receptor along the exocytic pathway via association with lipid rafts. *The Journal of Neuroscience, 22*, 8891–8901.

Marsh, D., & Barrantes, F. J. (1978). Immobilized lipid in acetylcholine receptor-rich membranes from Torpedo marmorata. *Proceedings of the National Academy of Sciences of the United States of America, 75*, 4329–4333.

Marsh, D., Watts, A., & Barrantes, F. J. (1981). Phospholipid chain immobilization and steroid rotational immobilization in acetylcholine receptor-rich membranes from Torpedo marmorata. *Biochimica et Biophysica Acta, 645*, 97–101.

Martin, M., Dotti, C. G., & Ledesma, M. D. (2010). Brain cholesterol in normal and pathological aging. *Biochimica et Biophysica Acta, 1801*, 934–944.

McCarthy, M. P., & Moore, M. A. (1992). Effects of lipids and detergents on the conformation of the nicotinic acetylcholine receptor from Torpedo californica. *The Journal of Biological Chemistry, 267*, 7655–7663.

Methot, N., Demers, C. N., & Baenziger, J. E. (1995). Structure of both the ligand- and lipid-dependent channel-inactive states of the nicotinic acetylcholine receptor probed by FTIR spectroscopy and hydrogen exchange. *Biochemistry, 34*, 15142–15149.

Methot, N., Ritchie, B. D., Blanton, M. P., & Baenziger, J. E. (2001). Structure of the pore-forming transmembrane domain of a ligand-gated ion channel. *The Journal of Biological Chemistry, 276*, 23726–23732.

Miller, P. S., & Aricescu, A. R. (2014). Crystal structure of a human GABAA receptor. *Nature, 512*, 270–275.

Mineur, Y. S., & Picciotto, M. R. (2008). Genetics of nicotinic acetylcholine receptors: Relevance to nicotine addiction. *Biochemical Pharmacology, 75*, 323–333.

Mitra, A., Bailey, T. D., & Auerbach, A. L. (2004). Structural dynamics of the M4 transmembrane segment during acetylcholine receptor gating. *Structure, 12*, 1909–1918.

Miyazawa, A., Fujiyoshi, Y., & Unwin, N. (2003). Structure and gating mechanism of the acetylcholine receptor pore. *Nature, 424*, 949–955.

Morales-Perez, C. L., Noviello, C. M., & Hibbs, R. E. (2016). X-ray structure of the human $\alpha4\beta2$ nicotinic receptor. *Nature, 538*, 411–415.

Mufson, E. J., Counts, S. E., Perez, S. E., & Ginsberg, S. D. (2008). Cholinergic system during the progression of Alzheimer's disease: Therapeutic implications. *Expert Review of Neurotherapeutics, 8*, 1703–1718.

Narayanaswami, V., & McNamee, M. G. (1993). Protein-lipid interactions and Torpedo californica nicotinic acetylcholine receptor function. 2. Membrane fluidity and ligand-mediated alteration in the accessibility of gamma subunit cysteine residues to cholesterol. *Biochemistry, 32*, 12420–12427.

Nashmi, R., Xiao, C., Deshpande, P., McKinney, S., Grady, S. R., Whiteaker, P., et al. (2007). Chronic nicotine cell specifically upregulates functional alpha 4* nicotinic receptors: Basis for both tolerance in midbrain and enhanced long-term potentiation in perforant path. *The Journal of Neuroscience, 27*, 8202–8218.

Nievas, G. A., Barrantes, F. J., & Antollini, S. S. (2007). Conformation-sensitive steroid and fatty acid sites in the transmembrane domain of the nicotinic acetylcholine receptor. *Biochemistry, 46*, 3503–3512.

Nury, H., Bocquet, N., Le Poupon, C., Raynal, B., Haouz, A., Corringer, P. J., et al. (2010). Crystal structure of the extracellular domain of a bacterial ligand-gated ion channel. *Journal of Molecular Biology, 395*, 1114–1127.

Nury, H., Van Renterghem, C., Weng, Y., Tran, A., Baaden, M., Dufresne, V., et al. (2011). X-ray structures of general anaesthetics bound to a pentameric ligand-gated ion channel. *Nature, 469*, 428–431.

Ochoa, E. L., Dalziel, A. W., & McNamee, M. G. (1983). Reconstitution of acetylcholine receptor function in lipid vesicles of defined composition. *Biochimica et Biophysica Acta*, *727*, 151–162.

Ortiz-Acevedo, A., Melendez, M., Asseo, A. M., Biaggi, N., Rojas, L. V., & Lasalde-Dominicci, J. A. (2004). Tryptophan scanning mutagenesis of the gammaM4 transmembrane domain of the acetylcholine receptor from Torpedo californica. *The Journal of Biological Chemistry*, *279*, 42250–42257.

Perillo, V. L., Penalva, D. A., Vitale, A. J., Barrantes, F. J., & Antollini, S. S. (2016). Trans-bilayer asymmetry and sphingomyelin composition modulate the preferential membrane partitioning of the nicotinic acetylcholine receptor in Lo domains. *Archives of Biochemistry and Biophysics*, *591*, 76–86.

Pons, S., Sallette, J., Bourgeois, J. P., Taly, A., Changeux, J. P., & Devillers-Thiery, A. (2004). Critical role of the C-terminal segment in the maturation and export to the cell surface of the homopentameric alpha 7-5HT3A receptor. *The European Journal of Neuroscience*, *20*, 2022–2030.

Popot, J. L., Demel, R. A., Sobel, A., Van Deenen, L. L., & Changeux, J. P. (1978). Interaction of the acetylcholine (nicotinic) receptor protein from Torpedo marmorata electric organ with monolayers of pure lipids. *European Journal of Biochemistry*, *85*, 27–42.

Rankin, S. E., Addona, G. H., Kloczewiak, M. A., Bugge, B., & Miller, K. W. (1997). The cholesterol dependence of activation and fast desensitization of the nicotinic acetylcholine receptor. *Biophysical Journal*, *73*, 2446–2455.

Reid, S. E., Moffat, D. J., & Baenziger, J. E. (1996). The selective enhancement and subsequent subtraction of atmospheric water vapour contributions from Fourier transform infrared spectra of proteins. *Spectrochimica Acta, Part A*, *52*, 1347–1356.

Reyes-Ruiz, J. M., Ochoa-de la Paz, L. D., Martinez-Torres, A., & Miledi, R. (2010). Functional impact of serial deletions at the C-terminus of the human GABArho1 receptor. *Biochimica et Biophysica Acta*, *1798*, 1002–1007.

Rotstein, N. P., Arias, H. R., Barrantes, F. J., & Aveldano, M. I. (1987). Composition of lipids in elasmobranch electric organ and acetylcholine receptor membranes. *Journal of Neurochemistry*, *49*, 1333–1340.

Ryan, S. E., Demers, C. N., Chew, J. P., & Baenziger, J. E. (1996). Structural effects of neutral and anionic lipids on the nicotinic acetylcholine receptor. An infrared difference spectroscopy study. *The Journal of Biological Chemistry*, *271*, 24590–24597.

Santiago, J., Guzman, G. R., Rojas, L. V., Marti, R., Asmar-Rovira, G. A., Santana, L. F., et al. (2001). Probing the effects of membrane cholesterol in the Torpedo californica acetylcholine receptor and the novel lipid-exposed mutation alpha C418W in Xenopus oocytes. *The Journal of Biological Chemistry*, *276*, 46523–46532.

Sauguet, L., Poitevin, F., Murail, S., Van Renterghem, C., Moraga-Cid, G., Malherbe, L., et al. (2013). Structural basis for ion permeation mechanism in pentameric ligand-gated ion channels. *The EMBO Journal*, *32*, 728–741.

Sauguet, L., Shahsavar, A., Poitevin, F., Huon, C., Menny, A., Nemecz, A., et al. (2014). Crystal structures of a pentameric ligand-gated ion channel provide a mechanism for activation. *Proceedings of the National Academy of Sciences of the United States of America*, *111*, 966–971.

Schaaf, C. P. (2014). Nicotinic acetylcholine receptors in human genetic disease. *Genetics in Medicine*, *16*, 649–656.

Schiebler, W., & Hucho, F. (1978). Membranes rich in acetylcholine receptor: Characterization and reconstitution to excitable membranes from exogenous lipids. *European Journal of Biochemistry*, *85*, 55–63.

Sebastiao, A. M., Colino-Oliveira, M., Assaife-Lopes, N., Dias, R. B., & Ribeiro, J. A. (2013). Lipid rafts, synaptic transmission and plasticity: Impact in age-related neurodegenerative diseases. *Neuropharmacology*, *64*, 97–107.

Shen, X. M., Deymeer, F., Sine, S. M., & Engel, A. G. (2006). Slow-channel mutation in acetylcholine receptor alphaM4 domain and its efficient knockdown. *Annals of Neurology, 60*, 128–136.
Simons, K., & Sampaio, J. L. (2011). Membrane organization and lipid rafts. *Cold Spring Harbor Perspectives in Biology, 3*, a004697.
Sturgeon, R. M., & Baenziger, J. E. (2010). Cations mediate interactions between the nicotinic acetylcholine receptor and anionic lipids. *Biophysical Journal, 98*, 989–998.
Sun, J., Comeau, J. F., & Baenziger, J. E. (2017). Probing the structure of the uncoupled nicotinic acetylcholine receptor. *Biochimica et Biophysica Acta, 1859*, 146–154.
Sunshine, C., & McNamee, M. G. (1992). Lipid modulation of nicotinic acetylcholine receptor function: The role of neutral and negatively charged lipids. *Biochimica et Biophysica Acta, 1108*, 240–246.
Sunshine, C., & McNamee, M. G. (1994). Lipid modulation of nicotinic acetylcholine receptor function: The role of membrane lipid composition and fluidity. *Biochimica et Biophysica Acta, 1191*, 59–64.
Taly, A., Corringer, P. J., Guedin, D., Lestage, P., & Changeux, J. P. (2009). Nicotinic receptors: Allosteric transitions and therapeutic targets in the nervous system. *Nature Reviews. Drug Discovery, 8*, 733–750.
Tamamizu, S., Guzman, G. R., Santiago, J., Rojas, L. V., McNamee, M. G., & Lasalde-Dominicci, J. A. (2000). Functional effects of periodic tryptophan substitutions in the alpha M4 transmembrane domain of the Torpedo californica nicotinic acetylcholine receptor. *Biochemistry, 39*, 4666–4673.
Tasneem, A., Iyer, L. M., Jakobsson, E., & Aravind, L. (2005). Identification of the prokaryotic ligand-gated ion channels and their implications for the mechanisms and origins of animal Cys-loop ion channels. *Genome Biology, 6*, R4.
Tobimatsu, T., Fujita, Y., Fukuda, K., Tanaka, K., Mori, Y., Konno, T., et al. (1987). Effects of substitution of putative transmembrane segments on nicotinic acetylcholine receptor function. *FEBS Letters, 222*, 56–62.
Tsetlin, V., Kuzmin, D., & Kasheverov, I. (2011). Assembly of nicotinic and other Cys-loop receptors. *Journal of Neurochemistry, 116*, 734–741.
Unwin, N. (2005). Refined structure of the nicotinic acetylcholine receptor at 4Å resolution. *Journal of Molecular Biology, 346*, 967–989.
Unwin, N., & Fujiyoshi, Y. (2012). Gating movement of acetylcholine receptor caught by plunge-freezing. *Journal of Molecular Biology, 422*, 617–634.
Vallejo, Y. F., Buisson, B., Bertrand, D., & Green, W. N. (2005). Chronic nicotine exposure upregulates nicotinic receptors by a novel mechanism. *The Journal of Neuroscience, 25*, 5563–5572.
Velisetty, P., & Chakrapani, S. (2012). Desensitization mechanism in prokaryotic ligand-gated ion channel. *The Journal of Biological Chemistry, 287*, 18467–18477.
Velisetty, P., Chalamalasetti, S. V., & Chakrapani, S. (2012). Conformational transitions underlying pore opening and desensitization in membrane-embedded GLIC. *The Journal of Biological Chemistry, 287*, 36864–36872.
Velisetty, P., Chalamalasetti, S. V., & Chakrapani, S. (2014). Structural basis for allosteric coupling at the membrane-protein interface in Gloeobacter violaceus ligand-gated ion channel (GLIC). *The Journal of Biological Chemistry, 289*, 3013–3025.
Vuong, N., Baenziger, J. E., & Johnston, L. J. (2010). Preparation of reconstituted acetylcholine receptor membranes suitable for AFM imaging of lipid-protein interactions. *Chemistry and Physics of Lipids, 163*, 117–126.
Wenz, J. J., & Barrantes, F. J. (2005). Nicotinic acetylcholine receptor induces lateral segregation of phosphatidic acid and phosphatidylcholine in reconstituted membranes. *Biochemistry, 44*, 398–410.

Xiao, C., Nashmi, R., McKinney, S., Cai, H., McIntosh, J. M., & Lester, H. A. (2009). Chronic nicotine selectively enhances alpha4beta2* nicotinic acetylcholine receptors in the nigrostriatal dopamine pathway. *The Journal of Neuroscience*, *29*, 12428–12439.

Xiu, X., Hanek, A. P., Wang, J., Lester, H. A., & Dougherty, D. A. (2005). A unified view of the role of electrostatic interactions in modulating the gating of Cys loop receptors. *The Journal of Biological Chemistry*, *280*, 41655–41666.

Xu, Y., Barrantes, F. J., Luo, X., Chen, K., Shen, J., & Jiang, H. (2005). Conformational dynamics of the nicotinic acetylcholine receptor channel: A 35-ns molecular dynamics simulation study. *Journal of the American Chemical Society*, *127*, 1291–1299.

Zhu, D., Xiong, W. C., & Mei, L. (2006). Lipid rafts serve as a signaling platform for nicotinic acetylcholine receptor clustering. *The Journal of Neuroscience*, *26*, 4841–4851.

Zimmermann, I., & Dutzler, R. (2011). Ligand activation of the prokaryotic pentameric ligand-gated ion channel ELIC. *PLoS Biology*, *9*, e1001101.

CHAPTER FIVE

Multiple Mechanisms of Regulation of Transient Receptor Potential Ion Channels by Cholesterol

Sara L. Morales-Lázaro[1], Tamara Rosenbaum[1]
Instituto de Fisiología Celular, Universidad Nacional Autónoma de México, Mexico City, Mexico
[1]Corresponding authors: e-mail address: saraluzm@ifc.unam.mx; trosenba@ifc.unam.mx

Contents

Abstract

The transient receptor potential (TRP) family of ion channels is constituted by several nonselective cation channels that are activated by diverse stimuli and that function as polymodal receptors. TRP ion channels are expressed in neural and nonneural tissues where they play important roles in cell physiology. The activation of these ion channels is achieved through changes in temperature, osmolarity, voltage, pH, pressure, and by some natural or synthetic chemical compounds that directly bind to these proteins to regulate their activity.

Other compounds that regulate TRP ion channel function are some endogenously synthetized lipid compounds. One such example of a compound of lipidic nature, commonly found in cells, is cholesterol. Cholesterol has been shown to exert positive and negative roles on TRP ion channel activity, albeit through different mechanisms which include a direct interaction of this molecule with specific amino acids located in the sequence of these channels or through the recruitment of the channels into specialized membrane microdomains (lipid rafts or caveolae). In this chapter, we will discuss important aspects of cholesterol as a regulator of some members of the TRP family of ion channels, and we will highlight the role of cholesterol on thermo-, chemo-, and osmosensation through regulation of the activity of these proteins.

Current Topics in Membranes, Volume 80
ISSN 1063-5823
http://dx.doi.org/10.1016/bs.ctm.2017.05.007

1. THE SUPER FAMILY OF TRANSIENT RECEPTOR POTENTIAL ION CHANNELS

Transient receptor potential (TRP) ion channels constitute a large family of ion channels involved in important physiological processes such as the perception of chemical, thermal, mechanic, and osmotic stimuli (Clapham, 2003). These receptors are mainly located at the cell surface but can also be found in membranes of intracellular organelles (Toro, Arias, & Brauchi, 2011). Due to their polymodal nature, they can be activated by several stimuli, allowing for the flow of ions across the cell membrane that results in changes of the membrane potential and cellular excitability (Clapham, 2003).

Mammalian TRP ion channels are classified into six subfamilies: TRPC (canonical), TRPM (melastatin), TRPML (mucolipin), TRPV (vanilloid), TRPA (ankyrin), and TRPP (polycystic) (Ramsey, Delling, & Clapham, 2006). All of them are constituted by four subunits (tetramers) where each subunit has six-transmembrane segments (S1–S6), with intracellularly localized N- and C-termini. Between the S5 and S6 segments are residues that, upon tetramerization, give rise to a central ion conducting pore that contains a selectivity filter that determines which ions can flow through the channel (Ramsey et al., 2006) (Fig. 1). The cytosolic regions contain ankyrin repeats (present in TRPA, TRPC, and TRPV subfamilies), calmodulin-binding sites, and PDZ domains that allow for the interaction with other proteins. Moreover, some of them contain an enzymatic domain (as what happens with TRPM6 and TRPM7 channels) (Montell, 2003; Schlingmann,

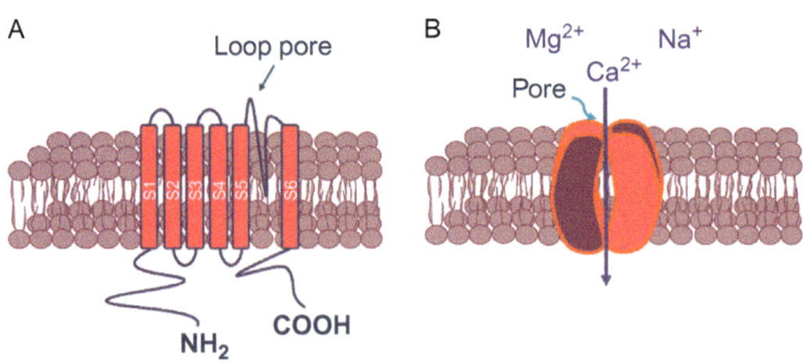

Fig. 1 Representation of the general TRP ion channel structure. (A) Each TRP subunit is composed of six-transmembrane segments (S1–S6), with amino (NH_2)- and carboxyl (COOH)-termini located intracellularly. The external pore loop is located between S5 and S6, and the pore is lined by the S6 domains. (B) A functional TRP ion channel is formed by four subunits that give rise to the functional channel.

Waldegger, Konrad, Chubanov, & Gudermann, 2007) or a large extracellular linker (between S1 and S2) with a polycystin motif (TRPP). These structural features impact on the function of TRP ion channels and their specific amino acid sequences represent essential targets for the binding of several ligands.

As mentioned earlier, different chemical and physical stimuli promote TRP channel activation. Some members of the TRP family are activated by temperature thus they function as sensors for cold, warm, and hot temperatures (TRPA1, TRPM3, TRPM8, TRPV1–4) and are referred to as thermoTRPs (Vriens, Nilius, & Voets, 2014). Most of these exquisite proteins are also regulated by distinct natural compounds found in plants (capsaicin, thymol, cinnamaldehyde, carvacrol, tetrahydrocannabinol, among others) or small endogenously produced molecules. Additionally, changes in intra- or extracellular pH and in the concentration of certain ions are also important chemical regulators of their activity (Ramsey et al., 2006).

Similar to what has been reported for other ion channels, regulators of TRP ion channels activation include molecules of a lipidic nature such as fatty acids, sphingolipids, phospholipids, and sterol compounds.

Sterols are a class of four-cyclic compounds with a cyclopentane perhydrophenanthrene nucleus, an hydroxyl group at position 3 of the A-ring, and a side chain located at carbon 17. These lipids are amphipathic molecules since the hydroxyl group provides a polar region, while the rest of the structure is nonpolar (Shieh, Hoard, & Nordman, 1977). These features allow for sterols to be important components of membrane structure with the hydroxyl group oriented toward the cytosolic face and the quite flat hydrophobic structure inserted in the lipid bilayer (Shieh et al., 1977). The most widely studied sterol in mammals is cholesterol (Fig. 2), not only

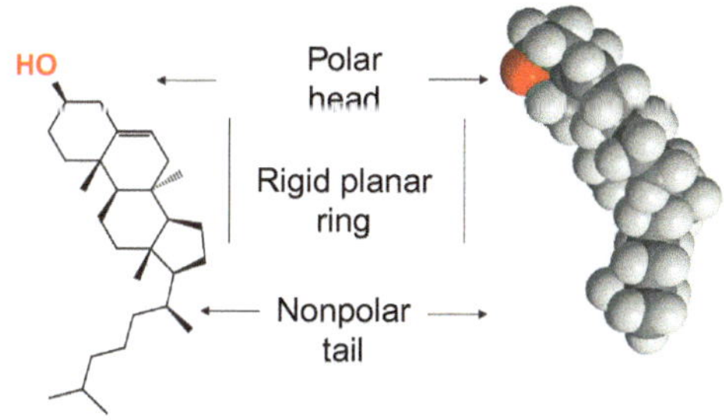

Fig. 2 Chemical structure of cholesterol. Cholesterol contains a hydrophilic hydroxyl group (polar head), fused rings which give rise to a rigid structure and a hydrophobic end constituted by a hydrocarbon tail. The chemical structure is shown on the *left* and a space-filled model of cholesterol is shown on the *right*.

because it is an important and abundant constituent of cell membranes but also because it actively modulates ion channel function.

Until now, only some members of three TRP channel subfamilies, TRPC, TRPM, and TRPV, have been shown to be modulated by cholesterol. The modulation of TRP ion channels by cholesterol can be achieved by two mechanisms: (1) by direct and specific interactions of cholesterol with the channel proteins or (2) by modifying the surrounding environment where these channels are located through changes in the lipid bilayer properties (Levitan, Fang, Rosenhouse-Dantsker, & Romanenko, 2010).

The following sections will discuss how TRP channel functions are regulated by cholesterol.

1.1 TRPC Ion Channels

The canonical subfamily of TRP ion channels is comprised of seven members (TRPC1–7), which are the closest mammalian homologues to the first described TRP channel in the *Drosophila* or fruit fly (Abramowitz & Birnbaumer, 2009). They are nonselective Ca^{2+}-permeable cation channels expressed in diverse tissues and in excitable and nonexcitable cells. These ion channels can be formed by four identical or by different subunits (homo- and heterotetramers, respectively) and the resulting combination of subunits ensues an increased diversity of their functionality in the cells where they are expressed (Abramowitz & Birnbaumer, 2009). The activation of these ion channels does not require a change in membrane potential; however, their opening can be regulated by voltage. Furthermore, TRPC channel activation is highly dependent on the phospholipase C (PLC) pathway since phosphatidylinositol 4,5-biphosphate (PIP2) hydrolysis produces phosphatidylinositol-3 (PI3), causing the release of Ca^{2+} and subsequent depletion of this ion from intracellular stores (Boulay et al., 1997), which in turn activates TRPC channels (Kiselyov et al., 1998; Putney, 2007; Yildirim, Kawasaki, & Birnbaumer, 2005). Additionally, diacylglycerol (DAG) produced by the actions of PLC can also directly activate some TRPC ion channels (TRPC3 and 6–7) (Hofmann et al., 1999). Other lipid compounds have also been reported as regulators of TRPC activity and among these are lysophospholipids, arachidonic acid (AA), and its metabolites, as well as cholesterol and its derivatives (Beech, 2012). These lipids have a positive role on TRPC regulation either by stimulating their activation or by promoting channel insertion into plasma membrane. In particular, cholesterol allows for the localization of TRPC into specialized microdomains in the plasma membrane, as will be discussed later.

1.1.1 TRPC1 Ion Channels Require Cholesterol for Their Localization in Caveolae

The TRPC1 ion channel is expressed in most human tissues with higher expression levels occurring in the brain, testis, ovaries, and in the cardiovascular system (Zhu, Chu, Peyton, & Birnbaumer, 1995). This protein is a nonvoltage-sensitive ion channel which is mainly activated by the depletion of Ca^{2+} from intracellular pools (Wes et al., 1995). It is mainly found in caveolae (specialized membrane cell invaginations enriched with cholesterol) where it has a relevant role for the physiology of human submandibular gland cells and vascular smooth muscles cells (Shi et al., 2012; Sun, Birnbaumer, & Singh, 2015). The TRPC1 ion channel directly interacts with caveolin-1, a tyrosine-phosphorylated-enriched protein in caveolae, which guarantees its correct localization in the cell membrane (Brazer, Singh, Liu, Swaim, & Ambudkar, 2003). The molecular interaction between TRPC1 and caveolin-1 allows for TRPC1 to be positioned correctly in the cell plasma membrane and to regulate the store-operated Ca^{2+} entry (SOCE) pathway at the caveolae (Fig. 3). The interaction between TRPC1 and caveolin-1 has been observed in rat endothelium-denuded caudal arteries where cholesterol depletion by methyl-β-cyclodextrin (MβCD) treatment (which scavenges cholesterol) reduces the interaction between TRPC1 and caveolin-1 leading to a decrease in SOCE and reducing the contractile effects that rely on the endothelin-1 pathway (Bergdahl et al., 2003). Similarly, it has been determined that there is an interaction between TRPC1 and caveolin-1 in lipid rafts from glioma cells, and the disruption of this specialized plasma membrane microdomains by MβCD impairs chemiotaxis induced by epidermal growth factor in these cells and also results in a decrease of the currents through TRPC1 (Bomben, Turner, Barclay, & Sontheimer, 2011). Considering that glioma is one of the more malignant and invasive type of tumors, the identification of novel regulators for the migration induced by growth factors is an important field of study that requires detailed analysis in order to understand the role of all the elements necessary for this cell process. Thus, it is not unwarranted that the TRPC1 ion channel could be a potential pharmacological target for the inhibition of this kind of pathology.

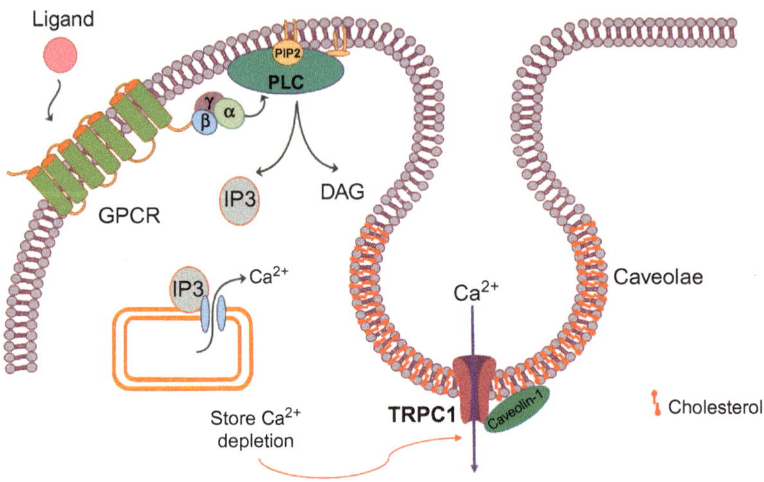

Fig. 3 Cholesterol is necessary for TRPC1 localization to caveolae. When a specific ligand binds to a G protein-coupled receptor (GPCR), this leads to the activation of the phospholipase C (PLC) pathway, resulting in PIP2 (phosphatidylinositol 4,5-biphosphate) hydrolysis to produce DAG (diacylglycerol) and IP3 (inositol 3 phosphate). The latter causes calcium store depletion, which activates TRPC1 ion channels located at caveolae. This localization is possible due to the interaction of TRPC1 with caveolin-1 and cholesterol.

1.1.2 The TRPC3 Ion Channel Is Positively Regulated by Cholesterol

The TRPC3 ion channel is highly expressed in brain and heart and it is a "lipid sensing" molecule since it is activated by DAG and modulated by other lipid compounds such as cholesterol. Similar to TRPC1 channels, for which the effects of cholesterol on its function are rather well defined, TRPC3 ion channels also need cholesterol for an adequate cell localization to occur. In this sense, it has been shown in HEK293 cells that TRPC3 expression is increased on the cell surface by cholesterol and is also the current dependent upon TRPC3 activation (Graziani et al., 2006). Other studies carried out using cerebral artery smooth muscle cells have shown that TRPC3 ion channels form a macrocomplex with caveolin-1 and inositol 1,4,5-triphosphate receptor type 1 (IP3-R1), which are located in caveolae. Here, TRPC3 is coupled to IP3 signaling leading to TRPC3 activation and promoting arterial vasoconstriction (Adebiyi, Narayanan, & Jaggar, 2011).

1.2 TRPM Ion Channels

The TRP melastatin subfamily (TRPM) contains eight members (M1–M8) described to date. This family was named after the first member identified, the TRPM1 which was isolated from a melanoma cDNA library

(Kraft & Harteneck, 2005). TRPM channels are structurally different to the channels from other TRP channel subfamilies since they lack ankyrin repeats and contain the largest amino (NH_2) region found in all of the TRP family of channels (Kraft & Harteneck, 2005). All of them are highly Ca^{2+} permeable and some members also allow for the flow of Mg^{2+} through their pores (TRPM6 and TRPM7) (Schmitz et al., 2003; Voets et al., 2004). The TRPM6 and TRPM7 proteins exhibit intrinsic enzymatic activity since they possess a serine/threonine kinase domain in their carboxyl (COOH) termini (Schlingmann & Gudermann, 2005).

These ion channels are activated by voltage, temperature, and specific ligands and some of them are regulated by cholesterol (e.g., TRPM3, TRPM7, and TRPM8).

1.2.1 TRPM3 Ion Channels Are Negatively Regulated by Cholesterol

TRPM3 expression has been identified in neuronal (sensory neurons) and nonneuronal tissues such as pancreatic beta cells and vascular smooth muscle cells (Naylor et al., 2010; Wagner et al., 2008). These channels are Ca^{2+} permeable and activated by heat and by specific ligands. The TRPM3 was one of the first TRP family channels to be identified as a target for the direct activation by steroids and shown to be activated by pregnenolone sulfate, highlighting an important role for this protein in insulin secretion (Wagner et al., 2008). TRPM3 expression in a subset of sensory neurons is important for noxious heat perception since this ion channel is activated by temperatures near 40°C (Vriens et al., 2011). Moreover, TRPM3 has an important role in the vascular system where its activation stimulates contraction and suppresses cytokine secretion from vascular smooth muscle cells (Naylor et al., 2010). Importantly, in these cells, it was reported that depletion of cholesterol by MβCD leads to TRPM3 activation even in the absence of the agonists of the channel and that cholesterol enrichment suppresses TRPM3 activation by pregnenolone sulfate (Naylor et al., 2010). These results demonstrate that endogenous cholesterol maintains TRPM3 partially inhibited and that a cholesterol overload, such as the one that prevails in atherosclerotic processes, could completely inhibit TRPM3 function in the cardiovascular system leading to inadequate cell contraction and allowing for proinflammatory cytokine secretion which result in a complicated scenario for the injured tissue (Naylor et al., 2010). The molecular mechanisms by which cholesterol suppresses TRPM3 activation remain unknown and represent an important field of future research that will allow for the complete understanding of the ways in which TRPM3 activation can be regulated.

1.2.2 The TRPM7 Channel as a Target for Positive Regulation by Cholesterol

The TRPM7 channel belongs to a group of proteins referred to as "chanzymes" due to their dual function as ion channels and serine/threonine kinases (Schmitz et al., 2003). It is important for adequate Mg^{2+} and Ca^{2+} homeostasis and it is activated by several stimuli such as pressure, shear stress, osmolarity, intracellular cations, and specific ligands. Interestingly, TRPM7 ion channels are specifically inhibited by intracellular Mg^{2+} nucleotides (Demeuse, Penner, & Fleig, 2006) and free intracellular Mg^{2+} (Nadler et al., 2001).

TRPM7's regulation by cholesterol has been shown to be achieved by means of two possible mechanisms. First, it was established that the bradykinin signaling pathway indirectly regulates TRPM7 function by allowing the suitable localization of TRPM7 into lipid rafts in vascular smooth muscle cells (Yogi, Callera, Tostes, & Touyz, 2009). This subcellular redistribution of TRPM7 ion channels in cholesterol-enriched microdomains induces a sustained increase in Mg^{2+} currents which could impact on the proinflammatory response of the cardiovascular system (Yogi et al., 2009). Second, TRPM7 is expressed in normal and cancerous prostate cells where its function is positively regulated by cholesterol resulting in cell proliferation and migration (Sun et al., 2014). The positive effects of cholesterol on TRPM7's activation were demonstrated by whole-cell patch-clamp experiments in prostate cells where acute and prolonged cholesterol treatment increased TRPM7 inward currents and stimulated cell proliferation and migration (Sun et al., 2014). These effects of cholesterol on TRPM7 activation could be blocked by statins that inhibit cholesterol synthesis. Thus, treatment with statins for the control of cholesterol levels could play protective roles in patients with prostate cancer.

As with TRPM3, the mechanism by which cholesterol modifies TRPM7 activity is unclear, and detailed studies are necessary in order to finely understand how cholesterol regulates the activity of these two members of the TRPM subfamily of ion channels.

1.2.3 TRPM8 Channel Activity Is Upregulated by Membrane Cholesterol Depletion

TRPM8 ion channel is found in primary afferent sensory neurons where it has a role as a sensor of low temperatures (threshold ~25°C). This ion channel is also activated by refreshing compounds such as menthol and icilin. Furthermore, TRPM8's activation by its agonists requires the presence of

PIP2 (Liu & Qin, 2005; Rohacs, Lopes, Michailidis, & Logothetis, 2005) and depletion of this phospholipid leads to channel desensitization.

TRPM8 regulation is sensitive to cholesterol content in the plasma membrane since this ion channel is segregated into lipid rafts, as determined for HEK293 cells transiently expressing this channel and in the endogenous expression system of neurons of the dorsal root ganglia (DRG) (Morenilla-Palao, Pertusa, Meseguer, Cabedo, & Viana, 2009). Lipid raft disruption (by cholesterol depletion with MβCD) modifies TRPM8 activation and enhances TRPM8 channel activation by menthol. Additionally, TRPM8 located outside of cholesterol-rich microdomains is linked to modifications of the thermal threshold for activation of TRPM8 since the depletion of cholesterol shifts TRPM8's activation curve toward warm temperatures (Morenilla-Palao et al., 2009). Interestingly, a post-translational modification in the TRPM8 ion channel, such as glycosylation of residue Asn934, is an important regulator of the localization of TRPM8 to lipid rafts. Moreover, mutation of this amino acid residue decreases the amount of channel found in these specialized domains without affecting expression levels of the protein in the plasma membrane. These results indicate that the lack of TRPM8 glycosylation is a way to positively regulate TRPM8 activation (Morenilla-Palao et al., 2009). Although the mechanism by which these cholesterol-rich microdomains modulate the activity of TRPM8 remains unknown, it is possible that the membrane environment affects the biophysical properties of the TRPM8 ion channel. Accordingly, there is a report showing that cholesterol reduction stabilizes TRPM8 in the plasma membrane, therefore increasing the number of channels in this region of the cell without altering its open probability or its unitary conductance (Veliz et al., 2010) (Fig. 4).

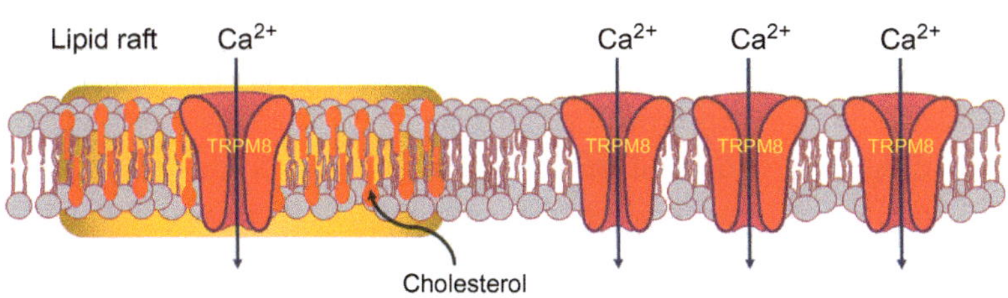

Fig. 4 TRPM8 ion channel levels on the cell surface are regulated by cholesterol. Cholesterol-enriched membrane microdomains (lipid rafts) contain low levels of TRPM8 ion channels. Cholesterol depletion increases the number of TRPM8 ion channels on the cell surface.

1.3 The TRPV Subfamily of Ion Channels

TRPV ion channels are the most extensively characterized subfamily of TRP proteins that is constituted by six members (TRPV1–6) (Gunthorpe, Benham, Randall, & Davis, 2002). They are named the vanilloid ion channels after the first member to be characterized, the TRPV1 channel which is activated by a vanilloid compound, capsaicin. TRPV1–4 are nonselective cation permeable channels (Vennekens, Owsianik, & Nilius, 2008) with preferred selectivity to Ca^{2+} ions, whereas TRPV5–6 only show high selectivity to Ca^{2+} (Vennekens et al., 2008). All of these channels exhibit the classical protein topology of TRP ion channels with six ankyrin repeats in their N-termini (Jin, Touhey, & Gaudet, 2006). TRPV1–V4 are thermosensors and nocisensors and have important roles in the transduction of thermal and nociceptive signals (Vay, Gu, & McNaughton, 2012). Conversely, TRPV5–6 are ion channels that are widely expressed in epithelial cells where they regulate Ca^{2+} homeostasis (Tomilin, Cherezova, Negulyaev, & Semenova, 2016). Most of the regulators of TRPV1–V4 activity are hydrophobic ligands and their function is also influenced by cholesterol.

1.3.1 Dual Roles for Cholesterol on TRPV1 Ion Channel Physiology

TRPV1 is a nonselective cationic ion channel and the better-studied protein of the TRP family of ion channels. It is mainly expressed in neurons of the DRG and trigeminal ganglia (Caterina et al., 1997), although its expression has been also detected in specific areas of the central nervous system (Mezey et al., 2000). TRPV1 has been widely associated to nociception since it can be activated by several noxious environmental stimuli, which results in signal transduction that is detected and interpreted by the brain as a painful message. This channel has a role as a thermosensor since it is activated by hot temperatures (≥43°C) (Caterina et al., 1997), changes in extra- and intracellular pH and pungent compounds (Caterina et al., 1997). To date, a large plethora of compounds that activate TRPV1 have been identified. Some of these are of natural origin: capsaicin, allicin, and resiniferatoxin (from *Capsicum* chili peppers, garlic, and from the cactus-like *Euphorbia resinifera*, respectively) (Caterina et al., 1997; Salazar et al., 2008; Szallasi & Blumberg, 1990). Furthermore, there are some TRPV1 activators which are endogenously produced such as anandamide, *N*-arachidonoyl-dopamine, metabolic products of lipoxygenase, lysophosphatidic acid, and among others (Morales-Lazaro, Simon, & Rosenbaum, 2013). These endogenous compounds are important pain mediators and act through the activation of this ion channel; thus, the search for inhibitors of TRPV1 function is a field that has received much attention. The discovery of analgesic molecules that influence TRPV1 activity is a much desired achievement. Despite large efforts to accomplish this, there are only few natural molecules identified as of yet as TRPV1 inhibitor. Examples of these are adenosine (Puntambekar, Van Buren, Raisinghani, Premkumar, & Ramkumar, 2004), resolvins (Xu et al., 2010), cholesterol (Picazo-Juarez et al., 2011), and oleic acid (Morales-Lázaro et al., 2016).

Initially, the role of cholesterol on TRPV1 regulation was resolved in experiments with DRG neurons (Liu, Huang, Wu, & Priestley, 2006; Szoke et al., 2010). The plasma membrane of these neurons was depleted of cholesterol by incubation with MβCD and whole-cell patch-clamp recordings determined that, TRPV1 currents evoked by capsaicin, were significantly reduced, as compared to untreated cells (Liu et al., 2006). Cholesterol depletion in DRG neurons abolished the presence of clusters of TRPV1 localized to the plasma membrane. This suggested that cholesterol is necessary for the proper localization of the TRPV1 channel in lipid rafts (Liu & Qin, 2005; Szoke et al., 2010) (Fig. 5).

Negative regulation of TRPV1 activation by cholesterol can also be achieved through a more direct and specific fashion. We evaluated the effects of cholesterol on TRPV1 function by using excised membranes patches of HEK293 cells transiently expressing rat TRPV1 channels. By promoting cholesterol enrichment of the membrane patches with a MβCD: cholesterol mix, we were able to determine the effects of cholesterol on TRPV1's activation by capsaicin. In this isolated and more controlled

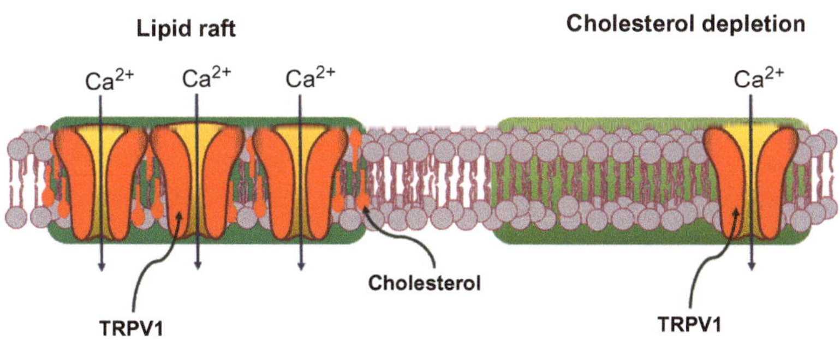

Fig. 5 TRPV1 ion channel needs cholesterol for its localization on lipid rafts. Cholesterol-enriched membrane microdomains (lipid rafts) contain high TRPV1 ion channel levels. Cholesterol depletion decreases TRPV1 ion channels located at the cell surface.

system, we could evaluate the direct effects of cholesterol on TRPV1 function and better exclude its role on trafficking-dependent processes. TRPV1 currents from excised membrane patches with cholesterol enrichment, exhibited a significant reduction in magnitude, as compared to control cells (Picazo-Juarez et al., 2011). The inhibition of TRPV1 activity by cholesterol addition under these experimental conditions produced no changes in the open probability or the conductance of the channel. The underlying mechanism for the reduction of TRPV1 currents by cholesterol was attributed to a decrease on the number of functional ion channels in the membrane patch, as determined from single channel experiments where TRPV1 channels were incubated in the presence of cholesterol (Picazo-Juarez et al., 2011).

Interestingly, inhibition of TRPV1 by cholesterol was stereospecific since enrichment of the membrane patches with epicholesterol (an epimeric form of cholesterol; Rog & Pasenkiewicz-Gierula, 2003) had not effects on capsaicin evoked currents; thus, specific structural requirements had to be met for TRPV1 to be inhibited by cholesterol.

For several ion channels, a direct interaction of cholesterol with a specific motif localized in their sequence has been described. This sequence is known as the CRAC motif (cholesterol recognition amino acid consensus) and classically this motif has the following sequence: L/V-(X_{1-5})-Y-(X_{1-5})-R/K, where X can be any amino acid (Fantini & Barrantes, 2013). Furthermore, it has been described that cholesterol can interact with proteins that exhibit a motif similar to the CRAC sequence but that is oriented in the opposite direction along the polypeptide chain, constituting an inverted CRAC sequence referred to as a CARC motif (Baier, Fantini, & Barrantes, 2011). Initially, we described the presence of a CRAC sequence on the fifth transmembranal segment (S5) of the rat TRPV1 channel. However, careful assessment of this first observation has led us to determine that this sequence is really oriented in an inverted order: $\mathbf{R}_{(579)}F_{(580)}M_{(581)}\mathbf{F}_{(582)}V_{(583)}Y_{(584)}\mathbf{L}_{(585)}$ (Fig. 6A). Moreover, it is worth noting that it does not present a tyrosine residue as what happens with the classical motif, leading us to propose that the rat TRPV1 ion channel has a CARC-like motif (Fig. 6A) and not a classical CRAC sequence, as initially thought. The functional characterization of the rat TRPV1 CARC motif revealed that mutations of the arginine (R579Q) or phenylalanine (F582Q) residues result in a partial reversal of the inhibitory effects of cholesterol on TRPV1's activation by capsaicin. Furthermore, the replacement of a leucine residue for isoleucine (L585I) completely abolished

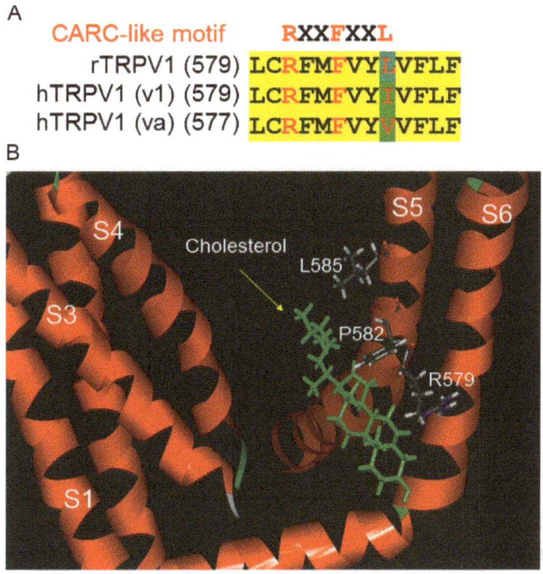

Fig. 6 The TRPV1 ion channel is directly regulated by cholesterol. (A) Consensus sequence for the CARC-like motif found in rat TRPV1 ion channels (NP_114188.1) and sequence alignment with the human TRPV1 variant 1 (v1) (NP_542435.2) and variant a (va) (ABA0665.1). (B) Docking simulation of cholesterol interaction with the CARC-like motif using the cryo-EM structure of TRPV1 in the closed state (PDB 3J5P, chain A). The cholesterol molecule is shown to be localized near the CARC-like motif found in the S5 of the TRPV1 ion cannel.

the inhibitory cholesterol effects on rat TRPV1 activation. In human, the expression of different TRPV1 variants (1–4, b, and a) has been described (Hayes et al., 2000). Interestingly, one of them, the human variant a (h-TRPV1 (va)), possesses a CARC-like motif in its S5: **R**FM**F**VY**V** (Fig. 6A) and capsaicin-evoked currents are partiality inhibited by the addition of cholesterol to membrane patches expressing this channel. In contrast, variant 1 (hTRPV1 (v1)) exhibits the motif **R**FM**F**VY**I** (Fig. 6A) and, in this case, cholesterol has no inhibitory effects on its activation (resembling the functional behavior of the rat-TRPV1 L585I mutant). However, if the isoleucine from hTRPV1 (v1) is substituted by a valine (such as what is found in the CARC-like motif of the hTRPV1 (va)), the activity of this mutated hTRPV1 (v1) (I585V) channel is only partially inhibit by cholesterol (Picazo-Juarez et al., 2011). Complete recovery from the inhibitory effects of cholesterol in this variant is achieved only when the isoleucine from hTRPV1 (v1) is substituted for a leucine (hTRPV1I585L). Together, these experiments demonstrate the importance of a single leucine residue for the inhibitory effects of cholesterol on TRPV1. Docking simulations that we performed suggest that, the leucine at position 585 together with other hydrophobic residues in the S5, give rise to an incipient cavity into which the aliphatic tail of cholesterol can introduce itself stabilizing the interaction with the channel, while the hydroxyl group of cholesterol electrostatically interacts with the arginine residue at position 579. Finally, the phenylalanine at position 582 allows for cholesterol stacking (Fig. 6B) (Picazo-Juarez et al., 2011).

In summary, cholesterol exerts a dual role on TRPV1 function since its depletion can either modify TRPV1 expression levels in the membrane or cholesterol can directly interact with a CARC-like motif localized to the S5 of the TRV1 ion channel causing a decrease in the number of functional ion channels that can respond to capsaicin.

To further complicate this picture, it has also been reported that cholesterol depletion modifies the permeability properties of the TRPV1 ion channel (Jansson et al., 2013). In a CHO (Chinese hamster ovary) cell line with inducible TRPV1 expression, it was shown, using whole-cell recordings and measuring the uptake rates of YO-PRO (a large fluorescent dye), that cholesterol depletion modifies the size of the ion conducting pathway of TRPV1. In this regard, some studies have shown that under hypocalcemic conditions, there is sustained TRPV1 activation by capsaicin and that this leads to an increase in the permeability of large cations (Chung, Guler, & Caterina, 2008). Interestingly, cholesterol depletion inhibits the permeability of large cations such as *N*-methyl-D-glucamine (NMDG), whereas permeability of Na^+ remains unchanged, indicating that cholesterol contents can modulate the ion selectivity properties of TRPV1 (Jansson et al., 2013). Thus, TRPV1 ion channels expressed in cells with high cholesterol content would be expected to be more "flexible" and allow for the flow of large cations, as compared to cells with low cholesterol content.

1.3.2 Positive Regulation of TRPV3 Channels by Cholesterol

The TRPV3 is a cation ion channel mainly expressed in keratinocytes at the epidermal layer and in hair follicles (Peier et al., 2002). It has been linked to skin physiology with an important role in thermal perception responses, hair development, and barrier function (Cheng et al., 2010; Moqrich et al., 2005). This channel is activated by warm and hot temperatures (33–45°C) (Peier et al., 2002; Xu et al., 2002) and by natural monoterpenoid compounds such as camphor (from Cinnamon camphora), carvacrol (from oregano), thymol (from thyme), and among others (Moqrich et al., 2005;

Vogt-Eisele et al., 2007). Unlike TRPV1, for which several endogenous activators have been identified, few of these molecules have been found to regulate TRPV3 and these have been mainly associated with inhibitory actions on channel activity. For example, while ATP serves as a positive regulator of TRPV1 activity, it has been shown that the application of this same molecule reduces sensitization of TRPV3 to its agonists (Phelps, Wang, Choo, & Gaudet, 2010). Furthermore, PIP2, a positive regulator for TRPV1 activation by capsaicin (Lukacs, Rives, Sun, Zakharian, & Rohacs, 2013), has a negative role on TRPV3 function since its hydrolysis by PLCβ potentiates TRPV3 currents and allows the TRPV3 activation by cold temperatures (Doerner, Hatt, & Ramsey, 2011). Other negative modulators of TRPV3 activation are resolvins, compounds which are endogenously synthesized from omega-3 fatty acids. Specifically, the 17-(*R*)-Resolvin D1 suppresses TRPV3 currents evoked by temperature (37°C), camphor and farnesyl pyrophosphate (FPP, a ligand of the LPA5 G protein-coupled receptor) (Bang, Yoo, Yang, Cho, & Hwang, 2012). The latter was the first identified specific endogenous agonist of the TRPV3 ion channel (Bang, Yoo, Yang, Cho, & Hwang, 2010). FPP is produced as an intermediate metabolite from the mevalonate pathway and interestingly, the intraplantar injection of FPP in mice induces a nociceptive behavior through TRPV3 activation (Bang et al., 2010).

The mevalonate pathway is a source for intermediary compounds which are used to produce isoprenoids such as cholesterol (Holstein & Hohl, 2004). Cholesterol as well as FPP (an intermediary of the same metabolic pathway) act as positive endogenous regulators of TRPV3 activation (Klein, Tannert, & Schaefer, 2014).

Positive regulation of TRPV3 by cholesterol was experimentally determined by measuring Ca^{2+} in a HEK293 cell line which stably expresses TRPV3. In this study, the cells were preloaded with MβCD or with a mix of MβCD:cholesterol in order to either extract or enrich the cholesterol content in the membrane, respectively. Enrichment with cholesterol modified the sensitivity of TRPV3 to its agonists (2-APB, carvacrol, and thymol) and shifted the EC50 for these to lower values (Klein et al., 2014). Whole-cell experiments confirmed that cholesterol supplementation altered TRPV3 current densities induced by 2-APB since cells treated with MβCD and cholesterol showed sixfold higher current densities than untreated cells or only treated with MβCD. Whole-cell patch-clamp experiments also showed that cholesterol enrichment increases TRPV3 current densities which respond to heat and that cholesterol shifts the temperature threshold for TRPV3 activation to lower values (Klein et al., 2014). This phenomenon of TRPV3 sensitization by cholesterol was also found in a native expression system of this ion channel of HaCaT cells, a human keratinocyte cell line (Klein et al., 2014).

As what we have discussed earlier, the mechanism underlying TRPV3 potentiation by cholesterol remains unclarified, albeit the fact that it is clear that the effect is not due to an increase in TRPV3 expression levels in the plasma membrane because cells enriched with cholesterol show a redistribution of TRPV3 localization mainly to intracellular compartments, an effect that is not observed in control or MβCD-treated cells where TRPV3 is bound to the plasma membrane (Klein et al., 2014). Another possible mechanism could be the presence of a CRAC sequence in TRPV3, as in the TRPV1 channel (Picazo-Juarez et al., 2011); however, this kind of motif is absent in the TRPV3 ion channel (Klein et al., 2014). Additionally, perturbation of membrane properties could also constitute a mechanism for the positive regulation roles of cholesterol on TRPV3 activation. In agreement with this notion, it has been determined that cells enriched with cholesterol were smaller in size and exhibited diminished basal current densities, as compared to control cells (Klein et al., 2014). Thus, it is possible that changes in the availability of this lipid in the membrane provide an environment that changes the expression and activation of the channel possibly due to alterations in fluidity of the membrane or even in their biophysical properties such as open probability and/or conductance. Nonetheless, it is interesting to note that cholesterol effects on TRPV3 sensitization are specific to this protein since enrichment of the membrane with cholesterol does not modify TRPV2 activation by heat, another close member of the vanilloid family (Klein et al., 2014) and, in the case of the TRPV1 channel, it has an opposite inhibitory role (Picazo-Juarez et al., 2011).

1.3.3 The TRPV4 Ion Channel and Its Interaction With Cholesterol

TRPV4 is an ion channel expressed in keratinocytes, DRG neurons, hippocampal neurons, and urothelial cells (Liedtke et al., 2000). Initially, this channel was characterized as an osmosensor because it can be activated by hypotonic osmotic stress (Liedtke et al., 2000). Presently, it is well known that TRPV4 also functions as a sensor of warm and hot temperatures (25–43°C) (Watanabe et al., 2002). TRPV4 activity is also regulated by lipidic compounds such as products of AA metabolism (Watanabe et al., 2003) and by pruritogenic compounds such as histamine and serotonin, pinpointing an important role for this channel in the itch response

(Cenac et al., 2010; Chen et al., 2016). Although TRPV4 has been localized to lipid rafts in hippocampal neurons from a cirrhotic rat model (Carreno, Ji, & Cunningham, 2009), its functional relevance has not been yet clarified. Recently, it was reported that TRPV4 has a CRAC motif confined to its S4 and that there is an inverted CRAC-like motif localized between loop 4 and the region between loop 4 and S5. These motifs are highly conserved among vertebrates and they are sites that allow for physical interaction with cholesterol and with its precursor, mevalonate, and their derivatives stigmasterol and aldosterone (Kumari et al., 2015). Interestingly, the TRPV4 ion channel colocalizes with caveolin-1, indicating that TRPV4 is present in caveolae, a situation that is highly dependent upon cholesterol content since the treatment with MβCD disrupts localization of TRPV4 in the plasma membrane channel (Kumari et al., 2015). Although it seems that cholesterol is important for a suitable cell localization of TRPV4 in the plasma membrane, which impacts on its functional roles, further studies are needed to explore the possibility of a direct interaction of cholesterol with CRAC and CRAC-like in the TRPV4 ion channel.

2. CONCLUDING REMARKS

Cholesterol regulates the physiology of TRP channels through diverse mechanisms. These include the localization of the channels to special membrane regions such as caveolae and lipid rafts which mostly regulate the function or expression of TRP channels in a positive fashion (e.g., TRPC channels). On the other hand, cholesterol has mixed effects on TRPM channel function, while it negatively regulates the activation of the TRPM3 protein by its agonist pregnenolone sulfate, it exerts positive effects on TRPM7 and TRPM8 channels by allowing for the localization of these channels to lipid rafts and promoting stabilization of the channels in the plasma membrane (TRPM7 and TRPM8) or by increasing the currents through these channels (e.g., TRPM7). For TRPV channels, contrasting effects have also been demonstrated. While some studies using experimental techniques where currents are measured from the whole cell show that cholesterol depletion from the membrane reduces the response of TRPV1 to capsaicin, other reports where the membrane patches containing these channels are isolated from the cell, suggest that TRPV1's activity is in fact diminished by addition of cholesterol through a direct interaction with a CARC-like motif. To further complicate this picture, it has also been shown (albeit using experimental approaches where the membrane integrity of the whole cell is intact) that cholesterol can produce changes in the pore. TRPV3 channels are positively regulated by cholesterol by unclarified mechanisms that could include relocation to different regions in the cell or a direct interaction with a CRAC motif in its sequence. Finally, TRPV4 channels it is known that they contain a CRAC motif but no definite interaction or functional consequence for this has been demonstrated. However, it has been shown for TRPV4 channels that cholesterol influences its localization to lipid rafts.

The regulation of ion channels by molecules of a lipidic nature is a field that has advanced considerably in the last decades. Nonetheless, there is a need to extend our knowledge on the fine mechanisms that intricately regulate the function of ion channels. For TRP channels, several efforts have led to the understanding of how lipid-related molecules modify the gating and expression of these proteins but we must still invest in pursuing further discoveries to shed light on how lipid molecules affect their activity and how this impacts in the physiology of the cell and of the whole organism.

ACKNOWLEDGMENTS

We thank Itzel Alejandra Llorente Gil, Ana Escalante, and Francisco Pérez at Instituto de Fisiología Celular, UNAM for expert technical support. This work was supported by grants from Dirección General de Asuntos del Personal Académico (DGAPA)—Programa de Apoyo a Proyectos de Investigación e Inovación Tecnológica (PAPIIT) IN200314 and IN200717, Consejo Nacional de Ciencia y Tecnología (CONACyT) CB-2014-01-238399, and a grant from Fronteras en la Ciencia No. 77 from CONACyT to T.R. and DGAPA-PAPIIT IA202717 to S.L.M.L.

REFERENCES

Abramowitz, J., & Birnbaumer, L. (2009). Physiology and pathophysiology of canonical transient receptor potential channels. *The FASEB Journal*, *23*(2), 297–328.

Adebiyi, A., Narayanan, D., & Jaggar, J. H. (2011). Caveolin-1 assembles type 1 inositol 1,4,5-trisphosphate receptors and canonical transient receptor potential 3 channels into a functional signaling complex in arterial smooth muscle cells. *The Journal of Biological Chemistry*, *286*(6), 4341–4348.

Baier, C. J., Fantini, J., & Barrantes, F. J. (2011). Disclosure of cholesterol recognition motifs in transmembrane domains of the human nicotinic acetylcholine receptor. *Scientific Reports*, *1*, 69.

Bang, S., Yoo, S., Yang, T. J., Cho, H., & Hwang, S. W. (2010). Farnesyl pyrophosphate is a novel pain-producing molecule via specific activation of TRPV3. *The Journal of Biological Chemistry*, *285*(25), 19362–19371.

Bang, S., Yoo, S., Yang, T. J., Cho, H., & Hwang, S. W. (2012). 17(R)-resolvin D1 specifically inhibits transient receptor potential ion channel vanilloid 3 leading to peripheral antinociception. *British Journal of Pharmacology*, *165*(3), 683–692.

Beech, D. J. (2012). Integration of transient receptor potential canonical channels with lipids. *Acta Physiologica*, *204*(2), 227–237.

Bergdahl, A., Gomez, M. F., Dreja, K., Xu, S. Z., Adner, M., Beech, D. J., et al. (2003). Cholesterol depletion impairs vascular reactivity to endothelin-1 by reducing store-operated Ca2+ entry dependent on TRPC1. *Circulation Research, 93*(9), 839–847.

Bomben, V. C., Turner, K. L., Barclay, T. T., & Sontheimer, H. (2011). Transient receptor potential canonical channels are essential for chemotactic migration of human malignant gliomas. *Journal of Cellular Physiology, 226*(7), 1879–1888.

Boulay, G., Zhu, X., Peyton, M., Jiang, M., Hurst, R., Stefani, E., et al. (1997). Cloning and expression of a novel mammalian homolog of Drosophila transient receptor potential (Trp) involved in calcium entry secondary to activation of receptors coupled by the Gq class of G protein. *The Journal of Biological Chemistry, 272*(47), 29672–29680.

Brazer, S. C., Singh, B. B., Liu, X., Swaim, W., & Ambudkar, I. S. (2003). Caveolin-1 contributes to assembly of store-operated Ca2+ influx channels by regulating plasma membrane localization of TRPC1. *The Journal of Biological Chemistry, 278*(29), 27208–27215.

Carreno, F. R., Ji, L. L., & Cunningham, J. T. (2009). Altered central TRPV4 expression and lipid raft association related to inappropriate vasopressin secretion in cirrhotic rats. *American Journal of Physiology. Regulatory, Integrative and Comparative Physiology, 296*(2), R454–R466.

Caterina, M. J., Schumacher, M. A., Tominaga, M., Rosen, T. A., Levine, J. D., & Julius, D. (1997). The capsaicin receptor: A heat-activated ion channel in the pain pathway. *Nature, 389*(6653), 816–824.

Cenac, N., Altier, C., Motta, J. P., d'Aldebert, E., Galeano, S., Zamponi, G. W., et al. (2010). Potentiation of TRPV4 signalling by histamine and serotonin: An important mechanism for visceral hypersensitivity. *Gut, 59*(4), 481–488.

Chen, Y., Fang, Q., Wang, Z., Zhang, J. Y., MacLeod, A. S., Hall, R. P., et al. (2016). Transient receptor potential vanilloid 4 ion channel functions as a pruriceptor in epidermal keratinocytes to evoke histaminergic itch. *The Journal of Biological Chemistry, 291*(19), 10252–10262.

Cheng, X., Jin, J., Hu, L., Shen, D., Dong, X. P., Samie, M. A., et al. (2010). TRP channel regulates EGFR signaling in hair morphogenesis and skin barrier formation. *Cell, 141*(2), 331–343.

Chung, M. K., Guler, A. D., & Caterina, M. J. (2008). TRPV1 shows dynamic ionic selectivity during agonist stimulation. *Nature Neuroscience, 11*(5), 555–564.

Clapham, D. E. (2003). TRP channels as cellular sensors. *Nature, 426*(6966), 517–524.

Demeuse, P., Penner, R., & Fleig, A. (2006). TRPM7 channel is regulated by magnesium nucleotides via its kinase domain. *The Journal of General Physiology, 127*(4), 421–434.

Doerner, J. F., Hatt, H., & Ramsey, I. S. (2011). Voltage- and temperature-dependent activation of TRPV3 channels is potentiated by receptor-mediated PI(4,5)P2 hydrolysis. *The Journal of General Physiology, 137*(3), 271–288.

Fantini, J., & Barrantes, F. J. (2013). How cholesterol interacts with membrane proteins: An exploration of cholesterol-binding sites including CRAC, CARC, and tilted domains. *Frontiers in Physiology, 4*, 31.

Graziani, A., Rosker, C., Kohlwein, S. D., Zhu, M. X., Romanin, C., Sattler, W., et al. (2006). Cellular cholesterol controls TRPC3 function: Evidence from a novel dominant-negative knockdown strategy. *The Biochemical Journal, 396*(1), 147–155.

Gunthorpe, M. J., Benham, C. D., Randall, A., & Davis, J. B. (2002). The diversity in the vanilloid (TRPV) receptor family of ion channels. *Trends in Pharmacological Sciences, 23*(4), 183–191.

Hayes, P., Meadows, H. J., Gunthorpe, M. J., Harries, M. H., Duckworth, D. M., Cairns, W., et al. (2000). Cloning and functional expression of a human orthologue of rat vanilloid receptor-1. *Pain, 88*(2), 205–215.

Hofmann, T., Obukhov, A. G., Schaefer, M., Harteneck, C., Gudermann, T., & Schultz, G. (1999). Direct activation of human TRPC6 and TRPC3 channels by diacylglycerol. *Nature, 397*(6716), 259–263.

Holstein, S. A., & Hohl, R. J. (2004). Isoprenoids: Remarkable diversity of form and function. *Lipids, 39*(4), 293–309.

Jansson, E. T., Trkulja, C. L., Ahemaiti, A., Millingen, M., Jeffries, G. D., Jardemark, K., et al. (2013). Effect of cholesterol depletion on the pore dilation of TRPV1. *Molecular Pain, 9*, 1.

Jin, X., Touhey, J., & Gaudet, R. (2006). Structure of the N-terminal ankyrin repeat domain of the TRPV2 ion channel. *The Journal of Biological Chemistry, 281*(35), 25006–25010.

Kiselyov, K., Xu, X., Mozhayeva, G., Kuo, T., Pessah, I., Mignery, G., et al. (1998). Functional interaction between InsP3 receptors and store-operated Htrp3 channels. *Nature, 396*(6710), 478–482.

Klein, A. S., Tannert, A., & Schaefer, M. (2014). Cholesterol sensitises the transient receptor potential channel TRPV3 to lower temperatures and activator concentrations. *Cell Calcium, 55*(1), 59–68.

Kraft, R., & Harteneck, C. (2005). The mammalian melastatin-related transient receptor potential cation channels: An overview. *Pflügers Archiv, 451*(1), 204–211.

Kumari, S., Kumar, A., Sardar, P., Yadav, M., Majhi, R. K., Kumar, A., et al. (2015). Influence of membrane cholesterol in the molecular evolution and functional regulation of TRPV4. *Biochemical and Biophysical Research Communications, 456*(1), 312–319.

Levitan, I., Fang, Y., Rosenhouse-Dantsker, A., & Romanenko, V. (2010). Cholesterol and ion channels. *Sub-Cellular Biochemistry, 51*, 509–549.

Liedtke, W., Choe, Y., Marti-Renom, M. A., Bell, A. M., Denis, C. S., Sali, A., et al. (2000). Vanilloid receptor-related osmotically activated channel (VR-OAC), a candidate vertebrate osmoreceptor. *Cell, 103*(3), 525–535.

Liu, M., Huang, W., Wu, D., & Priestley, J. V. (2006). TRPV1, but not P2X, requires cholesterol for its function and membrane expression in rat nociceptors. *The European Journal of Neuroscience, 24*(1), 1–6.

Liu, B., & Qin, F. (2005). Functional control of cold- and menthol-sensitive TRPM8 ion channels by phosphatidylinositol 4,5-bisphosphate. *The Journal of Neuroscience, 25*(7), 1674–1681.

Lukacs, V., Rives, J. M., Sun, X., Zakharian, E., & Rohacs, T. (2013). Promiscuous activation of transient receptor potential vanilloid 1 (TRPV1) channels by negatively charged intracellular lipids: The key role of endogenous phosphoinositides in maintaining channel activity. *The Journal of Biological Chemistry, 288*(49), 35003–35013.

Mezey, E., Toth, Z. E., Cortright, D. N., Arzubi, M. K., Krause, J. E., Elde, R., et al. (2000). Distribution of mRNA for vanilloid receptor subtype 1 (VR1), and VR1-like immunoreactivity, in the central nervous system of the rat and human. *Proceedings of the National Academy of Sciences of the United States of America, 97*(7), 3655–3660.

Montell, C. (2003). Mg^{2+} homeostasis: The Mg2+nificent TRPM chanzymes. *Current Biology, 13*(20), R799–R801.

Moqrich, A., Hwang, S. W., Earley, T. J., Petrus, M. J., Murray, A. N., Spencer, K. S., et al. (2005). Impaired thermosensation in mice lacking TRPV3, a heat and camphor sensor in the skin. *Science, 307*(5714), 1468–1472.

Morales-Lazaro, S. L., Simon, S. A., & Rosenbaum, T. (2013). The role of endogenous molecules in modulating pain through transient receptor potential vanilloid 1 (TRPV1). *The Journal of Physiology, 591*(13), 3109–3121.

Morales-Lázaro, S. L., Llorente, I., Sierra-Ramírez, F., López-Romero, A. E., Ortíz-Rentería, M., Serrano-Flores, B., et al. (2016). Inhibition of TRPV1 channels by a naturally occurring omega-9 fatty acid reduces pain and itch. *Nature Communications, 7*, 13092.

Morenilla-Palao, C., Pertusa, M., Meseguer, V., Cabedo, H., & Viana, F. (2009). Lipid raft segregation modulates TRPM8 channel activity. *The Journal of Biological Chemistry, 284*(14), 9215–9224.

Nadler, M. J., Hermosura, M. C., Inabe, K., Perraud, A. L., Zhu, Q., Stokes, A. J., et al. (2001). LTRPC7 is a Mg.ATP-regulated divalent cation channel required for cell viability. *Nature, 411*(6837), 590–595.

Naylor, J., Li, J., Milligan, C. J., Zeng, F., Sukumar, P., Hou, B., et al. (2010). Pregnenolone sulphate- and cholesterol-regulated TRPM3 channels coupled to vascular smooth muscle secretion and contraction. *Circulation Research, 106*(9), 1507–1515.

Peier, A. M., Reeve, A. J., Andersson, D. A., Moqrich, A., Earley, T. J., Hergarden, A. C., et al. (2002). A heat-sensitive TRP channel expressed in keratinocytes. *Science, 296*(5575), 2046–2049.

Phelps, C. B., Wang, R. R., Choo, S. S., & Gaudet, R. (2010). Differential regulation of TRPV1, TRPV3, and TRPV4 sensitivity through a conserved binding site on the ankyrin repeat domain. *The Journal of Biological Chemistry, 285*(1), 731–740.

Picazo-Juarez, G., Romero-Suarez, S., Nieto-Posadas, A., Llorente, I., Jara-Oseguera, A., Briggs, M., et al. (2011). Identification of a binding motif in the S5 helix that confers cholesterol sensitivity to the TRPV1 ion channel. *The Journal of Biological Chemistry, 286*(28), 24966–24976.

Puntambekar, P., Van Buren, J., Raisinghani, M., Premkumar, L. S., & Ramkumar, V. (2004). Direct interaction of adenosine with the TRPV1 channel protein. *The Journal of Neuroscience, 24*(14), 3663–3671.

Putney, J. W., Jr. (2007). Multiple mechanisms of TRPC activation. In W. B. Liedtke & S. Heller (Eds.), *TRP ion channel function in sensory transduction and cellular signaling cascades*. Boca Raton, FL: CRC Press.

Ramsey, I. S., Delling, M., & Clapham, D. E. (2006). An introduction to TRP channels. *Annual Review of Physiology, 68*, 619–647.

Rog, T., & Pasenkiewicz-Gierula, M. (2003). Effects of epicholesterol on the phosphatidylcholine bilayer: A molecular simulation study. *Biophysical Journal, 84*(3), 1818–1826.

Rohacs, T., Lopes, C. M., Michailidis, I., & Logothetis, D. E. (2005). PI(4,5)P2 regulates the activation and desensitization of TRPM8 channels through the TRP domain. *Nature Neuroscience, 8*(5), 626–634.

Salazar, H., Llorente, I., Jara-Oseguera, A., Garcia-Villegas, R., Munari, M., Gordon, S. E., et al. (2008). A single N-terminal cysteine in TRPV1 determines activation by pungent compounds from onion and garlic. *Nature Neuroscience, 11*(3), 255–261.

Schlingmann, K. P., & Gudermann, T. (2005). A critical role of TRPM channel-kinase for human magnesium transport. *The Journal of Physiology, 566*(Pt. 2), 301–308.

Schlingmann, K. P., Waldegger, S., Konrad, M., Chubanov, V., & Gudermann, T. (2007). TRPM6 and TRPM7—Gatekeepers of human magnesium metabolism. *Biochimica et Biophysica Acta, 1772*(8), 813–821.

Schmitz, C., Perraud, A. L., Johnson, C. O., Inabe, K., Smith, M. K., Penner, R., et al. (2003). Regulation of vertebrate cellular Mg2+ homeostasis by TRPM7. *Cell, 114*(2), 191–200.

Shi, J., Ju, M., Abramowitz, J., Large, W. A., Birnbaumer, L., & Albert, A. P. (2012). TRPC1 proteins confer PKC and phosphoinositol activation on native heteromeric TRPC1/C5 channels in vascular smooth muscle: Comparative study of wild-type and TRPC1 −/− mice. *The FASEB Journal, 26*(1), 409–419.

Shieh, H. S., Hoard, L. G., & Nordman, C. E. (1977). Crystal structure of anhydrous cholesterol. *Nature, 267*(5608), 287–289.

Sun, Y., Birnbaumer, L., & Singh, B. B. (2015). TRPC1 regulates calcium-activated chloride channels in salivary gland cells. *Journal of Cellular Physiology, 230*(11), 2848–2856.

Sun, Y., Sukumaran, P., Varma, A., Derry, S., Sahmoun, A. E., & Singh, B. B. (2014). Cholesterol-induced activation of TRPM7 regulates cell proliferation, migration, and viability of human prostate cells. *Biochimica et Biophysica Acta, 1843*(9), 1839–1850.

Szallasi, A., & Blumberg, P. M. (1990). Resiniferatoxin and its analogs provide novel insights into the pharmacology of the vanilloid (capsaicin) receptor. *Life Sciences, 47*(16), 1399–1408.

Szoke, E., Borzsei, R., Toth, D. M., Lengl, O., Helyes, Z., Sandor, Z., et al. (2010). Effect of lipid raft disruption on TRPV1 receptor activation of trigeminal sensory neurons and transfected cell line. *European Journal of Pharmacology, 628*(1–3), 67–74.

Tomilin, V. N., Cherezova, A. L., Negulyaev, Y. A., & Semenova, S. B. (2016). TRPV5/V6 channels mediate ca(2+) influx in Jurkat T cells under the control of extracellular pH. *Journal of Cellular Biochemistry, 117*(1), 197–206.

Toro, C. A., Arias, L. A., & Brauchi, S. (2011). Sub-cellular distribution and translocation of TRP channels. *Current Pharmaceutical Biotechnology, 12*(1), 12–23.

Vay, L., Gu, C., & McNaughton, P. A. (2012). The thermo-TRP ion channel family: Properties and therapeutic implications. *British Journal of Pharmacology, 165*(4), 787–801.

Veliz, L. A., Toro, C. A., Vivar, J. P., Arias, L. A., Villegas, J., Castro, M. A., et al. (2010). Near-membrane dynamics and capture of TRPM8 channels within transient confinement domains. *PloS One, 5*(10), e13290.

Vennekens, R., Owsianik, G., & Nilius, B. (2008). Vanilloid transient receptor potential cation channels: An overview. *Current Pharmaceutical Design, 14*(1), 18–31.

Voets, T., Nilius, B., Hoefs, S., van der Kemp, A. W., Droogmans, G., Bindels, R. J., et al. (2004). TRPM6 forms the Mg2+ influx channel involved in intestinal and renal Mg2+ absorption. *The Journal of Biological Chemistry, 279*(1), 19–25.

Vogt-Eisele, A. K., Weber, K., Sherkheli, M. A., Vielhaber, G., Panten, J., Gisselmann, G., et al. (2007). Monoterpenoid agonists of TRPV3. *British Journal of Pharmacology, 151*(4), 530–540.

Vriens, J., Nilius, B., & Voets, T. (2014). Peripheral thermosensation in mammals. *Nature Reviews. Neuroscience, 15*(9), 573–589.

Vriens, J., Owsianik, G., Hofmann, T., Philipp, S. E., Stab, J., Chen, X., et al. (2011). TRPM3 is a nociceptor channel involved in the detection of noxious heat. *Neuron, 70*(3), 482–494.

Wagner, T. F., Loch, S., Lambert, S., Straub, I., Mannebach, S., Mathar, I., et al. (2008). Transient receptor potential M3 channels are ionotropic steroid receptors in pancreatic beta cells. *Nature Cell Biology, 10*(12), 1421–1430.

Watanabe, H., Vriens, J., Prenen, J., Droogmans, G., Voets, T., & Nilius, B. (2003). Anandamide and arachidonic acid use epoxyeicosatrienoic acids to activate TRPV4 channels. *Nature, 424*(6947), 434–438.

Watanabe, H., Vriens, J., Suh, S. H., Benham, C. D., Droogmans, G., & Nilius, B. (2002). Heat-evoked activation of TRPV4 channels in a HEK293 cell expression system and in native mouse aorta endothelial cells. *The Journal of Biological Chemistry, 277*(49), 47044–47051.

Wes, P. D., Chevesich, J., Jeromin, A., Rosenberg, C., Stetten, G., & Montell, C. (1995). TRPC1, a human homolog of a Drosophila store-operated channel. *Proceedings of the National Academy of Sciences of the United States of America, 92*(21), 9652–9656.

Xu, H., Ramsey, I. S., Kotecha, S. A., Moran, M. M., Chong, J. A., Lawson, D., et al. (2002). TRPV3 is a calcium-permeable temperature-sensitive cation channel. *Nature, 418*(6894), 181–186.

Xu, Z. Z., Zhang, L., Liu, T., Park, J. Y., Berta, T., Yang, R., et al. (2010). Resolvins RvE1 and RvD1 attenuate inflammatory pain via central and peripheral actions. *Nature Medicine, 16*(5), 592–597. 591 p. following 597.

Yildirim, E., Kawasaki, B. T., & Birnbaumer, L. (2005). Molecular cloning of TRPC3a, an N-terminally extended, store-operated variant of the human C3 transient receptor potential channel. *Proceedings of the National Academy of Sciences of the United States of America*, *102*(9), 3307–3311.

Yogi, A., Callera, G. E., Tostes, R., & Touyz, R. M. (2009). Bradykinin regulates calpain and proinflammatory signaling through TRPM7-sensitive pathways in vascular smooth muscle cells. *American Journal of Physiology. Regulatory, Integrative and Comparative Physiology*, *296*(2), R201–R207.

Zhu, X., Chu, P. B., Peyton, M., & Birnbaumer, L. (1995). Molecular cloning of a widely expressed human homologue for the Drosophila trp gene. *FEBS Letters*, *373*(3), 193–198.

CHAPTER SIX

Direct Interactions of Cholesterol With Pentameric Ligand-Gated Ion Channels: Testable Hypotheses From Computational Predictions

Grace Brannigan[1]
Center for Computational and Integrative Biology, Rutgers University–Camden, Camden, NJ, United States
[1]Corresponding author: e-mail address: grace.brannigan@rutgers.edu

Contents

Abstract

Several essential ionotropic neurotransmitter receptors, including the nicotinic acetylcholine receptor (nAChR) and gamma-aminobutyric acid (GABA) type A receptor ($GABA_Ar$), belong to the family of pentameric ligand-gated ion channels (pLGICs). Function of these receptors is particularly sensitive to their lipid environment, including cholesterol and cholesterol-derived neurosteroids. Direct structural data investigating interactions between sterols and pLGICs, as well as their role in modulatory mechanisms, are largely unavailable. Physics-based computational approaches can serve a vital role in interpretation of more indirect data as well as hypothesis generation and experimental design. In this chapter, I report several examples in which computational approaches were used to predict direct binding interactions of steroids and pLGICs, evaluate the relative likelihood of possible interpretations of experimental data, and present rationally designed simple experiments. I conclude by offering several predictions that could be tested by future experiments.

Current Topics in Membranes, Volume 80
ISSN 1063-5823
http://dx.doi.org/10.1016/bs.ctm.2017.06.002

1. INTRODUCTION

Neuronal signaling relies heavily on ion channels, which are fundamental for the propagation of action potentials and are embedded in membranes with distinctive lipid compositions. These lipids have dual roles: collectively, they form a heterogeneous and viscoelastic lipid "solvent" as in other cell-types, but individual lipids can also have a particularly prominent role as signaling molecules in neuronal cells (Piomelli, Astarita, & Rapaka, 2007). This allows relatively rapid modulation and adaptation: lipid composition of the human brain is estimated to have evolved from that of ancestral primates about four times more quickly than nonneural tissues, with accelerating change in the human neocortex (Bozek et al., 2015). Furthermore, the differences in lipid composition are not restricted to lipids found at low concentrations; mammalian neuronal membranes are rich in polyunsaturated fatty acids (PUFAs), phosphatidylethanolamine (PE) headgroups, and cholesterol, with an overall composition less like other nonneuronal mammalian cells than like the electric organs of electrogenic fish (Table 1).

One of the earliest receptors to be purified, the nicotinic acetylcholine receptor (nAChR) was first extracted from the electric organ of rays from the genus *Torpedo*. In mammals, muscle-type nAChRs are found in the postsynaptic membrane at the neuromuscular junction and conduct cations upon binding of acetylcholine, depolarizing the membrane and promoting propagation of an action potential. Neuronal nAChRs are found in both the neuronal postsynaptic and presynaptic membrane, where they facilitate calcium regulation of neurotransmitter release (Garduño et al., 2012; Gotti, Zoli, & Clementi, 2006). nAChRs are members of the family of ionotropic receptors are known as pentameric ligand-gated ion channels (pLGICs). In humans, this includes excitatory channels nAChR and the 5HT-3 (serotonin) receptor, as well as inhibitory channels $GABA_Ar$ and the glycine receptor GlyR.

Many eukaryotic pLGICs are known to have requirements for the lipids in their surrounding membranes that are pronounced, yet remain unexplained and poorly defined. A steep drop in conductance may be observed by reducing the amount of membrane cholesterol below 15%–20%, but cholesterol can be replaced with several analogs with minimal effects (Criado, Eibl, & Barrantes, 1982; Dalziel, Rollins, & Mcnamee, 1980; Quesada et al., 2016). This is especially well established for the

Table 1 Lipid Composition of Plasma Membranes

	Torpedo[a]	Postsynaptic[b]	*Xenopus*[c]	Mammalian[d]
Phosphocholine (PC)	42	41	28	38
Phosphatidylethanolamine (PE)	34	35	17	22
PS + PI + PA[e]	16	18	10	12
Sphingomyelin (SM)	5	5	20	20
Other phospholipid	3	1	8	8
Saturated	42	49	44	53
Monounsaturated	26	15	39	20
Polyunsaturated	32	35	17	27
Cholesterol mole fraction	32	39	21	29

[a]*Torpedo californica* electric organ (Barrantes, 1989).
[b]Rat brain synaptosomes (Breckenridge, Gombos, & Morgan, 1972).
[c]*Xenopus* oocytes (Hill et al., 2005).
[d]"Idealized" (average) mammalian plasma membrane (Ingólfsson et al., 2014).
[e]Sum of total phosphatidylserine, phosphatidylinositol, and phosphatidic acid.
First group of rows indicates mole fraction of phospholipids among glycerophospholipid headgroups or sphingomyelin (SM); second group is distribution of fatty acyl chains. *Torpedo* and postsynaptic membranes have a higher abundance of PUFAs and also have PUFAs with longer chains and more double bonds. Final row has overall mole fraction of cholesterol, relative to the total number of phospholipids.

prototypical member of the family, the nAChR. For example, nAChR function in a POPE/DOPS/CHOL mixture is maximized at 30% cholesterol, with complete loss of function for lower concentrations and gradual loss of function for higher concentrations (Criado et al., 1982). Reconstitution of GABA$_A$rs similarly requires cholesterol within the mixture for native function (Bristow & Martin, 1987). Identifying the relevant properties of a functional lipid mixture has been challenging, with a recent extensive search identifying several distinct detergent-based mixtures yielding "optimal" function (Quesada et al., 2016).

The mode of interaction between cholesterol and nAChRs or GABA$_A$rs is unknown, but likely includes effects from (i) direct, specific interactions with lipids that may potentially bind quite deeply into the nAChR transmembrane domain, into sites inaccessible by phospholipids ("nonannular" sites) or remain in the first few lipidation shells around the protein ("annular" sites); and (ii) indirect effects via changing the lipid composition local to the protein or the membrane elasticity. Extensive and diverse experimental studies were undertaken, beginning in the 1980s (e.g., Dalziel et al., 1980,

Criado et al., 1982, Ochoa, Dalziel, & Mcnamee, 1983, Marsh & Barrantes, 1978, Ellena, Blazing, & Mcnamee, 1983, Middlemas & Raftery, 1987, Arias, Sankaram, Marsh, & Barrantes, 1990, Barrantes, Antollini, Blanton, & Prieto, 2000, DaCosta, Ogrel, Mccardy, Blanton, & Baenziger, 2002, Baenziger & Corringer, 2011, Baenziger, Hénault, Therien, & Sun, 2015, Corrie, Ogrel, Mccardy, Blanton, & Baenziger, 2002, DaCosta & Baenziger, 2009, Dacosta, Dey, Therien, & Baenziger, 2013, Barrantes, 1989, Barrantes, 2007, Barrantes et al., 2010, and others discussed elsewhere in this volume) to understand the underlying mechanism of nAChR lipid requirements, including identifying the likelihood of specific boundary lipids for which the nAChR has a higher affinity. Seemingly paradoxical trends suggest that several mechanisms are at work. Although addition of physiological amounts of cholesterol to a cholesterol-free reconstitution mixture can cause a fivefold increase in nAChR response, depletion (removal) of cholesterol by liposomes (Leibel, Firestone, Legler, Braswell, & Miller, 1987; Zabrecky & Raftery, 1985) or cyclodextrins (Borroni et al., 2007; Carlos et al., 2008; Santiago et al., 2001) can also cause gain-of-function. Many such experiments were carried out long before a detailed understanding of nAChR or pLGIC structure was available, and interpretation necessarily suffered, since even general estimates for the number of lipids found at the nAChR interface were unavailable.

Experiments were also carried out primarily with the working assumption of a randomly mixed membrane in the absence of protein, although lipid sorting by proteins was considered likely; there was little evidence available then that cholesterol by itself can induce nonrandom mixing and even domain formation. Formation of lipid microdomains, or "rafts," requires at least a ternary lipid mixture; one canonical "raft-mixture" consists of a saturated phospholipid, unsaturated phospholipid, and cholesterol. For certain compositions with similar numbers of each species, such a mixture will spontaneously phase separate into a liquid-ordered domain l_o which contains cholesterol and the saturated phospholipid, and a liquid-disordered domain l_{do} comprised of the unsaturated phospholipid (Janmey & Kinnunen, 2006; Simons & Gerl, 2010; Simons & Ikonen, 1997). Although we are now aware of the critical role of acyl chain unsaturation in this process, experiments have focused primarily on the role of cholesterol and phospholipid headgroup, and to some extent, acyl chain saturation or mono/di-unsaturation; they did not include the lipids with n-3 PUFA chains that are so abundant in both the fish electric organ and the postsynaptic membrane (Table 1), and in many cases did not include any unsaturated chains.

X-ray crystallography has recently yielded a diverse group of pLGIC structures, but still imposes two significant limitations: (1) X-ray structures of heteropentamers are still extremely uncommon; nearly all pLGICs crystallized with uniformly high resolution have been homomers, although heteromeric pLGICs serve critical roles in function of the vertebrate CNS. A very recent structure of a heteromeric $\alpha 4\beta 2$ nAChR receptor achieves 3.94–4.08 Å resolution (Morales-Perez, Noviello, & Hibbs, 2016). (2) pLGICs have only been successfully crystallized in highly artificial lipid–detergent environments. The medium resolution (nominally 4 Å, but without significantly discernable side-chain density in the TM domain, see Fig. 5A, inset) structure (PDB:2BG9) of the heteromeric $\alpha\beta\delta\alpha\gamma$ nAChR from the *Torpedo* electric ray, determined by cryo-electron microscopy in native membrane (Unwin, 2005), circumvents both these limitations, but introduces its own uncertainty due to its lower resolution. Recent advances in high-resolution electron microscopy allowed determination of an alternate structure for the glycine receptor (GlyR) (Du, Lü, Wu, Cheng, & Gouaux, 2015) and hold promise for heteropentamers common in the vertebrate CNS, in more realistic membranes.

As reported elsewhere in this volume, the current understanding of membrane organization combined with modern experimental approaches, particularly super-resolution light microscopy and cryo-electron microscopy, shows promise for decomposing the mechanism underlying cholesterol modulation, but these techniques are still far from being able to resolve individual lipids in a disordered membrane.

2. COMPUTATIONAL APPROACHES: POWER AND LIMITATIONS

Molecular dynamics (MD) simulation using rigorous physics-based algorithms is a technique whose development began almost immediately following the availability of scientific computers, and was first widely used in the 1960s to investigate atomic liquids (Verlet, 1967). The 2013 Nobel Prize in Chemistry was awarded to Karplus, Levitt, and Warshel for development of the widely used approach for realistic, atomic-resolution molecular dynamics (AA-MD) simulation of complex biomolecules. Nonetheless, a common disconnect between experimental and computational biophysicists has resulted in disparate understanding of the challenges, limitations, and power of such methods when applied to membrane-protein systems.

In principle, the MD "computational microscope" (Ingólfsson, Arnarez, Periole, & Marrink, 2016) has the potential to allow researchers to peer into the black box of the membrane surrounding a membrane protein: it allows dynamic sampling of a thermally active receptor in the complex fluid membrane, and naturally yields submolecular spatial resolution with no requirements for lipid order or receptor symmetry. For the purpose of observing spontaneous direct interactions, however, two major problems arise in practice:

Slow transition from bulk lipids to annular sites: Lipid disorder and entanglement among tails slows lipid diffusion, and lipids contacting the protein at the end of the simulation will mainly reflect those placed near the protein at the beginning of the simulation, not overall lipid composition or lipid–protein interactions.

Slow transition from annular to nonannular sites: Binding of lipids to nonannular sites typically encounters kinetic barriers that may be prohibitive in a typical simulation, depending on how buried the site is and how flexible the lipids are (acyl chains can penetrate grooves in pLGICs within a few nanoseconds, while reorientation and binding of cholesterol headgroup within the protein bundle can require hundreds of nanoseconds to microseconds).

One approach for overcoming these limitations requires placing lipids into putative sites (possibly identified by docking) and then running AA-MD to evaluate stability; kinetic barriers to unbinding can be further overcome using sophisticated free energy perturbation methods. We have used this approach numerous times, but visualization of spontaneous binding has largely been inaccessible.

A second widely used approach requires running MD on a coarse-grained version (CG-MD) of the hydrated membrane and transmembrane protein, to overcome time-scale limitations with equilibrating annular sites (and to some extent, nonannular sites). In MARTINI (Marrink, Risselada, Yefimov, Tieleman, & Vries, 2007) and similar CG forcefields, three to four atoms are grouped together in a single "bead," and interact via parameters that are calibrated to reproduce a range of properties determined in AA-MD simulations and in experiments. CG-MD can dramatically increase efficiency over AA-MD, through both the reduction in the number of particles and the removal of high-frequency vibrations, allowing each update of the system to represent a much larger incremental time step (20–30 fs for MARTINI vs 1–2 fs for typical AA forcefields). The CG-MD approach is suitable for nonspecific interactions but becomes more questionable for direct binding of lipids as ligands, due to the drastic smoothing of molecular shape. Reverse coarse-graining, in which CG lipids (and other molecules) are replaced with atomic coordinate representations, is required to realistically equilibrate a lipid in a nonannular site.

3. STEROLS AS BOUNDARY LIPIDS: NONANNULAR SITES

There are two natural locations for nonannular modulatory sites: at the interface between the subunits and within the subunit center. No high-resolution structures have provided conclusive evidence of embedded or nonannular sites for cholesterol in pLGICs, although the recent $\alpha4\beta2$ nAChR X-ray structure has included cholesteryl hemisuccinate in the purification mixture (Morales-Perez et al., 2016); intersubunit nonannular sites for POPC (with the headgroup buried in the TMD) have also been identified via crystallography (Althoff, Hibbs, Banerjee, & Gouaux, 2014).

Several studies preceding the era of feasible X-ray structures (Addona, Sandermann, Kloczewiak, Husain, & Miller, 1998; Corbin, Wang, & Blanton, 1998; Hamouda, Chiara, Sauls, Cohen, & Blanton, 2006) investigated the possibility of buried "nonannular" sites. The first medium resolution structure (PDBID:2BG9) (Unwin, 2005) used cryo-EM of the nAChR from *Torpedo* in a native membrane, revealing large gaps in protein density (Fig. 5A, inset). After observing the protein structure consistently expel water and collapse under MD simulation, we hypothesized that in the native system represented by the 2BG9 structure these sites were filled with lipids rather than water, namely cholesterol. Through docking, we found 15 potential cholesterol-binding sites within the receptor TM domain, in large hydrophobic gaps, including both intersubunit and intrasubunit sites (Fig. 1). MD simulations of the nAChR structure with embedded cholesterol significantly improved protein stability (Fig. 4) (Brannigan et al., 2008). The number of predicted buried nonannular sites was consistent with the previous estimate of 5–10 put forth by Jones and Mcnamee (1988) from fluorescence-quenching studies.

3.1 Intersubunit Sites

Docking algorithms placed multiple cholesterol molecules in sites fully embedded in the intersubunit cavities of the *Torpedo* nAChR TMD (2BG9), oriented parallel to the pore axis, with the hydroxyl located near the interface with the EC domain. In contrast, a later structure

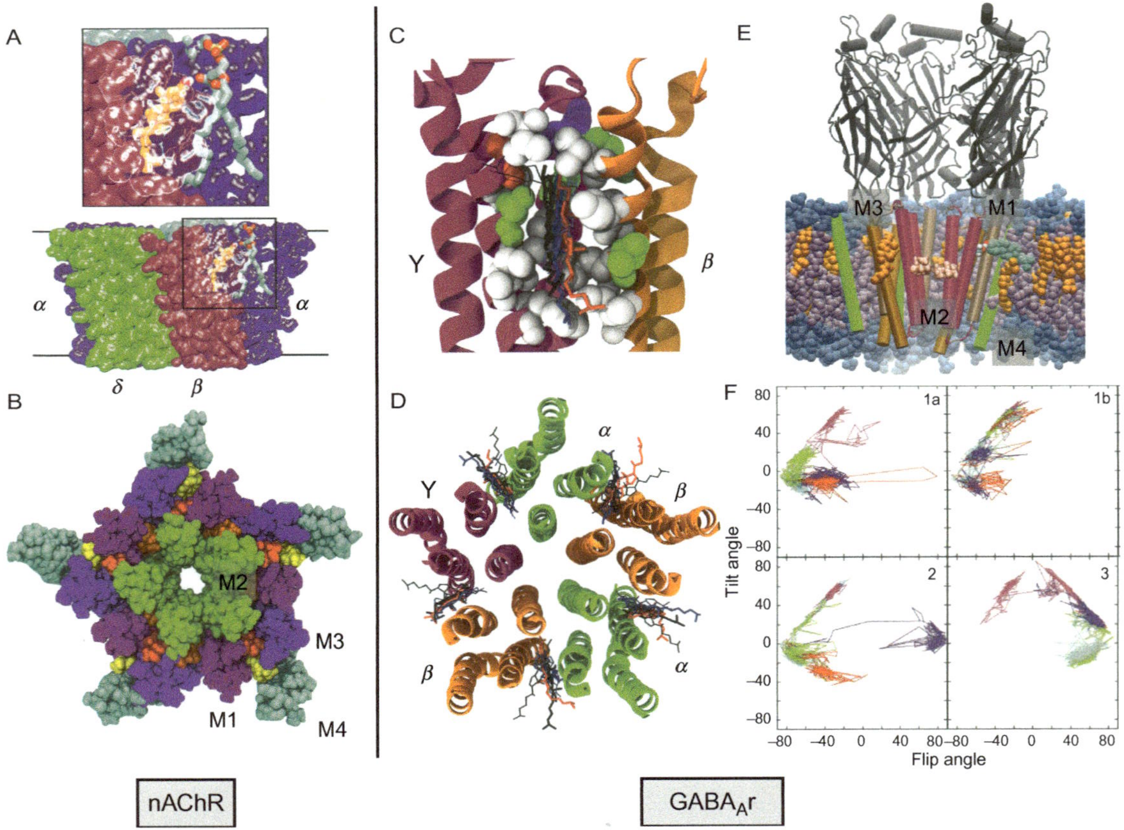

Fig. 1 See legend on opposite page.

(3RHW) for a pLGIC in complex with a TMD modulator, the Glutamate-gated ion channel from *Caenorhabditis elegans* with bound ivermectin (Hibbs & Gouaux, 2011), suggested a perpendicular orientation for an intersubunit-bound sterol molecule. Docking cholesterol to a homology model of $\alpha_1\beta_x\gamma_2$ GABA$_A$r based on 3RHW identified this orientation as by far the most favorable; the cholesterol hydroxyl contacted αS270, $\beta N/S265$, or γS280, all homologous residues to αS270 in the GluCl homomer. AA-MD of GABA$_A$ receptors in complex with cholesterol in these orientations were run over hundreds of nanoseconds using multiple groups of initial coordinates and models for the GABA$_A$ receptor (Fig. 1), one of which had very close correspondence to the GABA$_A$r β3 homomer released later (Miller & Aricescu, 2014). During these simulations cholesterol was observed to bind in one of the two main orientations, with the rough side facing either the M1 or M3 helices. Flipping between orientations was observed several times over the simulations, as was unbinding and spontaneous rebinding of cholesterol to the sites (Fig. 1).

The universality of nonannular interactions of cholesterol across pLGICs is unknown; both of these studies followed similar approaches and yielded

Fig. 1 Two potential orientations of cholesterol bound to intersubunit sites: fully embedded (A and B, Brannigan, Hénin, Law, Eckenhoff, & Klein, 2008) or partially embedded (C–F, Hénin, Salari, Murlidaran, & Brannigan, 2014). (A) Hydrogen bonding observed after 25 ns of AA-MD between intersubunit cholesterol (*orange*, except for OH group) with the acyl chain deeply buried in the nAChR (from structure 2BG9) and an external phosphatidylcholine molecule, *colored* by atom type. (B) Transmembrane portion of the nAChR in atomic detail shown in cross section parallel to the membrane and intersubunit (*orange*) and intrasubunit (*red* and *yellow*) cholesterol-binding modes resulting from automated docking (Brannigan et al., 2008). (C and D) Initial and final coordinates of bound cholesterol in MD simulations, viewed from within the membrane (C) and from outside the cell (D). *Blue* represents Autodock docking of cholesterol and *black* or *red* represents alignment of cholesterol to ivermectin in 3RHW. (E) Simulation system emphasizing hydrogen bond between cholesterol (bound to $\alpha-/\beta+$ interface) and β1:S265 (M2 S15′). Most cholesterol molecules are *orange*; hydrogen bonding partners (β1:S265 and one bound cholesterol) are *colored* by atom type, with most hydrogen atoms removed. The ring of leucines forming the hydrophobic gate is shown in *pink*. Water and ions not shown. (F) Reorientation of bound cholesterol molecules in four separate MD simulations. Each interface is *color-coded* (key and more details in Hénin et al. (2014)). The *lines* connect successive values of the tilt and flip angles for each bound cholesterol molecule. Tilt describes the angle between the cholesterol long axis and the membrane plane, whereas flip reflects the orientation of an axis orthogonal to the former, in the plane of the sterol nucleus. Simulations of models 1a, 1b, and 2 start from flip of −80 and tilt of −11, whereas initial conditions for model 3 are in the range flip 50–65 and tilt 10–35.

distinct predictions for cholesterol sites on the nAChR and $GABA_A$ receptors, with predicted sites on the $GABA_A$ receptor less deeply buried and more exposed to the lipid membrane. The two studies used structural templates with numerous differences, including experimental methodology, resolution, lipid environment, presence/absence of agonists and expected conformational state, in addition to the differences in sequence between nAChR and $GABA_A$ receptors. However, we note that the location of the backbone helices in the two templates (2BG9 and 3EAW) overlaps closely, so that the reduced protein density in the nAChR 2BG9 structure reflects the presence of smaller residues rather than a larger overall footprint of the TMD.

We recently carried out CG-MD simulations of nAChRs based on 2BG9 in domain-forming lipid mixtures, inspired by the *Torpedo* electroplax membrane and including cholesterol, DPPC, and DHA-PE, which is unusually abundant in both *Torpedo* electroplax and neuronal membranes. In CG-MD simulations, lipids may spontaneously bind to both inter and intrasubunit sites. We observe cholesterol binding to intersubunit sites in a partially embedded orientation similar to that predicted for $GABA_Ar$, with the semiflexible cholesterol tail exposed to the membrane. The remaining nAChR intersubunit cavity volume is filled instead with the extremely flexible DHA acyl chains. This combination is facilitated by partitioning of nAChR within the DHA-PE rich l_{do} phase, but separated from the cholesterol-rich l_o ("raft") phase by only one or two lipidation shells (Fig. 5). These simulations indicate that 2BG9 is not inconsistent with perpendicular intersubunit cholesterol (as we predicted for $GABA_Ar$).

Sites between subunits in the transmembrane domain of nAChR and $GABA_Ar$ have long been implicated in the actions of general anesthetics (Chiara, Dangott, Eckenhoff, & Cohen, 2003; Chiara et al., 2012; Jayakar, Dailey, Eckenhoff, & Cohen, 2013; Jayakar et al., 2014; Li, Chiara, Cohen, & Olsen, 2009; Nishikawa & Harrison, 2003; Stewart, Pierce, Hotta, Stern, & Forman, 2014) neurosteroids (Hosie, Wilkins, da Silva, & Smart, 2006; Westergard et al., 2015), and other modulators. The modulators that bind to these sites may (a) displace water, (b) displace lipid, and (c) bind via induced-fit, effectively opening a cavity that would be otherwise filled with amino acid density. Most binding mechanisms probably involve a combination of all three. It was initially assumed that both intersubunit and intrasubunit large gaps in the nAChR amino acid density found in the cryo-EM structure 2BG9 were filled with water that was then displaced by modulators (Miyazawa, Fujiyoshi, & Unwin, 2003), but we repeatedly observe most inter and intrasubunit cavities in the TMD expel water and become filled with acyl chains of the lipid tails (Brannigan et al., 2008; Hénin et al., 2014). Modulators binding to such sites may therefore need to displace acyl chains or cholesterol, rather than water.

Although crystal structures of pLGICs with modulators bound in the TMD are still rare, thus demanding time-intensive photolabeling efforts to detect binding sites for nonspecific modulators such as general anesthetics, we were able to build off information about intersubunit binding of ivermectin using strategically designed competition experiments. The thyroid hormone triiodothyronine is a dicyclic analog of the neurosteroids (Fig. 2) known to modulate $GABA_A$rs and is also a negative modulator of $GABA_A$r at low concentrations. Hypothesizing that it might bind at intersubunit sites, we tested for inhibition of activation by ivermectin or the neurosteroid allopregnanolone; using simple Schild analysis, we found that T3 competes with both ivermectin and allopregnanolone (Fig. 3). In MD

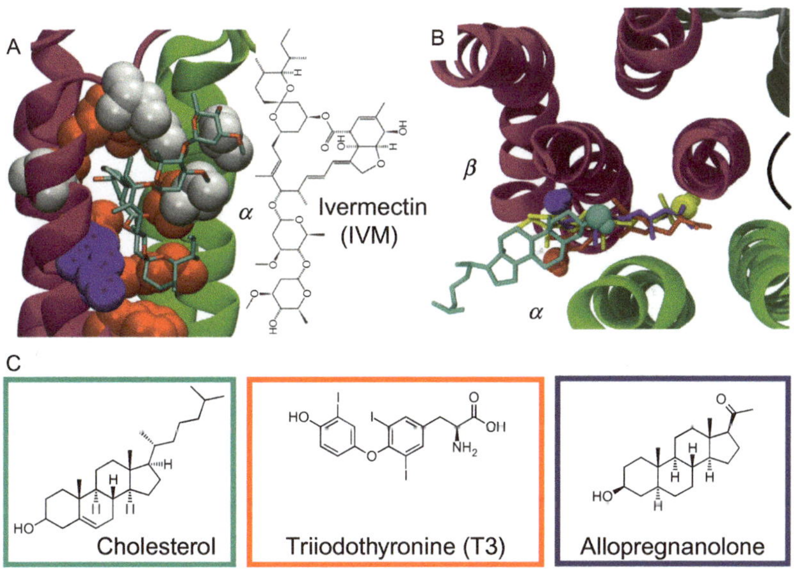

Fig. 2 Chemical analogy among steroid modulators and homology among receptors (Westergard, Salari, Martin, & Brannigan, 2015). (A) The GluCl IVM binding site. Residues identical to those at the $GABA_A$r $\alpha 1 - /\beta 1 +$ interface are *red*, similar residues are *white*, and residues with no similarity are *blue*. (B) Positive and negative modulators, *colored* as indicated by *border colors* in (C), except for pregnanolone sulfate (*yellow*), docked to intersubunit site in GluCl. Modulators are in licorice representation with hydroxyl head in space-filling representation. The central pore is represented by an arc. (C) Comparisons of structures of cholesterol, neurosteroid allopregnanolone, and thyroid hormone T3.

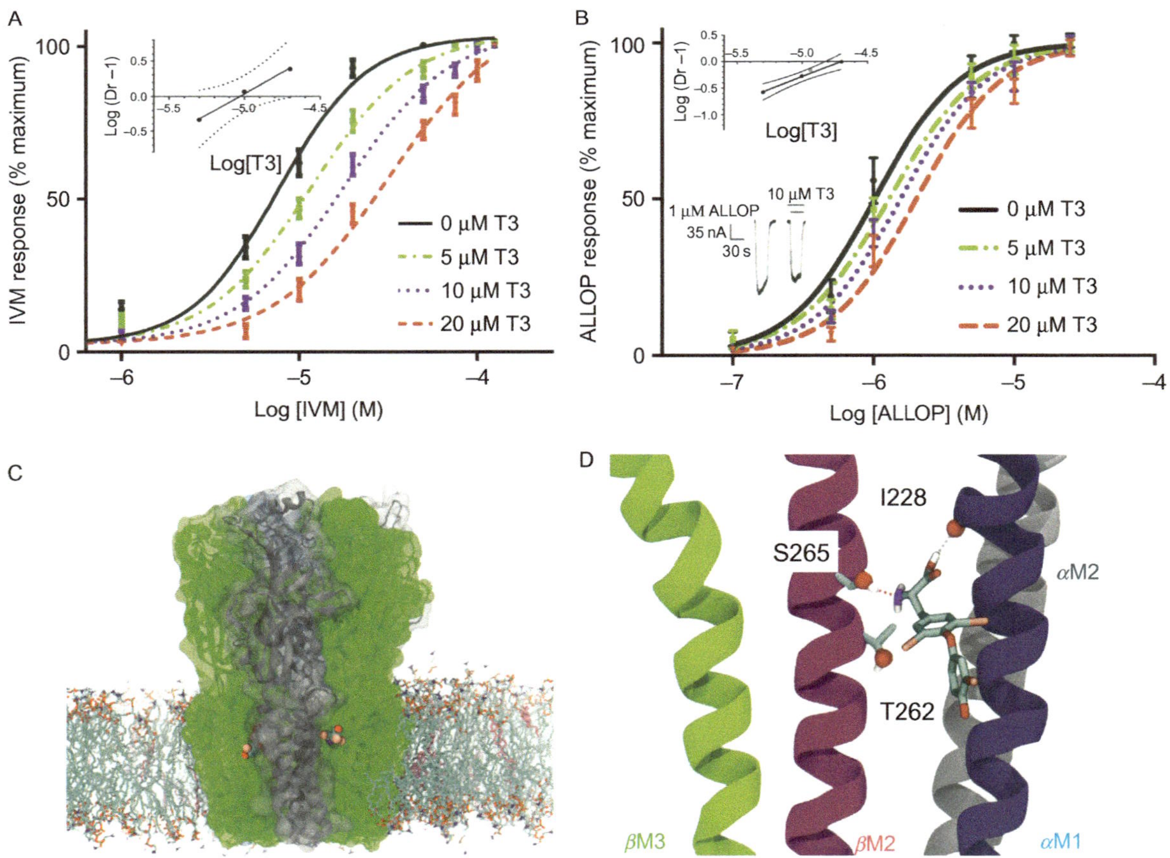

Fig. 3 See legend on opposite page.

simulations of T3 bound to $GABA_Ar$ intersubunit sites, we observed persistent hydrogen bonding of the T3 hydroxyl with the M1 helix backbone at a crystallographically observed bulge, suggesting a possible reason for why mutagenesis of intersubunit cavity residues has yielded largely inconclusive results.

These simulations do not provide insight into the mechanism by which binding to the intersubunit site causes its effects. Ivermectin is proposed to act via a wedge mechanism, where increasing the radius of the outer M1–M3 ring forces expansion of the inner pore-lining M2 helix ring. It is possible that other modulators, including neurosteroids, also act similarly; we saw slight increases in the M1–M3 radius with cholesterol bound to intersubunit sites in $GABA_Ar$ using one model and slight decreases using a different model; with the former model having the minimal RMSD with the later released β3 homomer structure (Miller & Aricescu, 2014).

3.2 Intrasubunit Sites

Docking algorithms placed multiple cholesterol molecules in sites fully embedded within the intrasubunit cavities of the *Torpedo* nAChR TMD (2BG9), also oriented parallel to the pore axis, with the hydroxyl located near the interface with the EC domain (Fig. 4A); docking algorithms did not identify intrasubunit cholesterol sites for any $GABA_Ar$. In nAChR, the K45–P280 contact anchors the γ EC domain to the γ TM domain. When it is broken, as at the end of the control simulation shown in Fig. 4C, the EC domain tilts away from its starting orientation and the TM

Fig. 3 Interactions of modulators binding to intersubunit sites (Westergard et al., 2015). (A and B) T3 inhibition of $GABA_Ar$ response to ivermectin (A) or allopregnanolone (B). Concentration–response curves for the effects of either agonist were constructed separately in the presence of 0, 5, 10, or 20 μM T3. The data are represented as means ± SEM for triplicate determinations. For each data point, $n = 3$–5. (*Inset*) Schild plot of the data from primary plot. "Dr" stands for dose-ratio. Neither Schild plot yielded a slope significantly different from unity according to 95% confidence intervals, consistent with competitive binding. (C and D) AA-MD simulations of $GABA_Ar$s with T3 bound to intersubunit sites. (C) $GABA_Ar$, in cartoon representation with surface overlay, viewed looking along the membrane. Lipids in the membrane are in licorice with POPC molecules colored by atom type and cholesterol molecules in *pink*. T3 molecules *colored* by atom type are shown at the interfaces between subunits. (D) Representative frame showing interactions between T3 at one of the $\alpha - /\beta +$ interfaces. T3 frequently formed simultaneous hydrogen bonds to α1M1 I228 (backbone at helical bulge) and β1M2 S265, similar to those we observed for propofol (Woll et al., 2016).

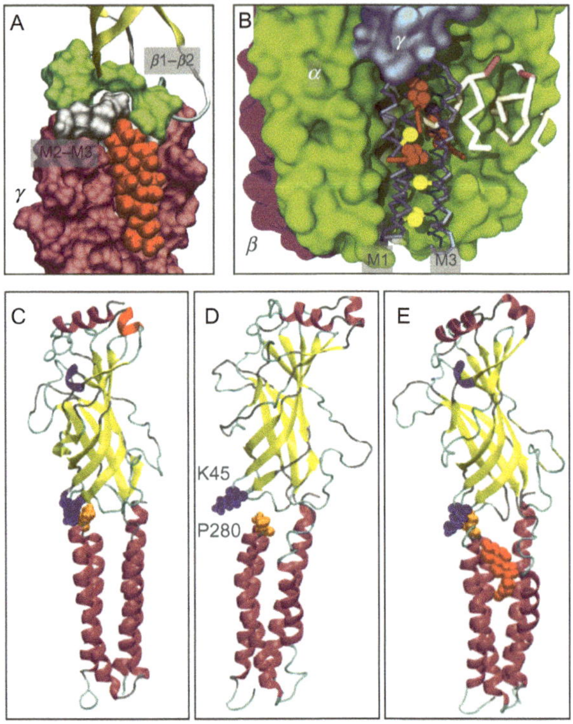

Fig. 4 Intrasubunit binding sites for cholesterol on nAChR. (A) Cholesterol (*red*) bound to nAChR γ subunit (*purple*) intrasubunit site, after brief 5 ns equilibration via AA-MD, bolstering contacts between M2 and M3 loop (*silver*) and β1–β2 loop (*green*). (B) Cholesterol after spontaneously partitioning to the center of the nAChR γ (*blue*) subunit during CG-MD (full system shown in Figs. 5 and 6) with experimentally identified accessible cysteine residues (Narayanaswami & McNamee, 1993) in *yellow*. (C–E) EC domain and TM domain of nAChR γ subunit are shown before AA-MD simulation (C), following 25 ns control simulation without cholesterol (D), and following 25 ns simulation with 15 molecules of cholesterol bound to intersubunit and intrasubunit sites as in Fig. 1B (E). *Panel (C)–(E) reprinted from Brannigan, G., Hénin, J., Law, R., Eckenhoff, R., & Klein, M. L. (2008). Embedded cholesterol in the nicotinic acetylcholine receptor.* Proceedings of the National Academy of Sciences of the United States of America, 105*(38), 14418–14423. http://dx.doi.org/10.1073/pnas.0803029105.*

domain, so that the only direct interaction between the two domains is through the flexible peptide chain linking M1 to the EC domain. Furthermore, β-sheets in the EC domain begin to unravel, reducing mechanical cohesion within the EC domain. Breaking of the K45–P280 contact is caused by the collapse of the loop containing P280 into the cholesterol-binding sites; occupation of these sites by cholesterol precludes such a collapse.

In CG-MD simulation, cholesterol molecules spontaneously bind to nAChR in orientations similar to those proposed in Brannigan et al. (2008), although they are likely to share the intrasubunit cavity with a polyunsaturated acyl chain. Different subunit species had different probabilities of being occupied by cholesterol vs phospholipid; the β and γ subunits both incorporated cholesterol molecules even when there was plentiful PUFA. Cholesterol occupying the intrasubunit cavity of the γ subunit contacted experimentally identified cysteine residues (Narayanaswami & McNamee, 1993) (Fig. 4).

4. STEROLS AS BOUNDARY LIPIDS: ANNULAR SITES

The lipids most commonly occupying annular binding sites will be affected by the protein's preferred membrane domain. Like cholesterol, unsaturated lipids have significant effects on organization and viscoelastic behavior of the surrounding membrane. The simplest lipid mixture shown to induce domain ("raft") formation consists of cholesterol and both a saturated and unsaturated species of phospholipid. In such mixtures, the liquid-ordered/raft phase is comprised of the saturated lipid and cholesterol, while the unsaturated lipid forms the liquid-disordered phase. Barrantes and colleagues (Wenz & Barrantes, 2005) found that the addition of nAChR to such a mixture increased the size of DPPC/Chol lipid-ordered domains, which (combined with additional FRET data) suggested that nAChR was embedded in liquid-ordered domains. One of the simplest, although not conclusive, technique for evaluating whether a transmembrane protein likely associates with a liquid-ordered domain involves testing for enrichment of the protein in the detergent resistant membrane (DRM); Bermudez, Antollini, Nievas, Aveldaño, and Barrantes (2010), however, found that nAChRs were not enriched in the DRM of a model, domain-forming mixture. Other experiments investigating whether nAChRs partition into lipid microdomains have been largely contradictory; Marchand, Devillers Thiéry, Pons, Changeux, and Cartaud (2002), Stetzkowski-Marden, Gaus, Recouvreur, Cartaud, and Cartaud (2006), and Willmann et al. (2006) suggest that nAChRs are associated with microdomains independent of stimulation by other proteins, while other studies (Campagna & Fallon, 2006; Zhu, Xiong, & Mei, 2006) found that nAChRs require stimulation by a protein such as agrin to partition into microdomains. Furthermore, formation and disassembly of the microdomains is highly sensitive to cholesterol concentration (Barrantes, 2007; Brusés, Chauvet, & Rutishauser, 2001;

Marchand et al., 2002; Pun et al., 2002; Zhu et al., 2006). Domain preferences are also sensitive to inclusion of sphingomyelin and adjustment of transbilayer asymmetry (Perillo, Peñalva, Vitale, Barrantes, & Antollini, 2016). This sensitivity may reflect an elastic effect: the cone-shaped pLGIC will impose a boundary condition on the membrane (Aranda-Espinoza, Berman, Dan, Pincus, & Safran, 1996) that is most easily accommodated by a flexible l_{do} phase, or l_o phases with high degrees of spontaneous curvature due to transbilayer asymmetry.

Computational equilibration of annular lipids in a nonrandom mixture, particularly one that involves phase separation, requires CG-simulations. We used a single nAChR from *Torpedo* in a range of ternary lipid mixtures, with each mixture containing DPPC, cholesterol, and a third lipid which could be one of twelve species of glycerophospholipid. The 12 species involved all combinations of two headgroups (PC or PE), three acyl chain saturation types (saturated, n-3 PUFA, n-6 PUFA), and two acyl chain lengths (16–18 CH_2 groups, or 20–22 CH_2 groups; MARTINI represents four CH_2 groups with a single bead, so it is not meaningful to distinguish between, e.g., 16 and 18 CH_2 groups).

In all systems in which domain formation occurred, we observed nAChR partition into cholesterol-poor l_{do} domains (Figs. 5 and 6); consequently, annular lipids were dominated by those lipids in the l_{do} domain, particularly PUFAs. However, as illustrated in Fig. 5, cholesterol still may occupy annular sites partially exposed to the cholesterol-poor l_{do} phase, although it is otherwise at very low concentration within the l_{do}. It is more likely to occupy annular sites on the l_o side of the nAChR and nonannular sites. Among phospholipids, n-3 PUFAs were far more likely to displace cholesterol from both annular and nonannular sites, relative to saturated or n-6 PUFA phospholipids. This likelihood was only weakly increased by replacing PC headgroups with PE and shorter acyl chains with longer acyl chains. This is consistent with the abundance of n-3 PUFAs(particularly DHA) in neuronal membranes, but no systematic experiments involving effect of n-3 PUFAs on nAChR function have been reported. While cholesterol bound to nonannular sites in both the subunit centers and interface with little specificity, nonannular binding of phospholipids was much more dependent upon subunit; for instance, DPPC was likely to occupy nonannular sites at the center of α subunits and between α and β subnits, but rarely occupied other sites. Correspondingly, comparison of partitioning of α_7 homomers with heteromers could indicate whether elastic effects or specific effects determine partitioning.

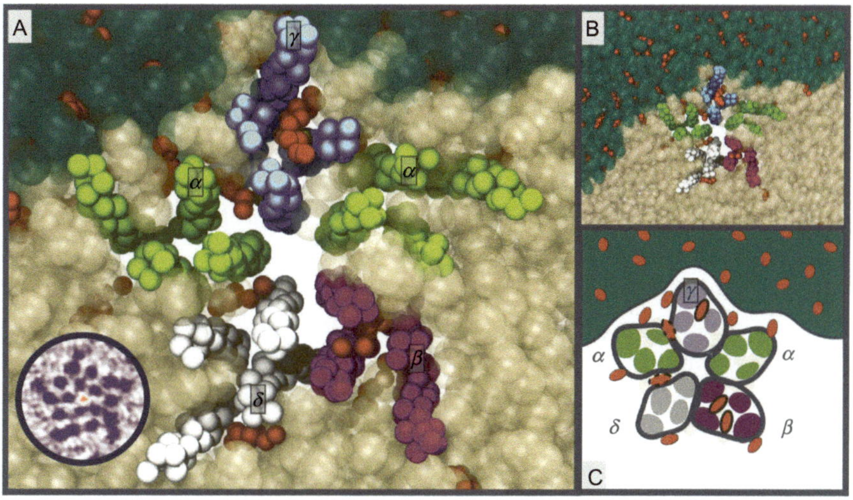

Fig. 5 Spontaneous binding of cholesterol to annular and nonannular sites on nAChR in a domain-forming membrane, observed via 2 μs of CG-MD. (A and B) nAChR (*colored* by subunit) in 2:2:1 DPPC:DHA-PE:CHOL, partitioned into l_{do} domain composed of *gray* DHA-PE, at the interface of l_o domain composed of *blue* DPPC and *red* CHOL. (*Inset*) EM density map used by Miyazawa et al. (2003) to derive a structure of the nAChR TM domain (PDB ID code 1OED). *Dark-blue circles* include protein backbone and side-chain density; material filling gaps between α-helices is unspecified. From the electron microscopy database, entry EMD-1044. (C) Cartoon illustrating partitioning of nAChR within l_{do} phase (*white*), near boundary between l_o (*blue*) phase with dominant binding mode for cholesterol molecules in simulation systems shown in (A).

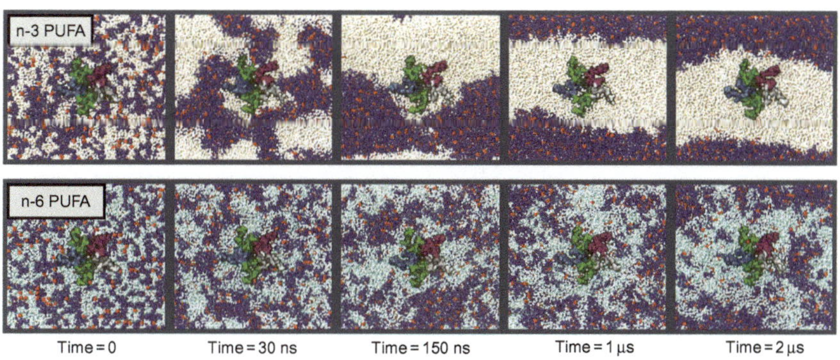

Fig. 6 Frames from 2 μs CG-MD simulation of nAChR in ternary mixture of DPPC (*navy*), CHOL (*red*), and (A) DHA-PE (*gray*, n-3 PUFA) or (B) DLoPC (*cyan*, n-6 PUFA).

5. PREDICTIONS

A long-standing barrier to characterizing lipid–protein interactions (including sterol–ion channel interactions) has been limited spatial resolution of most experimental approaches for systems with multiple lipid species or disorder over longer length scales. AA-MD simulation is a powerful method for investigating specific interactions of receptors with local lipids; however, due to nonrandom lipid mixing and slow diffusion of lipids over accessible time scales, realistic values for the local protein composition are almost totally unknown. As a result, investigation of the protein and its local environment cannot be fully decoupled from investigation of membrane organization over longer length scales.

A large number of experiments, ranging from the straightforward to the particularly sophisticated, have been carried out to investigate the mechanisms underlying cholesterol modulation of pLGICs. Unfortunately, approaches for hypothesis generation and data interpretation have lagged behind experimental techniques, and available systematic and physics-based techniques have been underutilized. The combination of results from AA-MD and CG-MD summarized here can be used to make numerous experimentally testable predictions, including

1. We predict function of nAChRs reconstituted in model membranes will be sensitive to replacement of even a small fraction of saturated or mono/diunsaturated acyl chains with n-6 and (especially) n-3 PUFAs. The trend may extend in a straightforward way to nAChRs in cell membranes because n-3 PUFAs are likely to compete successfully with native lipids for nAChR sites even at low concentrations; in contrast, depletion of cell membranes of cholesterol increases the abundance of other lipids which may now occupy nonannular sites and therefore be even more effective at increasing function.
2. If partitioning of pLGICs within domain-forming membranes is driven by membrane elasticity and the demand for a flexible membrane around the cone-shaped protein, partitioning will be strongly sensitive to changes in lipid composition that affect flexibility or spontaneous curvature of l_o and/or l_{do} phases, and only weakly sensitive to pLGIC sequence, since pLGICs are structurally conserved. If partitioning is instead driven by specific interactions with lipids, it will be far more sensitive to pLGIC sequence, particularly the presence of bulky vs small residues in the TMD.

3. The effects of membrane nonideality and domain formation on pLGIC function may be partially isolated by comparison of functional sensitivity to the total fraction of PUFA acyl chains vs the average number of lipids with two PUFA acyl chains; altering the latter while maintaining the former is likely to have significant effects if domain formation plays a substantial role in nAChR function, since lipids with one saturated chain and one PUFA chain will not be able to phase separate as effectively.

In CG-MD simulations with model membranes, we observed that affinities of lipid chains for nonannular (embedded) sites on nAChRs followed the trend n-3 PUFA > n-6 PUFA > CHOL ≫ satFA. Although competition of cholesterol and PUFAs for nonannular sites was initially surprising, we speculate that steroids and PUFAs have similar conformational entropy costs for binding to such sites. Steroids are rigid in the bulk membrane as well as annular and nonannular sites, while PUFAs are so flexible that available conformations are not significantly restricted by occupying nonannular sites among the tunnels within the nAChR TMD. In contrast, semiflexible saturated acyl chains may face a significant conformational entropic cost upon becoming embedded.

These CG-MD simulations involve both a lower resolution simulation model and an older pLGIC model, but serve as an important proof of principle, particularly in the context of high PUFA content in both native electroplax membranes and neurons. They further highlight that affinities for nonannular sites are determined not by the total number of lipid conformational degrees of freedom, but the number of new constraints imposed by binding. Entropic benefits of nonannular binding of lipids in general, while unintuitive, may be understood as allowing increased amino acid side-chain entropy relative to more dense, tightly packed proteins. Binding of membrane lipids in these gaps between hydrophobic residues precludes occupation by water, which is also entropically unfavorable as observed in the classical hydrophobic effect.

ACKNOWLEDGMENTS

I would like to thank Dr. Jérôme Hénin, Liam Sharp, Dr. Reza Salari, and Sruthi Murlidaran for their significant contributions to obtaining the computational results discussed within this chapter, as well as Thomas Westergard and Dr. Joseph Martin for carrying out the experiments involving effects of T3 on $GABA_A$r. Many of the results were obtained using computational resources from the National Science Foundation XSEDE program [MCB110149] as well as a local cluster funded by the National Science Foundation [DBI1126052]. Research was funded by grants from the National Science Foundation [MCB1330728], National Institutes of Health [P01GM55876], and Research Corporation for Scientific Advancement.

REFERENCES

Addona, G. H., Sandermann, H., Jr., Kloczewiak, M. A., Husain, S. S., & Miller, K. W. (1998). Where does cholesterol act during activation of the nicotinic acetylcholine receptor? *Biochimica et Biophysica Acta (BBA) - Biomembranes, 1370*(2), 299–309.

Althoff, T., Hibbs, R. E., Banerjee, S., & Gouaux, E. (2014). X-ray structures of GluCl in apo states reveal a gating mechanism of cys-loop receptors. *Nature, 512*(7514), 333–337. http://dx.doi.org/10.1038/nature13669.

Aranda-Espinoza, H., Berman, A., Dan, N., Pincus, P., & Safran, S. A. (1996). Interaction between inclusions embedded in membranes. *Biophysical Journal, 71*(2), 648–656. http://dx.doi.org/10.1016/S0006-3495(96)79265-2.

Arias, H., Sankaram, M., Marsh, D., & Barrantes, F. (1990). Effect of local anaesthetics on steroid-nicotinic acetylcholine receptor interactions in native membranes of Torpedo marmorata electric organ. *Biochimica et Biophysica Acta (BBA) - Biomembranes, 1027*(3), 287–294.

Baenziger, J. E., & Corringer, P.-J. (2011). 3D structure and allosteric modulation of the transmembrane domain of pentameric ligand-gated ion channels. *Neuropharmacology, 60*(1), 116–125. Retrieved from http://dx.doi.org/10.1016/j.neuropharm.2010.08.007. http://www.sciencedirect.com/science/article/pii/S0028390810002066%5Cnhttp://www.sciencedirect.com/science/article/pii/S0028390810002066/pdf?md5=c5c4609837a807a4c98555032c948225&pid=1-s2.0-S0028390810002066-main.pdf.

Baenziger, J. E., Hénault, C. M., Therien, J. P. D., & Sun, J. (2015). Nicotinic acetylcholine receptor-lipid interactions: Mechanistic insight and biological function. *Biochimica et Biophysica Acta - Biomembranes, 1848*(9), 1806–1817. http://dx.doi.org/10.1016/j.bbamem.2015.03.010.

Barrantes, F. J. (1989). The lipid environment of the nicotinic acetylcholine receptor in native and reconstituted membrane. *Critical Reviews in Biochemistry and Molecular Biology, 24*(5), 437–478.

Barrantes, F. J. (2007). Cholesterol effects on nicotinic acetylcholine receptor. *Journal of Neurochemistry, 103*(s1), 72–80.

Barrantes, F. J., Antollini, S. S., Blanton, M. P., & Prieto, M. (2000). Topography of nicotinic acetylcholine receptor membrane-embedded domains. *Journal of Biological Chemistry, 275*(48), 37333–37339.

Barrantes, F. J., Bermudez, V., Borroni, M. V., Antollini, S. S., Pediconi, M. F., Baier, J. C., et al. (2010). Boundary lipids in the nicotinic acetylcholine receptor microenvironment. *Journal of Molecular Neuroscience, 40*, 87–90. http://dx.doi.org/10.1007/s12031-009-9262-z.

Bermudez, V., Antollini, S. S., Nievas, G. A. F., Aveldaño, M. I., & Barrantes, F. J. (2010). Partition profile of the nicotinic acetylcholine receptor in lipid domains upon reconstitution. *Journal of Lipid Research, 51*(9), 2629–2641.

Borroni, V., Baier, C. J., Lang, T., Bonini, I., White, M. M., Garbus, I., et al. (2007). Cholesterol depletion activates rapid internalization of submicron-sized acetylcholine receptor domains at the cell membrane. *Molecular Membrane Biology, 24*(1), 1–15.

Bozek, K., Wei, Y., Yan, Z., Liu, X., Xiong, J., Sugimoto, M., et al. (2015). Organization and evolution of brain lipidome revealed by large-scale analysis of human, chimpanzee, macaque, and mouse tissues. *Neuron, 85*(4), 695–702. http://dx.doi.org/10.1016/j.neuron.2015.01.003.

Brannigan, G., Hénin, J., Law, R., Eckenhoff, R., & Klein, M. L. (2008). Embedded cholesterol in the nicotinic acetylcholine receptor. *Proceedings of the National Academy of Sciences of the United States of America, 105*(38), 14418–14423. http://dx.doi.org/10.1073/pnas.0803029105.

Breckenridge, W., Gombos, G., & Morgan, I. (1972). The lipid composition of adult rat brain synaptosomal plasma membranes. *Biochimica et Biophysica Acta (BBA) - Biomembranes*, *266*(3), 695–707. http://dx.doi.org/10.1016/0005-2736(72)90365-3.

Bristow, D. R., & Martin, I. L. (1987). Solubilisation of the γ-aminobutyric acid/benzodiazepine receptor from rat cerebellum: Optimal preservation of the modulatory responses by natural brain lipids. *Journal of Neurochemistry*, *49*(5), 1386–1393. http://dx.doi.org/10.1111/j.1471-4159.1987.tb01004.x.

Brusés, J. L., Chauvet, N., & Rutishauser, U. (2001). Membrane lipid rafts are necessary for the maintenance of the (alpha)7 nicotinic acetylcholine receptor in somatic spines of ciliary neurons. *The Journal of Neuroscience*, *21*(2), 504–512. (pii: 21/2/504).

Campagna, J., & Fallon, J. (2006). Lipid rafts are involved in c95 (4, 8) agrin fragment-induced acetylcholine receptor clustering. *Neuroscience*, *138*(1), 123–132.

Carlos, B.-P., Yaiza, M.-O., Jose, O.-C., Iris, S.-V., Guermarie, V., Alejandro, et al. (2008). Potential role of caveolin-1-positive domains in the regulation of the acetylcholine receptor's activatable pool: Implications in the pathogenesis of a novel congenital myasthenic syndrome. *Channels (Austin)*, *2*(3), 180–190.

Chiara, D., Dangott, L., Eckenhoff, R., & Cohen, J. (2003). Identification of nicotinic acetylcholine receptor amino acids photolabeled by the volatile anesthetic halothane. *Biochemistry*, *42*(46), 13457–13467.

Chiara, D. C., Dostalova, Z., Jayakar, S. S., Zhou, X., Miller, K. W., & Cohen, J. B. (2012). Mapping general anesthetic binding site(s) in human $\alpha 1\beta 3$ γ-aminobutyric acid type A receptors with [^{3}H]TDBzl-etomidate, a photoreactive etomidate analogue. *Biochemistry*, *51*(4), 836–847. Retrieved from http://www.ncbi.nlm.nih.gov/pubmed/22243422. http://dx.doi.org/10.1021/bi201772m.

Corbin, J., Wang, H. H., & Blanton, M. P. (1998). Identifying the cholesterol binding domain in the nicotinic acetylcholine receptor with [^{125}i] azido-cholesterol. *Biochimica et Biophysica Acta (BBA) - Biomembranes*, *1414*(1), 65–74.

Corrie, J., Ogrel, A. A., Mccardy, E. A., Blanton, M. P., & Baenziger, J. E. (2002). Lipid-protein interactions at the nicotinic acetylcholine receptor a functional coupling between nicotinic receptors and phosphatidic acid-containing lipid bilayers. *Journal of Biological Chemistry*, *277*(1), 201–208.

Criado, M., Eibl, H., & Barrantes, F. (1982). Effects of lipids on acetylcholine receptor. essential need of cholesterol for maintenance of agonist-induced state transitions in lipid vesicles. *Biochemistry*, *21*(15), 3622–3629.

DaCosta, C. J. B., & Baenziger, J. E. (2009). A lipid-dependent uncoupled conformation of the acetylcholine receptor. *Journal of Biological Chemistry*, *284*(26), 17819–17825. http://dx.doi.org/10.1074/jbc.M900030200.

Dacosta, C. J. B., Dey, L., Therien, J. P. D., & Baenziger, J. E. (2013). A distinct mechanism for activating uncoupled nicotinic acetylcholine receptors. *Nature Chemical Biology*, *9*(11), 701–707. Retrieved from http://www.ncbi.nlm.nih.gov/pubmed/24013278. http://dx.doi.org/10.1038/nchembio.1338.

DaCosta, C. J. B., Ogrel, A. A., Mccardy, E. A., Blanton, M. P., & Baenziger, J. E. (2002). Lipid-Protein Interactions at the Nicotinic Acetylcholine Receptor. *The Journal of Biological Chemistry*, *277*(1), 201–208. http://dx.doi.org/10.1074/jbc.M108341200.

Dalziel, A., Rollins, E., & Mcnamee, M. (1980). The effect of cholesterol on agonist-induced flux in reconstituted acetylcholine receptor vesicles. *FEBS Letters*, *122*(2), 193.

Du, J., Lü, W., Wu, S., Cheng, Y., & Gouaux, E. (2015). Glycine receptor mechanism elucidated by electron cryo-microscopy. *Nature*, *526*(7572), 224–229. http://dx.doi.org/10.1038/nature14853.

Ellena, J. F., Blazing, M. A., & Mcnamee, M. G. (1983). Lipid-protein interactions in reconstituted membranes containing acetylcholine receptor. *Biochemistry*, *22*(24), 5523–5535.

Garduño, J., Galindo-Charles, L., Jiménez-Rodríguez, J., Galarraga, E., Tapia, D., Mihailescu, S., et al. (2012). Presynaptic $\alpha 4\beta 2$ nicotinic acetylcholine receptors increase glutamate release and serotonin neuron excitability in the dorsal raphe nucleus. *Journal of Neuroscience*, *32*(43), 15148–15157.

Gotti, C., Zoli, M., & Clementi, F. (2006). Brain Nicotinic Acetylcholine Receptors: Native Subtypes and Their Relevance. *Trends in Pharmacological Sciences*, *27*(9), 482–491. http://dx.doi.org/10.1016/j.tips.2006.07.004.

Hamouda, A. K., Chiara, D. C., Sauls, D., Cohen, J. B., & Blanton, M. P. (2006). Cholesterol interacts with transmembrane α-helices m1, m3, and m4 of the Torpedo nicotinic acetylcholine receptor: Photolabeling studies using [3H] Azicholesterol. *Biochemistry*, *45*(3), 976–986.

Hénin, J., Salari, R., Murlidaran, S., & Brannigan, G. (2014). A predicted binding site for cholesterol on the GABAA receptor. *Biophysical Journal*, *106*(9), 1938–1949. http://dx.doi.org/10.1016/j.bpj.2014.03.024.

Hibbs, R. E., & Gouaux, E. (2011). Principles of activation and permeation in an anion-selective cys-loop receptor. *Nature*, *474*(7349), 54–60.

Hill, W. G., Southern, N. M., MacIver, B., Potter, E., Apodaca, G., Smith, C. P., et al. (2005). Isolation and characterization of the xenopus oocyte plasma membrane: A new method for studying activity of water and solute transporters. *American Journal of Physiology. Renal Physiology*, *289*(1), F217–F224. http://dx.doi.org/10.1152/ajprenal.00022.2005.

Hosie, A. M., Wilkins, M. E., da Silva, H. M. A., & Smart, T. G. (2006). Endogenous neurosteroids regulate GABAA receptors through two discrete transmembrane sites. *Nature*, *444*(7118), 486–489. http://dx.doi.org/10.1038/nature05324.

Ingólfsson, H. I., Arnarez, C., Periole, X., & Marrink, S. J. (2016). Computational microscopy of cellular membranes. *Journal of Cell Science*, *129*(2), 257–268.

Ingólfsson, H. I., Melo, M. N., Van Eerden, F. J., Arnarez, C., Lopez, C. A., Wassenaar, T. A., et al. (2014). Lipid organization of the plasma membrane. *Journal of the American Chemical Society*, *136*(41), 14554–14559. http://dx.doi.org/10.1021/ja507832e.

Janmey, P. A., & Kinnunen, P. K. J. (2006). Biophysical properties of lipids and dynamic membranes. *Trends in Cell Biology*, *16*(10), 538–546. http://dx.doi.org/10.1016/j.tcb.2006.08.009.

Jayakar, S. S., Dailey, W. P., Eckenhoff, R. G., & Cohen, J. B. (2013). Identification of propofol binding sites in a nicotinic acetylcholine receptor with a photoreactive propofol analog. *Journal of Biological Chemistry*, *288*(9), 6178–6189. http://dx.doi.org/10.1074/jbc.M112.435909.

Jayakar, S. S., Zhou, X., Chiara, D. C., Dostalova, Z., Savechenkov, P. Y., Bruzik, K. S., et al. (2014). Multiple propofol-binding sites in a γ-aminobutyric acid type a receptor (GABAAR) identified using a photoreactive propofol analog. *Journal of Biological Chemistry*, *289*(40), 27456–27468. http://dx.doi.org/10.1074/jbc.M114.581728.

Jones, O. T., & Mcnamee, M. G. (1988). Annular and nonannular binding sites for cholesterol associated with the nicotinic acetylcholine receptor. *Biochemistry*, *27*(7), 2364–2374.

Leibel, W. S., Firestone, L. L., Legler, D. C., Braswell, L. M., & Miller, K. W. (1987). Two pools of cholesterol in acetylcholine receptor-rich membranes from Torpedo. *Biochimica et Biophysica Acta (BBA) - Biomembranes*, *897*(2), 249–260.

Li, G. D., Chiara, D. C., Cohen, J. B., & Olsen, R. W. (2009). Neurosteroids allosterically modulate binding of the anesthetic etomidate to γ-aminobutyric acid type A receptors. *Journal of Biological Chemistry*, *284*(18), 11771–11775. http://dx.doi.org/10.1074/jbc.C900016200.

Marchand, S., Devillers-Thiéry, A., Pons, S., Changeux, J.-P., & Cartaud, J. (2002). Rapsyn escorts the nicotinic acetylcholine receptor along the exocytic pathway via association with lipid rafts. *The Journal of Neuroscience: The Official Journal of the Society for Neuroscience, 22*(20), 8891–8901.

Marrink, S. J., Risselada, H. J., Yefimov, S., Tieleman, D. P., & Vries, A. H. D. (2007). The MARTINI force field: Coarse grained model for biomolecular simulations. *The Journal of Physical Chemistry B, 111*(27), 7812–7824.

Marsh, D., & Barrantes, F. (1978). Immobilized lipid in acetylcholine receptor-rich membranes from Torpedo marmorata. *Proceedings of the National Academy of Sciences, 75*(9), 4329–4333.

Middlemas, D. S., & Raftery, M. A. (1987). Identification of subunits of acetylcholine receptor that interact with a cholesterol photoaffinity probe. *Biochemistry, 26*(5), 1219–1223.

Miller, P. S., & Aricescu, A. R. (2014). Crystal structure of a human GABAA receptor. *Nature, 512*(7514), 270–275. http://dx.doi.org/10.1038/nature13293.

Miyazawa, A., Fujiyoshi, Y., & Unwin, N. (2003). Structure and gating mechanism of the acetylcholine receptor pore. *Nature, 423*(6943), 949–955. http://dx.doi.org/10.1038/nature01748.

Morales-Perez, C. L., Noviello, C. M., & Hibbs, R. E. (2016). X-ray structure of the human $\alpha4\beta2$ nicotinic receptor. *Nature, 538*(7625), 411–415. http://dx.doi.org/10.1038/nature19785.

Narayanaswami, V., & McNamee, M. G. (1993). Protein-lipid interactions and Torpedo californica nicotinic acetylcholine receptor function. 2. membrane fluidity and ligand-mediated alteration in the accessibility of gamma subunit cysteine residues to cholesterol. *Biochemistry, 32*(46), 12420–12427.

Nishikawa, K., & Harrison, N. L. (2003). The actions of sevoflurane and desflurane on the gamma-aminobutyric acid receptor type A: Effects of TM2 mutations in the alpha and beta subunits. *Anesthesiology, 99*(3), 678–684. http://dx.doi.org/10.1097/00000542-200309000-00024.

Ochoa, E. L., Dalziel, A. W., & Mcnamee, M. G. (1983). Reconstitution of acetylcholine receptor function in lipid vesicles of defined composition. *Biochimica et Biophysica Acta (BBA) - Biomembranes, 727*(1), 151–162.

Perillo, V. L., Peñalva, D. A., Vitale, A. J., Barrantes, F. J., & Antollini, S. S. (2016). Transbilayer asymmetry and sphingomyelin composition modulate the preferential membrane partitioning of the nicotinic acetylcholine receptor in Lo domains. *Archives of Biochemistry and Biophysics, 591*, 76–86. http://dx.doi.org/10.1016/j.abb.2015.12.003.

Piomelli, D., Astarita, G., & Rapaka, R. (2007). A neuroscientist's guide to lipidomics. *Nature Reviews. Neuroscience, 8*(10), 743–754. http://dx.doi.org/10.1038/nrn2233.

Pun, S., Sigrist, M., Santos, A. F., Ruegg, M. A., Sanes, J. R., Jessell, T. M., et al. (2002). An intrinsic distinction in neuromuscular junction assembly and maintenance in different skeletal muscles. *Neuron, 34*(3), 357–370.

Quesada, O., González-Freire, C., Ferrer, M. C., Colón -Sáez, J. O., Fernández-García, E., Mercado, J., et al. (2016). Uncovering the lipidic basis for the preparation of functional nicotinic acetylcholine receptor detergent complexes for structural studies. *Scientific Reports, 6*, 32766. Retrieved from http://www.nature.com/articles/srep32766. http://dx.doi.org/10.1038/srep32766.

Santiago, J., Guzmán, G. R., Rojas, L. V., Marti, R., Asmar-Rovira, G. A., Santana, L. F., et al. (2001). Probing the effects of membrane cholesterol in the Torpedo californica acetylcholine receptor and the novel lipid-exposed mutation αC418W in Xenopus oocytes. *Journal of Biological Chemistry, 276*(49), 46523–46532.

Simons, K., & Gerl, M. J. (2010). Revitalizing membrane rafts: New tools and insights. *Nature Reviews. Molecular cell Biology, 11*(10), 688–699. http://dx.doi.org/10.1038/nrm2977.

Simons, K., & Ikonen, E. (1997). Functional rafts in cell membranes. *Nature, 387*(6633), 569–572. Retrieved from http://www.ncbi.nlm.nih.gov/pubmed/9177342. http://dx.doi.org/10.1038/42408.

Stetzkowski-Marden, F., Gaus, K., Recouvreur, M., Cartaud, A., & Cartaud, J. (2006). Agrin elicits membrane lipid condensation at sites of acetylcholine receptor clusters in C2C12 myotubes. *Journal of Lipid Research*, *47*(10), 2121–2133.

Stewart, D. S., Pierce, D. W., Hotta, M., Stern, A. T., & Forman, S. A. (2014). Mutations at beta N265 in γ-aminobutyric acid type a receptors alter both binding affinity and efficacy of potent anesthetics. *PLoS ONE*, *9*(10), e111470. http://dx.doi.org/10.1371/journal.pone.0111470.

Unwin, N. (2005). Refined structure of the nicotinic acetylcholine receptor at 4A resolution. *Journal of Molecular Biology*, *346*(4), 967–989.

Verlet, L. (1967). Computer "Experiments" on Classical Fluids. I. Thermodynamical properties of Lennard-Jones molecules. *Physical Review*, *159*(1), 98–103. http://dx.doi.org/10.1103/PhysRev.159.98.

Wenz, J. J., & Barrantes, F. J. (2005). Nicotinic acetylcholine receptor induces lateral segregation of phosphatidic acid and phosphatidylcholine in reconstituted membranes. *Biochemistry*, *44*(1), 398–410.

Westergard, T., Salari, R., Martin, J. V., & Brannigan, G. (2015). Interactions of l-3,5,3'-triiodothyronine, allopregnanolone, and ivermectin with the GABAA receptor: Evidence for overlapping intersubunit binding modes. *PLoS One*, *10*(9), e0139072. http://dx.doi.org/10.1371/journal.pone.0139072.

Willmann, R., Pun, S., Stallmach, L., Sadasivam, G., Santos, A. F., Caroni, P., et al. (2006). Cholesterol and lipid microdomains stabilize the postsynapse at the neuromuscular junction. *The EMBO Journal*, *25*(17), 4050–4060.

Woll, K. A., Murlidaran, S., Pinch, B. J., Hénin, J., Wang, X., Salari, R., et al. (2016). A novel bifunctional alkylphenol anesthetic allows characterization of GABAA receptor subunit binding selectivity in synaptosomes. *The Journal of Biological Chemistry*, *291*, 20473–20486. http://dx.doi.org/10.1074/jbc.M116.736975.

Zabrecky, J. R., & Raftery, M. A. (1985). The role of lipids in the function of the acetylcholine receptor. *Journal of Receptors and Signal Transduction*, *5*(5–6), 397–417.

Zhu, D., Xiong, W. C., & Mei, L. (2006). Lipid rafts serve as a signaling platform for nicotinic acetylcholine receptor clustering. *The Journal of Neuroscience: The Official Journal of the Society for Neuroscience*, *26*(18), 4841–4851. http://dx.doi.org/10.1523/JNEUROSCI.2807-05.2006.

CHAPTER SEVEN

Insights Into the Molecular Requirements for Cholesterol Binding to Ion Channels

Avia Rosenhouse-Dantsker[1]
University of Illinois at Chicago, Chicago, IL, United States
[1]Corresponding author: e-mail address: dantsker@uic.edu

Contents

Abstract

The concept that cholesterol binds to proteins via specific binding motifs, and thereby modulates their function, has emerged two decades ago. When we recently embarked on studies to uncover the putative binding region(s) of cholesterol in the Kir2.1 channel, we carried out an unbiased approach that combines computational and experimental methods. This approach resulted in the identification of novel cholesterol-binding regions distinct from known cholesterol-binding motifs. In recent years, a plethora of structures of proteins complexed with cholesterol have been determined revealing variegated cholesterol-binding regions that can provide invaluable insights into the prerequisites for cholesterol binding. Thus, using this database of structures, the goal of this chapter is to present a comprehensive analysis of representative cholesterol-binding regions, and thereby determine the molecular requirements for cholesterol binding. The analysis demonstrates that the primary requirement for cholesterol binding is a highly hydrophobic environment, and that the interaction with the cholesterol molecule can be stabilized by stacking interactions between its ring structure and hydrophobic aromatic residues, and by hydrogen bonding between its hydroxyl group and a variety of protein residues. This general requirement suggests that the known cholesterol-binding motifs describe a subset of cholesterol-binding regions, and provides a framework for expanding the search for novel cholesterol-binding regions in ion channels.

Current Topics in Membranes, Volume 80
ISSN 1063-5823
http://dx.doi.org/10.1016/bs.ctm.2017.05.003

1. CHOLESTEROL-BINDING MOTIFS: FIRST INSIGHTS INTO SPECIFIC INTERACTIONS BETWEEN MEMBRANE PROTEINS AND CHOLESTEROL

During the last decade, it has been demonstrated that a variety of ion channels are sensitive to the level of membrane cholesterol (reviewed by Maguy, Hebert, & Nattel, 2006; Martens, O'Connell, & Tamkun, 2004; Rosenhouse-Dantsker, Mehta, & Levitan, 2012). In most cases, cholesterol has been found to suppress channel function. This effect was observed in the majority of inwardly rectifying K^+ channels (Romanenko, Fang, et al., 2004; Rosenhouse-Dantsker, Leal-Pinto, Logothetis, & Levitan, 2010), Ca^{2+}-sensitive K^+ channels (Bukiya, Belani, Rychnovsky, & Dopico, 2011; Crowley, Treistman, & Dopico, 2003), voltage-gated K^+ channels (Abi-Char et al., 2007; Hajdú, Varga, Pieri, Panyi, & Gáspár, 2003), voltage-gated Na^+ and Ca^{2+} channels (Lundbaek et al., 2004), volume-regulated anion channels (Romanenko, Rothblat, & Levitan, 2004), vanilloid transient receptor potential (TrpV) channels (Picazo-Juarez et al., 2011), and the nicotinic acetylcholine receptor (nAChR) (Barrantes, 2004, 2007). In contrast, only a few channels were enhanced by cholesterol enrichment and/or suppressed by cholesterol depletion. Examples include atrial and neuronal G-protein inwardly rectifying potassium channels (Bukiya, Durdagi, Noskov, & Rosenhouse-Dantsker, 2017; Deng et al., 2012), epithelial Na^+ channels (Awayda et al., 2011; Shlyonsky, Mies, & Sariban-Sohraby, 2003), specific subtypes of transient receptor potential (TRP) channels (Lockwich et al., 2000), and Ca^{2+}-permeable stretch-activated cation channels (Chubinskiy-Nadezhdin, Negulyaev, & Morachevskaya, 2011).

Two mechanisms have been proposed to explicate the molecular basis for the regulation of ion channels by cholesterol: an indirect mechanism that can be traced to the rigidity of the fused ring system of cholesterol and a direct mechanism that can be traced to the type of interactions that the cholesterol molecule can form with different amino acids.

More specifically, due to the rigidity of the ring system of cholesterol (Fig. 1), changes in cholesterol levels lead to changes in the physical properties of lipid bilayers including their rigidity, fluidity, and thickness (reviewed by Rosenhouse-Dantsker et al., 2012). Such changes in the physical properties of lipid bilayers result in a hydrophobic mismatch between the transmembrane domains and the lipid bilayer (Lundbaek & Andersen, 1999;

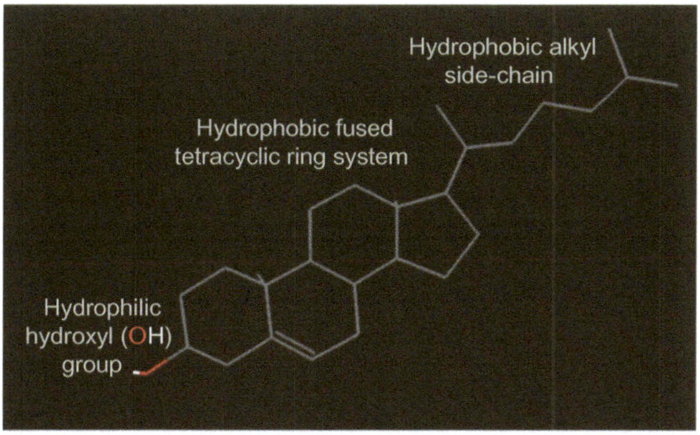

Fig. 1 Structure of the cholesterol molecule. Stick presentation of the cholesterol molecule based on PDB ID 1LRI. Cholesterol is a tetracyclic molecule with a fused ring system that has a polar hydrophilic hydroxyl group on one end and a hydrophobic eight-carbon alkyl group on its other end.

Lundbaek, Birn, Hansen, & Andersen, 1996). This, in turn, affects the function of membrane proteins, thereby providing an indirect mechanism for cholesterol regulation of ion channels.

Alternatively, a growing number of studies have demonstrated that cholesterol may also play a direct role in the regulation of ion channels via specific cholesterol–protein interactions (Epand, 2006; Levitan, Singh, & Rosenhouse-Dantsker, 2014; Marsh & Barrantes, 1978; Rosenhouse-Dantsker et al., 2012). In particular, three cholesterol-binding motifs have been proposed to delineate the molecular requirements for cholesterol binding.

The first and most established cholesterol-binding motif proposed was the cholesterol recognition amino acid consensus (CRAC) motif, which was defined as $(L/V)X_{1-5}YX_{1-5}(R/K)$ where X_{1-5} represents 1–5 residues of any amino acid (Epand, 2006; Li & Papadopoulos, 1998). This motif was proposed in studies of the peripheral-type benzodiazepine receptor, a transmembrane protein that mediates the translocation of cholesterol (Li & Papadopoulos, 1998). Subsequently, the CRAC motif was also identified in several ion channels including the large conductance Ca^{2+} and voltage-gated K^+ (BK) channels (Singh et al., 2012) and the nAChR (Fantini & Barrantes, 2013).

More recently, an inverted CRAC sequence, the CARC motif, was proposed (Fantini & Barrantes, 2013). The CARC motif, which was defined as $(R/K)X_{1-5}(Y/F)X_{1-5}(L/V)$, has been suggested to be responsible for cholesterol interactions with the nAChR and was shown to be more consistent in predicting cholesterol recognition motifs in integral membrane proteins (Fantini & Barrantes, 2013). In accordance with this notion, the CARC motif identified in the transmembrane domains of TRPV1 (Picazo-Juarez et al., 2011) was found to be critical for cholesterol sensitivity of the channel.

The third established cholesterol-binding motif is the cholesterol consensus motif (CCM) (Hanson et al., 2008). This motif was based on the complexed crystal structure of the β-adrenergic receptor with cholesterol, and included residues on adjacent helices of the G protein-coupled receptor: (W/Y)(I/V/L)(K/R) on one helix and (F/Y/R) on a second helix (Cherezov et al., 2007; Hanson et al., 2008). The notion that cholesterol may bind to membrane proteins in between different α-helices within the transmembrane domain was further supported by the similar characteristics of the binding sites that were identified in a variety of G protein-coupled receptors (Yeagle, 2014) including, for example, the proton pumping Rhodopsin receptor ARII (Wada et al., 2011), the μ-opiod receptor (Manglik et al., 2012), the A2A adenosine receptor (Liu et al., 2012), the dopamine receptor (Penmatsa, Wang, & Gouaux, 2013), and the 5-HT2B/ERG receptor (Wacker et al., 2013).

The common characteristic of the three motifs is that they all include specific types of residues that are expected to interact with the three distinct parts of the cholesterol molecule. The majority of the cholesterol molecule is hydrophobic, consisting of a tetracyclic structure with an eight-carbon alkyl group attached on one end, and as such is expected to interact with hydrophobic residues. In addition, however, it also has a polar hydroxyl group on its other end that can form hydrogen bonding with a variety of residues (Fig. 1). The overarching principle that underlies all three cholesterol-binding motifs is that each of the three parts of the cholesterol molecule interacts with a specific kind of amino acids via a particular type of interaction. The types of interactions included in all three motifs are (1) hydrophobic interactions between the alkyl group of cholesterol and aliphatic hydrophobic amino acids such as isoleucine, leucine, and valine; (2) stacking interactions between the ring structure of the cholesterol molecule and aromatic hydrophobic amino acids such as phenylalanine, tyrosine, and tryptophan; (3) hydrogen bonding between the hydroxyl group of cholesterol and the side-chain of a positively charged amino acid such as arginine and lysine.

However, major open questions are whether these three cholesterol-binding motifs account for all the possible cholesterol–protein interactions, and whether cholesterol may interact with membrane proteins via other types of residues. For example, whereas hydrophobic residues can be expected to form an important part of the binding surface of the hydrophobic cholesterol molecule, there are several hydrophobic amino acids that may contribute to a hydrophobic environment in addition to isoleucine, leucine, and valine. Another question is whether the presence of a positively charged amino acid is critical for the formation of a cholesterol-binding site. Aside from interacting with arginine or lysine, the polar hydroxyl group of the cholesterol molecule could hypothetically form hydrogen bonding with the side-chains of a variety of polar or charged amino acids or with the backbone carbonyl or amino groups of any residue.

To address these questions, we present two independent studies. First, we describe our earlier study of the putative cholesterol-binding regions in the inwardly rectifying potassium channel, Kir2.1, using an unbiased computational–experimental approach (Rosenhouse-Dantsker, Noskov, Durdagi, Logothetis, & Levitan, 2013). Second, based on an extensive database of structures of a variety of proteins complexed with cholesterol, we determine the molecular prerequisites for cholesterol binding to proteins.

2. CHOLESTEROL-BINDING REGIONS IN KIR2.1: BEYOND KNOWN CHOLESTEROL-BINDING MOTIFS

Prior to employing the computational–experimental approach to identify putative cholesterol-binding regions in the Kir2.1 channel, we first searched for the known cholesterol-binding motifs. A CRAC motif found in the inwardly rectifying potassium channel Kir2.1 was located in a cytosolic segment outside the transmembrane domain, rendering it highly energetically unfavorable for cholesterol binding (Rosenhouse-Dantsker et al., 2013).

CARC motifs were also identified in both the transmembrane and cytosolic domains of Kir2.1 (Rosenhouse-Dantsker et al., 2013). Whereas the CARC motifs located in the cytosolic domain were unfavorable for cholesterol binding, the two CARC motifs located in the transmembrane domain at its interface with the cytosolic domain provided a more favorable putative cholesterol-binding site (Rosenhouse-Dantsker et al., 2013). One transmembrane CARC motif involved residues R67—F73—V77 and the second involved residues R82—F88—L90. In the first CARC motif, the role of R67 could not be tested because its mutation resulted in a nonfunctional channel. The V77I mutation, on the other hand, resulted in loss of cholesterol sensitivity. F73 was shown to interact with the cholesterol molecule via stacking interactions in molecular dynamics simulations. Thus, although its mutation also resulted in a nonfunctional channel, it too may play a critical role in the sensitivity of the channel to cholesterol. In contrast, mutations of all three residues of the second CARC motif did not affect the sensitivity of the channel to cholesterol, suggesting that it does not describe a cholesterol-binding region in these channels. This also suggests that the existence of a cholesterol-binding motif does not necessarily imply that cholesterol would bind to the protein at the corresponding region.

We thus proceeded to determine where cholesterol binds to the channel protein employing an unbiased approach that combined computational and experimental techniques (Rosenhouse-Dantsker et al., 2013). Specifically, to identify possible cholesterol-binding sites in Kir2.1, we employed a combination of docking studies, all-atom molecular dynamics simulations, site-directed mutagenesis, and two electrode voltage clamp using the *Xenopus* oocyte heterologous expression system. This approach resulted in the identification of two novel putative cholesterol-binding regions (Rosenhouse-Dantsker et al., 2013). The energetically more favorable binding region was located in the center of the transmembrane domain of Kir2.1 (region 1). A second binding region, which may represent a transient site with weak and possibly short-lived cholesterol binding, was located at the interface between the transmembrane and cytosolic domains of the channel (region 2).

Within region 1, we identified eight residues whose mutation abrogated or significantly decreased the sensitivity of the channel to cholesterol (Fig. 2A–D). These eight residues were not in one continuous segment of the channel protein as in the CRAC/CARC motifs, but rather distributed among the α-helices of two adjacent channel subunits indicating that the cholesterol molecule would bind in between these α-helices similarly to the case in the CCM. Out of the eight residues whose mutation affected the sensitivity of the channel to cholesterol, six were the same type of aliphatic hydrophobic residues that are included in the cholesterol-binding motifs, namely, isoleucine, leucine, and valine (L69, L85, V93, I166, V167, I175). In contrast, the other two residues were residues that are not included in any of the three known cholesterol-binding motifs. One residue was a hydrophobic alanine (A70), and the other was a polar uncharged serine (S95). Mutations of several aromatic residues (W81F, F88L, F92Y, F98L/A, F163V, F174V) and of a positively charged arginine residue (R82T) located in this region did not affect the sensitivity of the channel to cholesterol. Mutations of two other aromatic residues (F73L, F99L) and an arginine (R67Q/W) rendered the channel nonfunctional, and their role in cholesterol binding could not be determined experimentally. However, based on a 50 ns all-atom full-membrane molecular dynamics simulation,

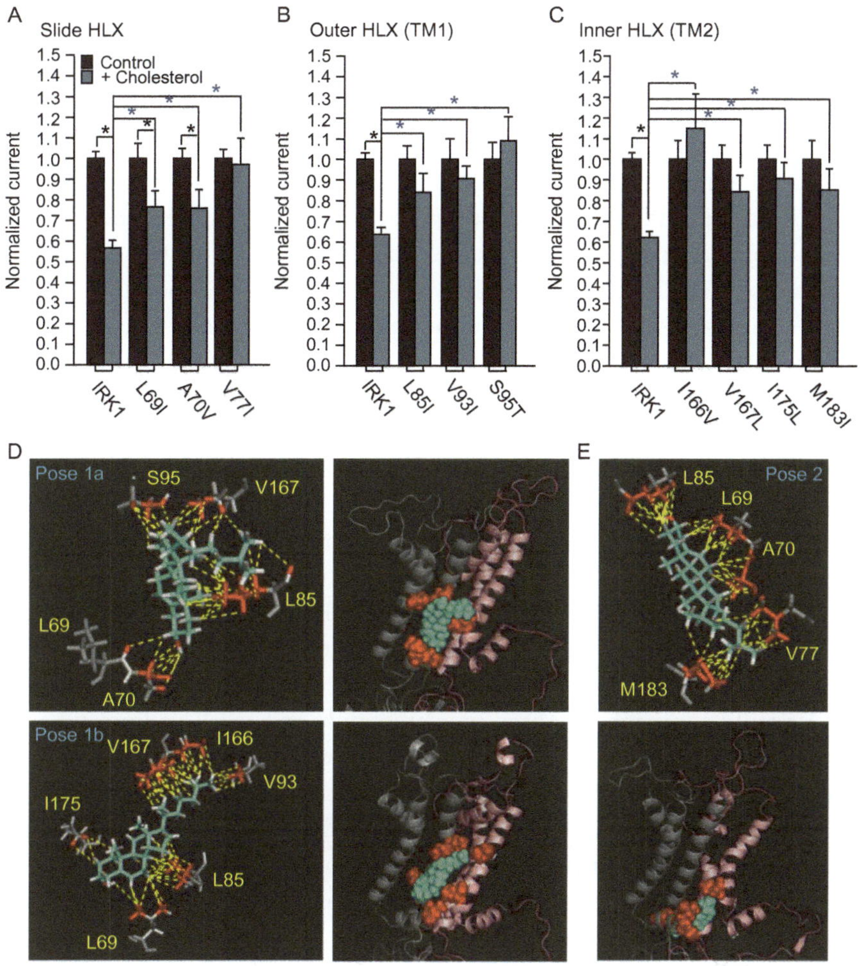

Fig. 2 Cholesterol recognition residues in the two putative transmembrane binding regions in Kir2.1. Whole-cell basal currents recorded in *Xenopus* oocytes at −80 mV showing the effect of cholesterol enrichment on Kir2.1 and (A) L69I, A70V, and V77I of the slide helix; (B) L85I, V93I, and S95T of the outer helix; and (C) I166V, V167L, I175L, and M183I of the inner helix ($n = 9$–90). Significant difference is indicated by an *asterisk* (*$P \leq 0.05$). A *black asterisk* indicates significant difference between whole-cell currents obtained for same construct following cholesterol enrichment and in the absence of treatment (control). A *blue asterisk* indicates significant difference between the effect of cholesterol enrichment on a mutant and the wild-type Kir2.1 channel. (D–E) Stick and ball presentations of the cholesterol recognition residues that surround the cholesterol molecule in each of the two putative binding regions after 50 ns of all-atom full-membrane molecular dynamics simulations. Two representative poses in region 1 are shown in (D) and one representative pose in region 2 is shown in (E). The relative position of each binding region in the transmembrane domain of the channel is shown in the ball presentation that includes the two adjacent subunits of the channel that interact with the cholesterol molecule in a cartoon presentation. *From Rosenhouse-Dantsker, A., Noskov, S., Durdagi, S., Logothetis, D. E., & Levitan, I. (2013). Identification of novel cholesterol-binding regions in Kir2 channels. Journal of Biological Chemistry, 288(43), 31154–31164.*

both W81 and F73 may interact with the rigid ring structure of the cholesterol molecule, suggesting that the very mild W81F mutation may not sufficiently affect the interaction of the channel by cholesterol, and that F73 may also be important for cholesterol binding.

Within region 2, we identified five hydrophobic residues whose mutation abrogated or significantly decreased the sensitivity of the channel to cholesterol (Fig. 2A–C, E). Similarly to the cholesterol binding pocket in region 1, these five residues were located in two adjacent subunits of the channel. Accordingly, also in this region, cholesterol would bind in between α-helices. Three of these residues were the hydrophobic residues leucine and valine (L69, V77, L85) that are included in all three cholesterol-binding motifs. The other two residues, an alanine (A70) and a methionine (M183), are not included in any cholesterol-binding motif. Mutations of several aromatic residues (W81F, F88L, F174V) and of positively charged lysine residues (K182Q, K185Q, K187Q) located in this region did not have an effect on the sensitivity of the channel to cholesterol. In this region, the effect of only one nonfunctional mutation of an aromatic residue that was shared with region 1 (F73I) could not be determined.

A similar approach was used in a later study of Kir2.1 leading to a very similar conclusion (Fürst, Nichols, Lamoureux, & D'Avanzo, 2014). Thus, whereas similarly to known cholesterol-binding motifs, both cholesterol-binding regions in Kir2.1 include aliphatic hydrophobic residues and possibly aromatic residues, it is not clear whether any positively charged residue (arginine or lysine) plays any role in cholesterol binding to either regions of the channel (Rosenhouse-Dantsker et al., 2013). Moreover, in both cholesterol-binding regions, mutations of residues that are not included in any of the three cholesterol-binding motifs affected the sensitivity of the channel to cholesterol. These included an alanine that was shared by both regions, a serine in region 1, and a methionine in region 2 (Rosenhouse-Dantsker et al., 2013). Alanine is an example of a hydrophobic residue that may contribute to forming a hydrophobic environment, and serine is an example of a residue that has a polar side-chain that could form hydrogen bonding with the hydroxyl group of the cholesterol molecule.

3. CHOLESTEROL-BINDING REGIONS IN PROTEINS: INSIGHTS FROM A DATABASE OF STRUCTURES

We next turned to the RCSB Protein Data Bank for representative structures of different proteins complexed with cholesterol. A search for structures that included the ligand CLR (cholesterol) resulted in 92 structures. Within this database of structures, there were some sets of structures in which the protein was crystallized in different conditions (e.g., different ligands, different agonists, etc.). In such cases, one structure was randomly selected for analysis. Also, some proteins were crystallized with cholesterol in different occasions. If the residues in the cholesterol-binding regions were very similar (>90% identity), only one of the structures was included in the analysis. This process resulted in a database of 24 structures. Since some of the structures included more than one cholesterol molecule, all unique (nonduplicate) binding regions were included in the analysis resulting in the 38 cholesterol-binding regions depicted in Fig. 3A–X. The structures are ordered according to their release date in the RCSB Protein Data Bank.

In the molecular dynamics simulations described above, cholesterol molecules bound to Kir2.1 exhibited significant flexibility, continuously exploring a considerable conformational space within each binding region, and interacting with different sets of residues in these two regions (Rosenhouse-Dantsker et al., 2013). Thus, to determine the types of residues that form cholesterol-binding regions in this database of 24 structures while accounting for this potential flexibility, we plotted the types of residues located within 5 Å (4 ± 1 Å) from each cholesterol molecule. The results are depicted in Fig. 4. For each cholesterol-binding region, the amino acids are shown as bars, organized from left to right according to the order and color in the key. The height of each bar indicates the number of amino acids of the corresponding type that are within 5 Å from the cholesterol molecule. Further details are included in Table 1 that lists all the residues located in proximity to the cholesterol molecule according to type as defined in the three known cholesterol-binding motifs.

Examination of this information leads to the following observations:

A. The vast majority of cholesterol-binding regions include multiple aliphatic hydrophobic amino acids that are among the residues in the known cholesterol-binding motifs (37 of the 38 binding regions) (Table 1). On the average, within the database of 38 cholesterol-binding regions, there were 1.79 isoleucine residues, 2.11 leucine residues, and 1.18 valine residues totaling in 5.08 aliphatic hydrophobic residues of this type per binding region. This indicates that this type of residues is critical for forming an adequate hydrophobic environment for cholesterol binding.

B. The majority of cholesterol-binding regions include hydrophobic aromatic residues that appear in the known cholesterol-binding motifs

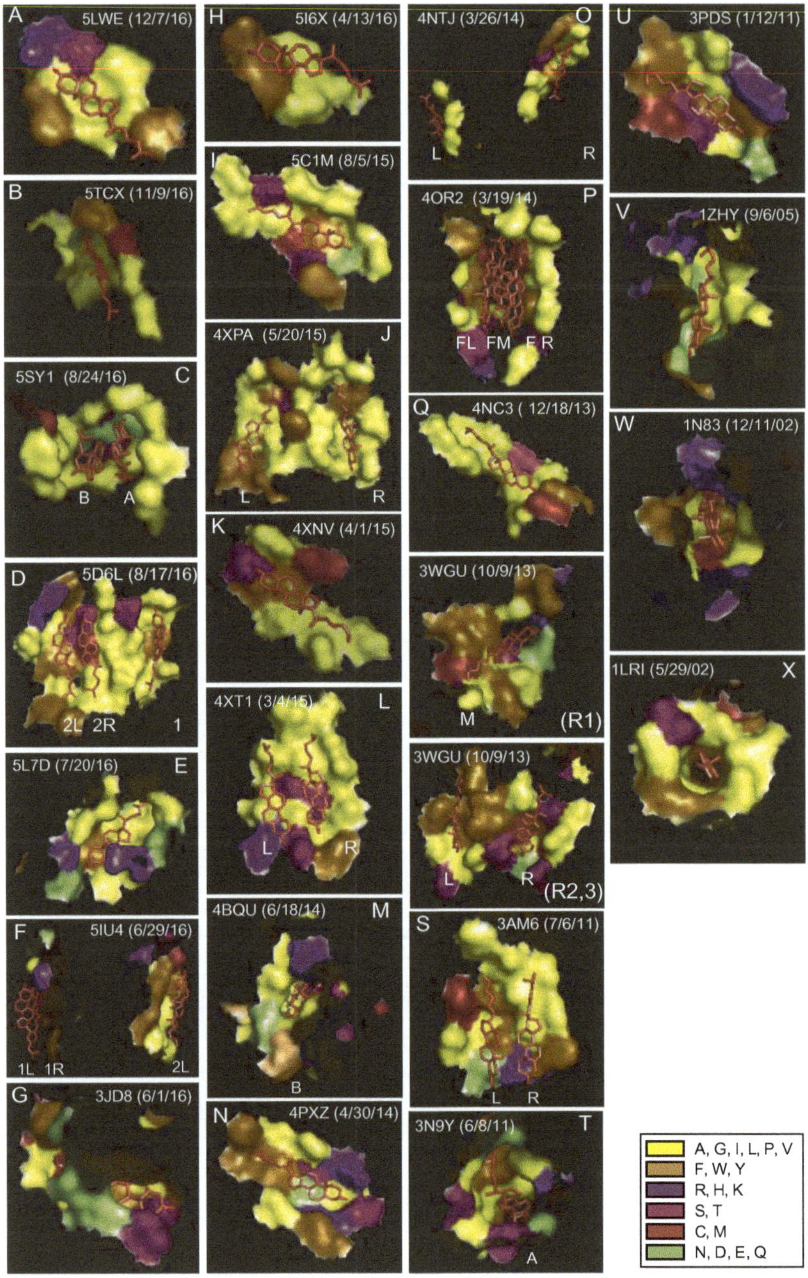

Fig. 3 Representative set of structures of cholesterol-binding regions. Surface representation of the binding regions and the residues that participate in the interactions with cholesterol in a representative set of 24 structures of proteins complexed with cholesterol. Both the PDB accession number and the release date of the structure are included in each figure. The amino acids in the binding regions are colored according to the color code in the key based on residue type as follows: (1) hydrophobic amino acids: A, G, I, L, P, V; (2) aromatic amino acids: F, W, Y; (3) hydrophilic, positively charged amino acids with basic side chains: R, H, K; (4) amino acids with aliphatic hydroxyl-containing side chains: S, T; (5) amino acids with sulfur-containing side chains: C, M; (6) amino acids with acidic side chains and their derivatives: N, D, E, Q.

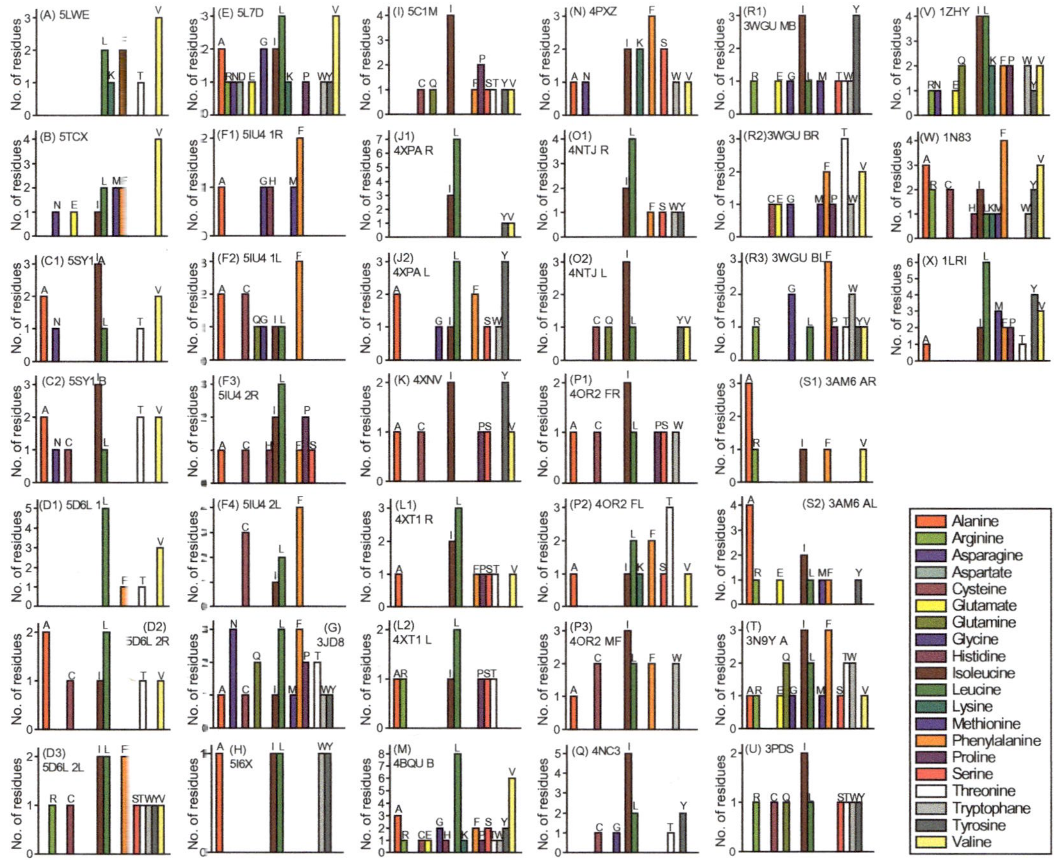

Fig. 4 Types of residues involved in cholesterol-binding regions in a representative database of structures of proteins complexed with cholesterol. For each cholesterol-binding region depicted in Fig. 3, the amino acids located within 5 Å from the cholesterol molecule are shown as bars, organized from left to right according to the order and color in the key. A separate bar graph is plotted for each cholesterol-binding region shown in Fig. 3. The PDB accession number of the corresponding structure is shown on each bar graph. The height of each bar indicates the number of amino acids of the corresponding type that are involved in the binding region.

Table 1 Cholesterol Interacting Residues in a Representative Set of Structures of Cholesterol-Binding Regions

PDB ID (Release Date); Protein Name	Nonpolar Aliphatic Residues (I/L/V)	Nonpolar Aromatic Residues (F/Y/W)	Positively Charged Residues (R/K)	Other Residues	Most Proximal Atom(s) to O(CLR)
(A) 5LWE (12/7/16); CCR9 chemokine receptor	V262, L312, V315, L316, V320	F308, F319	K254	T258	NZ(K254): 4.4 Å
(B) 5TCX (11/9/16); Tetraspanin CD81	I64, V68, V71, V75, L98, L101, V212, I215	F21, F94	None	N18, M72, M216, E219	OE1(E219): 2.6 Å OE2(E219): 3.0 Å ND2(N18): 3.0 Å
(C1) 5SY1 A (8/24/16); STRA6 receptor	A: I416, L419, V420, I423, I520, B: V512	None	None	A: A516, N519, B: T515, A516	OD1(N519): 2.5 Å OG1(T515): 3.2 Å
(C2) 5SY1B	A: V512, A516, B: I416, L419, V420, I423, I520	None	None	A: T515, A516, B: C424, T515, A516, N519	OD1(N519): 2.5 Å OG1(T515): 3.4 Å
(D1) 5D6L 1 (8/17/16); β2 adrenergic receptor	V44, L45, V48, V52, L53, L84, L339, L340	F49	None	T56	O(L339): 3.2 Å
(D2) 5D6L 2R	I55, L80, V81, L84	None	None	A59, T73, C77, A85	O(CLR 2 L): 5.0 Å
(D3) 5D6L 2L	T73, V81, I112, L115, I154, L155	Y70, F108, W158, F166	R151	T73, S74, C77	NE(R151): 4.1 Å NH2(R151): 5.1 Å
(E) 5L7D (7/20/16); Smoothened homolog	L108, L112, I156, V157, V210, V488, L489, I496	W109, Y130	K105, R161	D95, G111, N114, A115, E160, G162, P164, A492	OD2(D95): 2.8 Å

(F1) 5IU4 1R (6/29/16); A2A adenosine receptor	None	F79, F133	None	A73, H75, G76, M140	N(G76): 4.4 Å
(F2) 5IU4 1L	L58, I80	F62, F70, F79	None	C71, A72, A73, G76, C77, Q163	O(H_2O 2526): 2.6 Å N(A73): 2.9 Å
(F3) 5IU4 2R	L244, L247, I251, I252, L269	F255	None	P248, C262, S263, H264, A265, P266	O(S263): 2.6 Å O(H_2O 2614): 2.9 Å N(S263): 3.5 Å
(F4) 5IU4 2L	L187, L247, I251	F182, F183, F255, F258	None	C254, C259, C262	O(H_2O 2589): 2.9 Å
(G) 3JD8 (6/1/16); NPC1	L83, L111, L176	W27, F108, Y192, F194	None	N41, Q79, T82, N86, P90, T112, C177, M193, N198, Q200, A201, P202	ND2(N41): 2.8 Å NE2(Q79): 2.9 Å
(H) 5I6X (4/13/16); ts3 serotonin transporter	I576, L577	Y572, W573	None	A580	OH(Y572): 5.5 Å
(I) 5C1M (8/5/15); μ-opioid receptor	I290, V291, I298, I302, I308	Y299, F313	None	T294, P295, P309, Q314, S317, C321	OE1(Q314): 3.7 Å
(J1) 4XPA R (5/20/15); Dopamine transporter	V34, L37, L38, I41, L270, L276, L277, L280, L347, I351, I358	Y273	None	None	N(V34): 5.3 Å
(J2) 4XPA L	L66, L73, I78, L346	Y69, Y342, F350, F357, Y518, W519	None	G70, A74, S353, A354	NE1(W519): 4.6 Å OH(Y342): 4.7 Å
(K) 4XNV (4/1/15); P2Y1 receptor	I178, V181, I216	Y189, Y217	None	A182, P185, S213, C220	OG(S213): 2.8 Å

Continued

Table 1 Cholesterol Interacting Residues in a Representative Set of Structures of Cholesterol-Binding Regions—cont'd

PDB ID (Release Date); Protein Name	Nonpolar Aliphatic Residues (I/L/V)	Nonpolar Aromatic Residues (F/Y/W)	Positively Charged Residues (R/K)	Other Residues	Most Proximal Atom(s) to O(CLR)
(L1) 4XT1 R (3/4/15); Viral GPCR bound to chemokine CX3CL1	I238, I239, L242, L246, V250, L267	F265	None	P243, T247, S268, A272	N(L269): 4.2 Å O(S268): 4.5 Å O(F265): 4.8 Å
(L2) 4XT1 L	I239, L275, L279	None	R271	P243, S268, A272, T276	OG(S268): 3.4 Å
(M) 4BQU B (6/18/14); Japanin	L12, L14, L21, V26, V29, L45, V65, L74, V76, V82, L86, L102, L104, V144	F22, Y47, F63, Y92, W117	R43, K89	B: P5, A6, H17, S18, E23, G27, T31, A78, G87, A88, C114, S115	N(E23): 3.0 Å O(H_2O 2012): 3.4 Å
(N) 4PXZ (4/30/14); P2Y12 receptor	I61, I68, V146	F51, F106, W149, F153	K64, K142	S55, N65, S145, A150	OG(S55): 2.7 Å
(O1) 4NTJ R (3/26/14); P2Y12 receptor	I23, L27, L31, I247, L286, L289	F28, Y278, W285	None	S282	O(H_2O 1310): 4.8 Å OH(Y278): 5.1 Å
(O2) 4NTJ L	I120, I200, L203, I204, V207	Y123	None	Q124, C208	NE2(Q124): 3.6 Å
(P1) 4OR2 FR (3/19/14); Metabotropic glutamate receptor 1	A: I594, I597, L602	A: W588	None	A: S589, A598, C601, B: P1056	OG(S589): 4.3 Å O(CLR MF): 4.3 Å
(P2) 4OR2 FL	B: I604, V643, L659	B: F639, F646	B: K651	B: T647, A650, T653, T655, S656	OG1(T653): 2.7 Å

(P3) 4OR2 MF	A: I597, I604, L605, B: I604, L605	A: W588, F646, B: W588, F646	None	A: C601, B: C601, A650	O(CLR FR): 4.3 Å O(CLR MB): 4.6 Å
(Q) 4NC3 (12/18/13); 5-HT2B receptor	I61, I65, I66, I69, L74, L77, I395	Y394, Y399	None	G70, T73, C397	O(Y394): 2.5 Å OH(Y399): 2.8 Å
(R1) 3WGU MB (10/9/13); Na,K-ATPase	C: I777, L784, I843, D: I35	C: Y847, D: W32, Y39, Y43	D: R27	C: T781, E840, S844, G848, M852	NH1(R27): 2.9 Å OE1(E840): 3.2 Å
(R2) 3WGU BR	C: V864, D: V56	C: F860, W980, F982	None	C: E868, T979, C983, P986, D: G52, T53, M57, T60	OE2(E868): 2.7 Å
(R3) 3WGU BL	C: L976, E: V26	C: W924, Y963, W981, F982, F985, E: F33	E: R27	C: P978, E: T25, G29, G30	O(V26): 2.7 Å
(S1) 3AM6 AR (7/6/11); Rhodopsin AR2	I25, V29	F33	R34	A22, A26, A30	NH1(R34): 4.1 Å
(S2) 3AM6 AL	I19, L219, I220	F27, Y42	R34	A22, A23, A26, A30, M216, E223	NH2(R34): 4.4 Å OE1(E223): 4.6 Å
(T) 3N9Y A (6/8/11); CYP11A1	I84, L101, I351, V353, L460, I461	F82, W87, F202, W231, F458	R81	M201, E283, A286, G287, T291, S352, T354, Q356, Q377	O(H_2O 573): 2.6 Å O(H_2O 538): 2.8 Å
(U) 3PDS (1/12/11); β2 Adrenoceptor	I58, I154, L155	Y70, W158	R151	Q65, T73, S74, C77	OH(Y70): 3.2 Å

Continued

Table 1 Cholesterol Interacting Residues in a Representative Set of Structures of Cholesterol-Binding Regions—cont'd

PDB ID (Release Date); Protein Name	Nonpolar Aliphatic Residues (I/L/V)	Nonpolar Aromatic Residues (F/Y/W)	Positively Charged Residues (R/K)	Other Residues	Most Proximal Atom(s) to O(CLR)
(V) 1ZHY (9/6/05); Yeast oxysterol binding protein Osh4	L24, I33, L39, I167, L177, V179, L201, I203, I206, V213	W10, F13, F42, W46, Y97	R100, K108, K109	Q96, E107, P110, N165, Q181, P211	NE2(Q96): 2.7 Å O(H_2O 2003): 2.8 Å O(H_2O 2004): 2.8 Å
(W) 1N83 (12/11/02); Orphan Nuclear Hormone Receptor ROR(α)-LBD	I327, V364, V379, L394, I400, V403	W320, F365, Y380, F381, F391, F399, Y507	K326, R367, R370	C323, A324, A330, M368, A371, C396, H484	O(H_2O 1008): 2.8 Å O(H_2O 1005): 2.9 Å O(H_2O 1020): 3.0 Å
(X) 1LRI (5/29/02); β-cryptogein	L15, V16, L19, L36, L41, I60, I63, V75, L80, L82, V84	Y12, F24, Y33, Y47, Y87, F91	None	M35, P42, M50, M59, P76, T77, A88	OH(Y47): 2.6 Å O(H_2O 105): 2.7 Å

For each cholesterol-binding region depicted in Fig. 3, the residues located within 5 Å from the cholesterol molecule are listed according to type as defined in the three known cholesterol-binding motifs, CRAC, CARC, and CCM. Residues within 4 Å from the cholesterol molecule are *underlined*. Also shown are the most proximal atom(s) to the oxygen atom of the hydroxyl group of the cholesterol molecule in each cholesterol-binding region.

(34 of the 38 binding regions) (Table 1). On the average, there were 1.37 phenylalanine residues, 0.55 tryptophan residues, and 0.79 tyrosine residues totaling in 2.71 aromatic hydrophobic residues of this type per binding region. This suggests that also this type of residue is important for cholesterol binding. The roles that these aromatic residues can play to stabilize the interactions between the membrane protein and cholesterol are multifold. Aside from forming stacking interactions with the rigid ring structure of the cholesterol molecule, these hydrophobic residues can also contribute to the hydrophobic environment necessary for binding to the mostly hydrophobic cholesterol molecule. Furthermore, two of these amino acids, tyrosine and tryptophan, are amphipathic, and their side-chains include a hydrophilic group (an ionizable hydroxyl group on the phenyl ring of the tyrosine and a nitrogen in the aromatic indole ring of the tryptophan). As such, these two residues can interact with the hydroxyl group of cholesterol via their side-chains as evident in several of the crystal structures (Table 1).

C. The majority of cholesterol-binding regions (23 of the 38 binding regions) do not include an arginine or a lysine (Table 1). These positively charged residues that were included in all cholesterol-binding motifs, and are expected to interact with the hydroxyl group of cholesterol, are replaced in the majority of the binding regions in the complexed structures by other types residues. The latter interact with the cholesterol hydroxyl group either via their side-chain or backbone functional groups. Among the residues that may interact with the hydroxyl group via their side-chains were asparagine, aspartate, glutamate, glutamine, serine, threonine, tryptophan, and tyrosine. Residues observed to potentially interact with the hydroxyl group of cholesterol via their backbone carboxyl or amino groups include alanine, asparagine, leucine, valine, serine, and phenylalanine. Even in cases in which there was an arginine or a lysine within 5 Å from the cholesterol molecule, it did not necessarily interact with the hydroxyl group of the cholesterol molecule (see Table 1). In some cases, the interaction of the hydroxyl group with a positively charged amino acid was replaced by an interaction with surrounding water molecules or with the hydroxyl group of an adjacent cholesterol molecule. This suggests that positively charged residues such as arginine or lysine are not a prerequisite for cholesterol binding.

D. In many structures, additional types of residues could also be found in cholesterol-binding regions. The most common ones in the database of

38 cholesterol-binding regions that we examined were alanine (an average of 1.00 residues per binding region), threonine (an average of 0.68 residues per binding region), cysteine (an average of 0.58 residues per binding region), proline (an average of 0.47 residues per binding region), serine (an average of 0.45 residues per binding region), glycine (an average of 0.34 residues per binding region), and methionine (an average of 0.32 residues per binding region). Interestingly, this group of residues includes four hydrophobic residues (alanine, proline, glycine, and methionine) and three polar residues (threonine, cysteine, and serine). Whereas the hydrophobic residues can contribute to the formation of a favorable hydrophobic environment for cholesterol binding, the side-chains of the threonine and serine may interact with the hydroxyl group of the cholesterol molecules as evident in several of the complexed structures (Table 1).

4. CHOLESTEROL-BINDING REGIONS IN PROTEINS: CONCLUSIONS AND FUTURE DIRECTIONS

In this chapter, we have attempted to delineate the molecular characteristics of cholesterol binding to proteins based on the clues that emerge from 38 cholesterol-binding regions identified in a database of 24 representative protein structures complexed with one or more cholesterol molecules. Comparison of the residues involved in cholesterol binding in this database of structures suggests that the three known cholesterol-binding motifs, CRAC, CARC, and CCM, describe a subset of cholesterol-binding regions. The critical factor for cholesterol binding seems be a requirement for a highly hydrophobic environment in which cholesterol binding may be stabilized by stacking interactions with hydrophobic aromatic residues and hydrogen bonding between the hydroxyl of the cholesterol molecule and protein residue(s) (Fig. 5 and Table 1). Per the known cholesterol-binding motifs, the latter is expected to occur via an interaction with an arginine or a lysine side-chain. In the database of complexed structures, however, the hydroxyl of the cholesterol molecule interacted with side-chains of a variety of polar or charged residues and/or backbone functional groups of both polar and nonpolar residues (Table 1).

Whereas an ever-growing list of ion channels has been shown to be modulated by cholesterol, the question of whether this occurs via an indirect effect or via direct binding of cholesterol to the channel protein remains

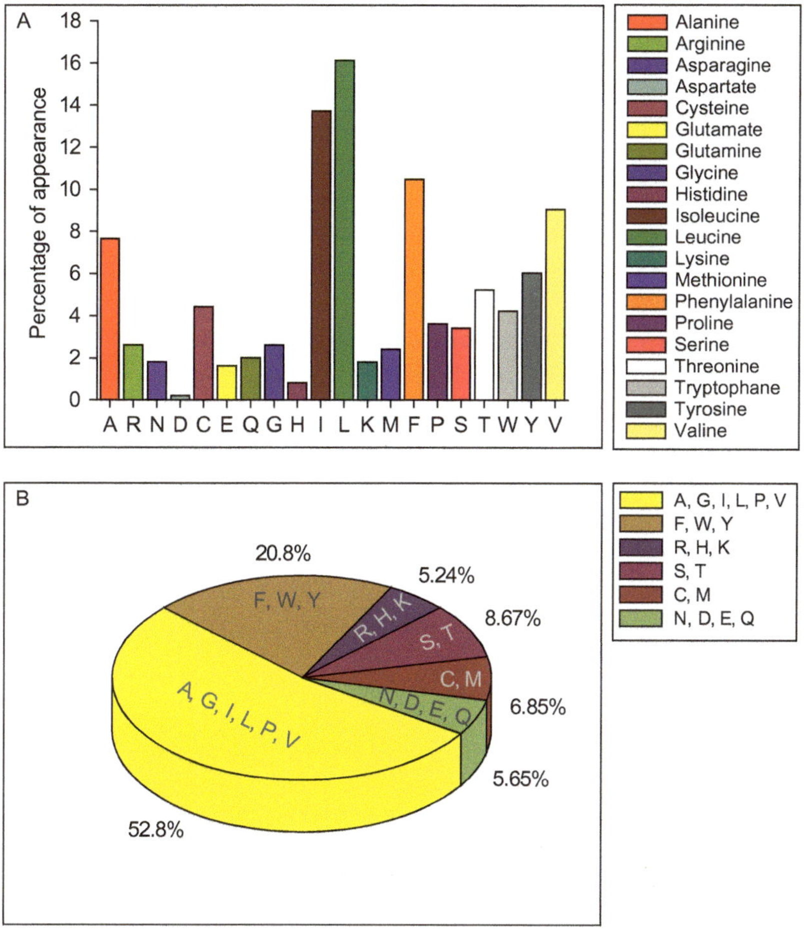

Fig. 5 Distribution of residues involved in cholesterol binding. Summary based on all cholesterol-binding regions in the representative database of structures of proteins complexed with cholesterol (Figs. 3–4 and Table 1) of (A) the percentage of appearance of each of the 20 amino acids and (B) the combined percentage of appearance of different types of amino acids. In (A) the bars are organized from left to right according the same color key used in Fig. 4. The amino acids in (B) were grouped as described in Fig. 3. The fraction of appearance of each residue is calculated by dividing the total number of appearances of the residue in all cholesterol-binding regions analyzed by 38, the total number of cholesterol-binding regions.

open for the majority of channels. The molecular requirements for cholesterol binding that emerge from the analysis of the complexed structures database can provide a framework for expanding the search for cholesterol-binding regions in ion channels, and may result in the identification of additional ion channels that are modulated by cholesterol via direct binding.

In particular, our computational–experimental combined approach identified putative cholesterol-binding regions in the Kir2.1 channel that differed from the known cholesterol-binding motifs (Rosenhouse-Dantsker et al., 2013). These binding regions fit the general requirements for cholesterol binding according to the database of structures of proteins complexed with cholesterol. They included multiple aliphatic hydrophobic residues (isoleucine, leucine, valine), potential interactions with aromatic hydrophobic residues (phenylalanine, tryptophan), and other residues that were commonly found in the cholesterol-binding regions in the database of complexed structures (alanine, serine, methionine). This suggests that an unbiased approach is required to identify cholesterol-binding regions in membrane proteins in general, and ion channels in particular.

Further analysis of the database of complexed structures to determine the actual manner (position and orientation) in which cholesterol binds to proteins will provide further understanding of the molecular characteristics of cholesterol binding and their functional consequences.

REFERENCES

Abi-Char, J. l., Maguy, A., Coulombe, A., Balse, E., Ratajczak, P., Samuel, J.-L., et al. (2007). Membrane cholesterol modulates Kv1.5 potassium channel distribution and function in rat cardiomyocytes. *The Journal of Physiology*, *582*(3), 1205–1217.

Awayda, M. S., Awayda, K. L., Pochynyuk, O., Bugaj, V., Stockand, J. D., & Ortiz, R. M. (2011). Acute cholesterol-induced anti-natriuretic effects: Role of epithelial Na^+ channel activity, protein levels, and processing. *Journal of Biological Chemistry*, *286*(3), 1683–1695.

Barrantes, F. J. (2004). Structural basis for lipid modulation of nicotinic acetylcholine receptor function. *Brain Research Reviews*, *47*(1–3), 71–95.

Barrantes, F. J. (2007). Cholesterol effects on nicotinic acetylcholine receptor. *Journal of Neurochemistry*, *103*(Suppl. 1), 72–78.

Bukiya, A. N., Belani, J. D., Rychnovsky, S., & Dopico, A. M. (2011). Specificity of cholesterol and analogs to modulate BK channels points to direct sterol-channel protein interactions. *The Journal of General Physiology*, *137*(1), 93–110.

Bukiya, A. N., Durdagi, S., Noskov, S., & Rosenhouse-Dantsker, A. (2017). Cholesterol up-regulates neuronal G protein-gated inwardly rectifying potassium (GIRK) channel activity in the hippocampus. *Journal of Biological Chemistry*, *292*, 6135–6147.

Cherezov, V., Rosenbaum, D. M., Hanson, M. A., Rasmussen, S. G., Thian, F. S., Kobilka, T. S., et al. (2007). High-resolution crystal structure of an engineered human β2-adrenergic G protein-coupled receptor. *Science*, *318*(5854), 1258–1265.

Chubinskiy-Nadezhdin, V. I., Negulyaev, Y. A., & Morachevskaya, E. A. (2011). Cholesterol depletion-induced inhibition of stretch-activated channels is mediated via actin rearrangement. *Biochemical and Biophysical Research Communications*, *412*(1), 80–85.

Crowley, J. J., Treistman, S. N., & Dopico, A. M. (2003). Cholesterol antagonizes ethanol potentiation of human brain BKCa channels reconstituted into phospholipid bilayers. *Molecular Pharmacology*, *64*(2), 365–372.

Deng, W., Bukiya, A. N., Rodríguez-Menchaca, A. A., Zhang, Z., Baumgarten, C. M., Logothetis, D. E., et al. (2012). Hypercholesterolemia induces up-regulation of K_{ACh} cardiac currents via a mechanism independent of phosphatidylinositol 4,5-bisphosphate and Gβγ. *Journal of Biological Chemistry*, *287*(7), 4925–4935.

Epand, R. M. (2006). Cholesterol and the interaction of proteins with membrane domains. *Progress in Lipid Research*, *45*(4), 279–294.

Fantini, J., & Barrantes, F. J. (2013). How cholesterol interacts with membrane proteins: An exploration of cholesterol-binding sites including CRAC, CARC and tilted domains. *Frontiers in Physiology*, *4*, 31.

Fürst, O., Nichols, N. G., Lamoureux, G., & D'Avanzo, N. (2014). Identification of a cholesterol-binding pocket in inward rectifier K(+) (Kir) channels. *Biophysical Journal*, *107*(12), 2786–2796.

Hajdú, P., Varga, Z., Pieri, C., Panyi, G., & Gáspár, R. J. (2003). Cholesterol modifies the gating of Kv1.3 in human T lymphocytes. *Pflügers Archiv*, *445*(6), 674–682.

Hanson, M. A., Cherezov, V., Griffith, M. T., Roth, C. B., Jaakola, V. P., Chien, E. Y. T., et al. (2008). A specific cholesterol binding site is established by the 2.8Å structure of the human β-adrenergic receptor. *Structure*, *16*(6), 897–905.

Levitan, I., Singh, D. K., & Rosenhouse-Dantsker, A. (2014). Cholesterol binding to ion channels. *Frontiers in Physiology*, *5*, 65.

Li, H., & Papadopoulos, V. (1998). Peripheral-type benzodiazepine receptor function in cholesterol transport. Identification of a putative cholesterol recognition/interaction amino acid sequence and consensus pattern. *Endocrinology*, *139*(12), 4991–4997.

Liu, W., Chun, E., Thompson, A. A., Chubukov, P., Xu, F., Katritch, V., et al. (2012). Structural basis for allosteric regulation of GPCRs by sodium ions. *Science*, *337*(6091), 232–236.

Lockwich, T. P., Liu, X., Singh, B. B., Jadlowiec, J., Weiland, S., & Ambudkar, I. S. (2000). Assembly of Trp1 in a signaling complex associated with caveolin-scaffolding lipid raft domains. *Journal of Biological Chemistry*, *275*(16), 11934–11942.

Lundbaek, J. A., & Andersen, O. S. (1999). Spring constants for channel-induced lipid bilayer deformations estimates using gramicidin channels. *Biophysical Journal*, *76*(2), 889–895.

Lundbaek, J. A., Birn, P., Hansen, A. J., & Andersen, O. S. (1996). Membrane stiffness and channel function. *Biochemistry*, *35*(12), 3825–3830.

Lundbaek, J. A., Birn, P., Hansen, A. J., Sogaard, R., Nielsen, C., Girshman, J., et al. (2004). Regulation of sodium channel function by bilayer elasticity: The importance of hydrophobic coupling. Effects of micelle-forming Amphiphiles and cholesterol. *The Journal of General Physiology*, *123*(5), 599–621.

Maguy, A., Hebert, T. E., & Nattel, S. (2006). Involvement of lipid rafts and caveolae in cardiac ion channel function. *Cardiovascular Research*, *69*(4), 798–807.

Manglik, A., Kruse, A. C., Kobilka, T. S., Thian, F. S., Mathiesen, J. M., Sunahara, R. K., et al. (2012). Crystal structure of the micro-opioid receptor bound to a morphinan antagonist. *Nature*, *485*(7398), 321–326.

Marsh, D., & Barrantes, F. J. (1978). Immobilized lipid in acetylcholine receptor-rich membranes from *Torpedo marmorata*. *Proceedings of the National Academy of Sciences of the United States of America*, *75*(9), 4329–4333.

Martens, J. R., O'Connell, K., & Tamkun, M. (2004). Targeting of ion channels to membrane microdomains: Localization of KV channels to lipid rafts. *Trends in Pharmacological Sciences*, *25*(1), 16–21.

Penmatsa, A., Wang, K. H., & Gouaux, E. (2013). X-ray structure of dopamine transporter elucidates antidepressant mechanism. *Nature*, *503*(7474), 85–90.

Picazo-Juarez, G., Romero-Suarez, S., Nieto-Posadas, A., Llorente, I., Jara-Oseguera, A., Briggs, M., et al. (2011). Identification of a binding motif in the S5 helix that confers cholesterol-sensitivity to TRPV1. *Journal of Biological Chemistry*, *286*, 24966–24976.

Romanenko, V. G., Fang, Y., Byfield, F., Travis, A. J., Vandenberg, C. A., Rothblat, G. H., et al. (2004a). Cholesterol sensitivity and lipid raft targeting of Kir2.1 channels. *Biophysical Journal*, *87*(6), 3850–3861.

Romanenko, V. G., Rothblat, G. H., & Levitan, I. (2004b). Sensitivity of volume-regulated anion current to cholesterol structural analogues. *The Journal of General Physiology*, *123*(1), 77–88.

Rosenhouse-Dantsker, A., Leal-Pinto, E., Logothetis, D. E., & Levitan, I. (2010). Comparative analysis of cholesterol sensitivity of Kir channels: Role of the CD loop. *Channels*, *4*(1), 63–66.

Rosenhouse-Dantsker, A., Mehta, D., & Levitan, I. (2012). Regulation of ion channels by membrane lipids. *Comprehensive Physiology*, *2*(1), 31–68.

Rosenhouse-Dantsker, A., Noskov, S., Durdagi, S., Logothetis, D. E., & Levitan, I. (2013). Identification of novel cholesterol-binding regions in Kir2 channels. *Journal of Biological Chemistry*, *288*(43), 31154–31164.

Shlyonsky, V. G., Mies, F., & Sariban-Sohraby, S. (2003). Epithelial sodium channel activity in detergent-resistant membrane microdomains. *American Journal of Physiology. Renal Physiology*, *284*(1), F182–F188.

Singh, A. K., McMillan, J., Bukiya, A. N., Burton, B., Parrill, A. L., & Dopico, A. M. (2012). Multiple cholesterol recognition/interaction amino acid consensus (CRAC) motifs in cytosolic C tail of Slo1 subunit determine cholesterol sensitivity of Ca2+– and voltage-gated K+ (BK) channels. *Journal of Biological Chemistry*, *287*(24), 20509–20521.

Wacker, D., Wang, C., Katritch, V., Han, G. W., Huang, X. P., Vardy, E., et al. (2013). Structural features for functional selectivity at serotonin receptors. *Science*, *340*(6132), 615–619.

Wada, T., Shimono, K., Kikukawa, T., Hato, M., Shinya, N., Kim, S. Y., et al. (2011). Crystal structure of the eukaryotic light-driven proton-pumping rhodopsin, Acetabularia rhodopsin II, from marine alga. *Journal of Molecular Biology*, *411*(5), 986–998.

Yeagle, P. L. (2014). Non-covalent binding of membrane lipids to membrane proteins. *Biochimica et Biophysica Acta*, *1838*(6), 1548–1559.

SECTION 3

Emerging Topics of Cholesterol Regulation of Ion Channels

CHAPTER EIGHT

Regulation of P2X Purinergic Receptor Signaling by Cholesterol

Ruth D. Murrell-Lagnado[1]
Sussex Neuroscience, School of Life Sciences, University of Sussex, Brighton, United Kingdom
[1]Corresponding author: e-mail address: r.murrell-lagnado@sussex.ac.uk

Contents

Abstract

P2X receptors are cation-selective channels that are activated by the binding of extracellular ATP. They have a high permeability to Ca^{2+}, Na^+, and K^+ and are expressed widely throughout the nervous, immune, cardiovascular, skeletal, gastrointestinal, respiratory, and endocrine systems. Seven mammalian subtypes of P2X receptor subunits have been identified, P2X1–7, and those that function as homotrimeric receptors (P2X1, 2, 3, 4, and 7) are targeted to lipid rafts, although they show limited resistance to solubilization by Triton X-100. Recent crystal structures of P2X3 and P2X4 receptors have provided considerable high-resolution information about the architecture of this family of receptors and yet the molecular details of how they are regulated by cholesterol are unknown. Currents mediated by the P2X1–4 receptors are either inhibited or relatively insensitive to cholesterol depletion, but there is no clear evidence to support the direct binding of cholesterol to these receptors. In contrast, the activity of the

Current Topics in Membranes, Volume 80
ISSN 1063-5823
http://dx.doi.org/10.1016/bs.ctm.2017.05.004

low-affinity, proinflammatory P2X7 receptor is potentiated by cholesterol depletion and regions within the proximal C-terminus play an important role in coupling changes in cholesterol to the gating of the pore. Based upon our understanding of the lipid signaling events that are triggered downstream of P2X7 receptor activation, a change in the levels of cholesterol may contribute to the sensitization of receptor currents and the dilation of the pore that occurs following prolonged, high-level stimulation. This chapter focuses on the regulation of P2X7 receptor signaling by cholesterol and our current understanding of the mechanisms that underlie this.

1. INTRODUCTION

P2X receptors have a very widespread distribution throughout the human body and purinergic signaling mediated by these receptors takes place in all tissues (Khakh & North, 2006). All cells release ATP by a combination of mechanisms including vesicular release, ion channel-mediated release, and cell lysis (Burnstock & Verkhratsky, 2012). Under normal physiological conditions, the levels of extracellular ATP are very low because of the highly active, membrane-bound ectonucleotidases that rapidly break down ATP to ADP, AMP, and adenosine. However, levels of extracellular ATP are much higher following injury or in diseases that are associated with inflammation. These include cancer, neurodegenerative, and metabolic diseases, in addition to the well-characterized chronic inflammatory diseases such as arthritis and atherosclerosis. High ATP concentrations are recognized as a danger signal, the effects of which are mediated by the uniquely low-affinity P2X7 receptor (Khakh & North, 2006). Conditions that lead to enhanced P2X7 receptor signaling are also conditions that result in perturbations of membrane lipids and remodeling of rafts (Fessler & Parks, 2011).

Lipid rafts are specialized, liquid-ordered microdomains in the plasma membrane enriched in cholesterol, sphingolipids, and saturated phospholipids. They can be identified biochemically by their low buoyant density on a sucrose density gradient, their resistance to solubilization in detergent, and the presence of protein markers such as caveolin (Lingwood & Simons, 2010; Pike, 2004). The precise nature of rafts can differ with respect to their size and lipid/protein composition. For example, caveolae are a special type of lipid raft that form small invaginations of the plasma membrane and contain caveolin, an integral membrane protein. Some other proteins, such as glycosylphosphatidylinositol-anchored proteins, are targeted to rafts, but others are excluded, resulting in compartmentalization and organization of signaling events. Dynamic remodeling of rafts is important in signaling and the control of cholesterol trafficking through rafts by environmental factors and signaling cascades can be an important determinant of the cellular response (Fessler & Parks, 2011).

2. P2X RECEPTORS FORM A STRUCTURALLY DISTINCT CLASS OF LIGAND-GATED ION CHANNELS

P2X receptors assemble as homo- or heterotrimeric complexes. The seven identified subtypes share the same general structure, possessing two transmembrane regions connected via a large extracellular loop and intracellular N- and C-termini. The seven P2X subunits are 40%–50% identical in their amino acid sequence, but they range in size because of a very divergent C-terminal domain which is short in P2X1, P2X3, P2X4, and P2X6 (~30–50 residues) longer in P2X2 and P2X5 and considerably longer in P2X7(North, 2002). P2X7 is composed of 595 amino acids, 242 of which are within the cytoplasmic C-terminus.

In 2009, Gouaux's group published the crystal structure of the zebrafish P2X4 receptor in its closed state, at 3.1 Å resolution but without the cytoplasmic N- and C-termini (Kawate, Michel, Birdsong, & Gouaux, 2009). This structure showed the subunits arranged into chalice-shaped trimeric receptors, with each subunit possessing a large extracellular domain and two transmembrane α-helices, the second of which (TM2) contributes to the central conduction pore. Three years later, the same group published a second structure of the zebrafish P2X4 receptor, again without the cytoplasmic regions, but this time in an ATP-bound state and with an open pore. This confirmed the location of three ATP-binding sites lying at the interfaces between the subunits and approximately 40 Å from the transmembrane domains (TMDs) (Hattori & Gouaux, 2012). It showed that upon binding ATP, the upper regions of the extracellular domain close together while the lower parts flex outward. In this structure, the pore-lining helices were surprisingly widely spaced and it was postulated that the gaps in between would be occupied by the membrane lipids.

More recently, crystal structures were described for the human P2X3 receptor in closed, open, and desensitized states (Mansoor et al., 2016). An important advance over the previous structures was the inclusion of more of the cytoplasmic N- and C-terminal regions, with only 5 residues deleted at the N-terminus (14 less than the previous structures) and 33 at the C-terminus (16 less than the previous structures) (Fig. 1A). In the

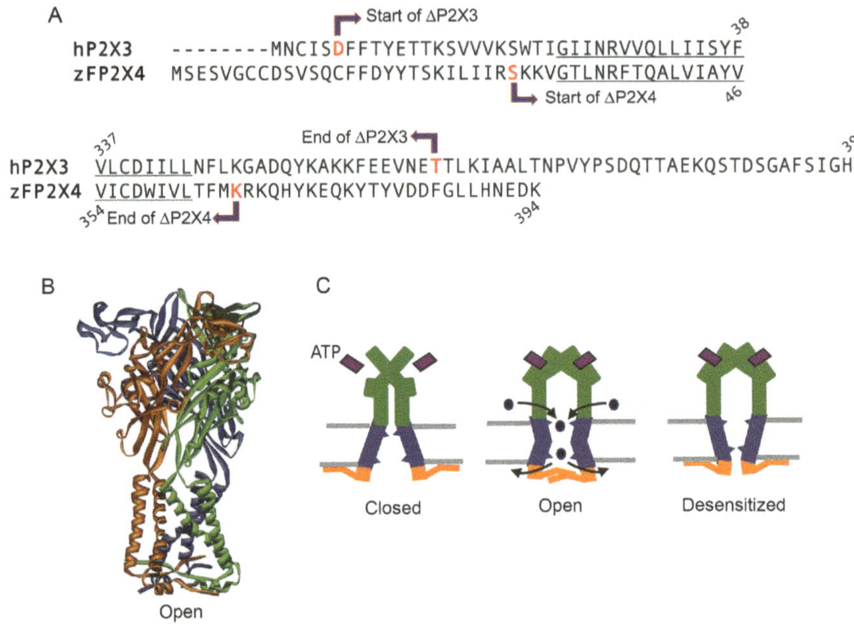

Fig. 1 The trimeric structure of P2X receptors. (A) A comparison of the cytoplasmic N- and C-terminal sequences of hP2X3 and zFP2X4 receptors, upstream of the first TMD (*top*) and downstream of the second TMD (*bottom*). The residues highlighted in *red* represent the start and end of the receptor constructs used for the crystallization studies. The hP2X3 receptor has only 5 amino acids deleted from the N-terminus and 33 from the C-terminus. (B) The architecture of the hP2X3 receptor in the open state shows the cytoplasmic cap formed from the N- and C-terminal regions (PDB ID: 5SVK, structure at 2.77 Å resolution). (C) A cartoon to summarize the gating cycle and to highlight the importance of the cytoplasmic cap in this process.

new open state structure, the membrane proximal N- and C-terminal regions are shown to form a domain that was termed the "cytoplasmic cap" (Fig. 1B). Each subunit contributes two sequential β-strands from the N-terminus and a β-strand from the C-terminus which assemble to form the cap. The cytoplasmic cap sits below the TMDs, and its transient formation and stability are thought to have a central role in receptor gating. During channel opening, TM2 rotates to open the pore and the cytoplasmic cap is formed to anchor the cytoplasmic surfaces of the TM domains. In the transition from the open to the desensitized state, the cytoplasmic half of TM2 moves upward and the cytoplasmic cap is destabilized (Fig. 1C). The crucial role of these cytoplasmic regions in P2X receptor channel gating is relevant to understanding their regulation by cholesterol and other lipids. As detailed later, mutations within these regions modulate cholesterol sensitivity and contribute to the binding site for the cofactor, phosphatidylinositol 4,5 bisphosphate (PIP_2) (Zhao, Yang, Ting, & Logothetis, 2007).

3. P2X RECEPTORS ARE LOCALIZED TO LIPID RAFTS

Several members of the P2X family are detected in the buoyant lipid raft fractions including P2X1, P2X2, P2X3, P2X4, and P2X7 (Allsopp, Lalo, & Evans, 2010; Barth et al., 2008, 2007; Gnanasekaran, Sundukova, van den Maagdenberg, Fabbretti, & Nistri, 2011; Gonnord et al., 2009; Vacca, Amadio, Sancesario, Bernardi, & Volonte, 2004; Vial & Evans, 2005; Vial, Fung, Goodall, Mahaut-Smith, & Evans, 2006). The other two subtypes, P2X5 and P2X6, have very limited or no function as homomeric receptors, and their association with rafts has not been reported. The presence of P2X receptors within lipid rafts is dependent upon the method for isolating rafts, and this has been shown for both recombinant and native receptors. They are associated with rafts prepared in the absence of detergent or with very low concentrations of Triton X-100 (0.05%–0.1%), but the association decreases at higher concentrations of Triton X (Allsopp et al., 2010; Gonnord et al., 2009). There is some subtype dependency to this with P2X2 appearing to be more strongly associated with rafts than the other subtypes. There is also some dependence on cell type, with regard to the proportion of receptors within raft vs nonraft fractions. Triton X depletes inner leaflet membrane lipids, whereas these are retained in detergent-free conditions, which suggests that these lipids might play a role in targeting P2X receptors to rafts.

4. REGULATION OF P2X1–4 RECEPTORS BY THEIR ASSOCIATION WITH LIPID RAFTS

4.1 P2X1 Receptors

The P2X1 receptor is expressed in smooth muscle where it mediates contraction of arteries, resistant arterioles, vas deferens, and bladder (Vial & Evans, 2000). Similar to the recombinant receptor expressed in HEK293 cells, the native receptor in smooth muscle is predominantly within lipid rafts (Vial & Evans, 2005; Vial et al., 2006). P2X1 receptor currents are profoundly sensitive to disruption of lipid rafts, with the current mediated by the recombinant receptor expressed in HEK293 cells inhibited by 90% following treatment with the cholesterol-depleting agent, methyl β-cyclodextrin (meβCD) (Allsopp et al., 2010). Similarly, in rat tail artery, P2X1 receptor-mediated contractions are reduced by approximately 50% (Vial & Evans, 2005). In platelets, only 20% of the P2X1 receptor population is present in raft fractions, but cholesterol depletion still reduces the receptor-mediated Ca^{2+} response by 80%, suggesting that the receptors within rafts are the main functional pool (Vial et al., 2006).

The mechanism that underlies the cholesterol sensitivity of P2X1 was investigated by taking advantage of the cholesterol insensitivity of the P2X2 receptor currents. Chimeras were generated between P2X1 and P2X2, and the region of P2X1 responsible for conferring meβCD-sensitivity to receptor currents was identified as the amino-terminal region in close proximity to the first TMD (Allsopp et al., 2010). Point mutations highlighted two clusters of residues within this region, 20–23 and 27–29, that abolished the inhibitory effect of cholesterol depletion on P2X1 receptor currents when they were substituted with the corresponding P2X2 residues. Crucially, this same region underlies the differential sensitivity of P2X1 and P2X2 receptor currents to the intact actin cytoskeleton, suggesting an association between the dependence on lipid rafts and the dependence on the cytoskeleton. P2X1 interacts with a range of cytoskeletal proteins and disruption of the actin cytoskeleton with cytochalasin D abolishes P2X1 but not P2X2 receptor currents (Lalo, Roberts, & Evans, 2011). Stabilization of the cytoskeleton with jasplakinolide (a macrocyclic peptide that promotes actin polymerization) not only blocks the inhibitory effects of cytochalasin D on P2X1 receptor currents but also blocks the inhibitory effects of meβCD treatment. These results suggest that depletion of cholesterol inhibits P2X1 receptor function not via a direct effect on the receptor itself but via an effect on the actin cytoskeleton. The actin cytoskeleton is sensitive to a loss of raft integrity because it is intimately associated with membrane lipid rafts by virtue of its association with cholesterol-binding proteins (Head, Patel, & Insel, 2014). Thus targeting of P2X1 receptors to lipid rafts is an important determinant of their function, but cholesterol does not directly affect the gating of the pore.

4.2 P2X3 Receptors

P2X3 receptors are present within lipid rafts in trigeminal sensory neurons and in dorsal root ganglion, where they play a key role in sensory transmission (Gnanasekaran et al., 2011; Vacca et al., 2004). P2X3 receptor currents are relatively insensitive to cholesterol depletion, although a small level of inhibition was reported in trigeminal sensory neurons, suggesting that there is no direct effect of sterols on the gating of the pore (Allsopp et al., 2010; Gnanasekaran et al., 2011; Vacca et al., 2004). Signaling downstream of P2X3 does however shows a dependence upon rafts because receptor endocytosis triggered by the binding of agonist occurs within rafts and P2X3-mediated signaling continues within the endosome (Chen et al., 2012; Gnanasekaran et al., 2013). P2X3-positive endosomes identified postagonist treatment were shown to colocalize with caveolin-1 and depletion of cholesterol inhibited both agonist-stimulated endocytosis and the retrograde transport of P2X3-positive endosomes along the axon of sensory neurons (Chen et al., 2012). In addition to playing a role in receptor endocytosis, caveolin-1 also appears to be involved in the agonist-induced delivery of P2X3 receptors to the plasma membrane, a process that contributes to receptor sensitization (Chen, Zhu, Wang, Zhang, & Bao, 2014).

4.3 P2X4 Receptors

P2X4 receptor currents are reported to be insensitive to changes in cholesterol (Allsopp et al., 2010) although others observed an inhibition of receptor-mediated currents following cholesterol depletion (Li & Fountain, 2012). In THP-1 monocytes, where P2X4 exerts a pro-inflammatory role, calcium responses mediated by P2X4 were reduced by acute depletion of cholesterol with meβCD and by chronic depletion of cellular cholesterol with fluvastatin (Li & Fountain, 2012). The reason for these different results and the underlying mechanisms involved in the cholesterol sensitivity have not been established. Unlike P2X1, P2X4 receptor currents are insensitive to disruption of the cytoskeleton, suggesting that changes in the cytoskeleton do not contribute to P2X4 cholesterol sensitivity.

5. P2X7 RECEPTORS AND THE MECHANISMS BY WHICH THEY ASSOCIATE WITH RAFTS

P2X7 receptors are highly expressed in immune cells, including mast cells, monocytes, macrophages, microglia, and T-lymphocytes, but are also found in other cells including epithelia, vascular endothelial cells, fibroblasts, osteoclasts, and osteoblasts (Surprenant & North, 2009). The nature of lipid rafts differs in these cells, with caveolae being well described in fibroblasts and epithelial cells, but less well characterized in macrophages (Fessler & Parks, 2011). Thus the regulation of P2X7 receptor signaling by lipid rafts is likely to vary in a cell type-dependent manner. For example, activation of the P2X7 receptor in macrophages plays a key role in the innate immune response, in particular, in triggering the release of the proinflammatory cytokine IL-1β (Wiley, Sluyter, Gu, Stokes, & Fuller, 2011). It is also involved in the killing of intracellular pathogens such as *Mycobacterium tuberculosis* (Fairbairn, Stober, Kumararatne, & Lammas, 2001; Shemon et al., 2006). When strongly activated, P2X7 receptors promote membrane permeabilization and cell death. While this can be beneficial in the killing of infected macrophages, it can also contribute to tissue damage and disease progression in other pathological conditions that are associated with chronic inflammation such as fibrosis in the kidney, lung, liver, and pancreas (Gentile, Natale, Lazzerini, Capecchi, & Laghi-Pasini, 2015). In other contexts, such as in cancer cells, the P2X7 receptor has a more trophic role, promoting cell proliferation, cell migration, and invasiveness (Adinolfi et al., 2011; Jelassi et al., 2011).

The association of the P2X7 receptor with cholesterol-enriched microdomains is reported in a number of different cell types including the submandibular gland, peritoneal macrophages, osteoblasts, and lung alveolar cells, as well as heterologously expressed receptors in HEK293 cells (Barth et al., 2008, 2007; Gangadharan, Nohe, Caplan, Czymmek, & Duncan, 2015; Garcia-Marcos et al., 2006; Gonnord et al., 2009). In general, sucrose gradient density centrifugation suggests that there are two discrete populations, the raft and nonraft pools. Two mechanisms have been proposed for the targeting of P2X7 to lipid rafts, one involving an interaction with caveolin-1 and another involving palmitoylation of the receptor. Both the relative importance and the impact of these two mechanisms remain unclear and it might vary depending upon the cell type.

5.1 A Role for Caveolin-1 in Targeting P2X7 Receptors to Lipid Rafts

Evidence in support of a role for caveolin-1 was obtained using an airway epithelial cell line, in which immunogold labeling of P2X7 showed gold particles mostly associated with caveolae (Barth et al., 2007). Also in these cells, the transient knockdown of caveolin-1 by siRNA caused a reduction in P2X7 receptor expression. Similarly, in caveolin-1-deficient mice, the expression of P2X7 in both lung epithelia and pulmonary endothelia was reduced compared to wild-type mice. Barth and colleagues went on to show the coimmunoprecipitation of P2X7 and caveolin-1 from airway epithelial cells and the isolation of a complex, under nondenaturing conditions, that comprised P2X4, P2X7, and caveolin-1 (Barth et al., 2008). In contrast to these findings, a study carried out using mouse osteoblast cells used a similar coimmunoprecipitation approach, but were unable to detect P2X7 and caveolin-1 within the same complex (Gangadharan et al., 2015). They also reported an increase in P2X7 receptor currents following knockdown of caveolin-1, which appears to be inconsistent with the results in airway epithelia but does support the idea that the lipid raft environment influences P2X7 receptor function. We investigated the possibility of an interaction between P2X7 and caveolin-1 following the coexpression of the recombinant proteins in HEK293 cells. Similar to Gangadharan et al. (2015), we were not able to coimunoprecipitate the two proteins, although we did observe an increase in cell toxicity when the proteins were coexpressed that was not seen when they were expressed individually (unpublished results). Thus, we conclude likely that the association of P2X7 with caveolae is probably not mediated via a direct interaction with caveolin-1 but instead involves additional proteins that are within the caveolae but not ubiquitously expressed.

5.2 The Role of Palmitoylation in Targeting the P2X7 Receptor to Lipid Rafts

The role of palmitoylation in targeting P2X7 receptors to lipid rafts was investigated in HEK293 cells expressing the recombinant receptor (Gonnord et al., 2009). Palmitate is a16-carbon fatty acid, which is linked via a thioester bond to cysteine residues. Palmitoylation is a reversible process and is controlled by a family of palmitoyltransferases and the acyl protein thioesterases. Gonnord et al. (2009) used ^{3}H-palmitate to show that the receptor is palmitoylated and that this pool appears in the light fraction after sucrose density centrifugation. Disruption to this process using the palmitic acid analogue 2-bromopalmitate (2-BP) reduced the appearance of the receptor in the lipid raft fraction. The P2X7 receptor has 16 conserved intracellular cysteines within the N- and C-terminal tail that could potentially be sites of palmitoylation. These were systematically mutated to alanines in groups of between two and four in an attempt to identify the target cysteines (Gonnord et al., 2009). Mutations within four of the seven groups, located within the proximal, middle, and distal C-terminus, either significantly reduced or abolished palmitoylation, as measured by incorporation of ^{3}H-palmitate. These results are difficult to interpret in terms of identifying a site of palmitoylation. There may be more than one site, but it is unclear why mutations within one group would abolish palmitoylation of another group unless the mutations disrupted the overall structure of the C-terminal domain. These mutations also inhibited the plasma membrane expression of the receptor, consistent with the misfolding and/or misassembly of the mutated protein. An alternative possibility is that palmitoylation of P2X7 has a more direct role in regulating its trafficking. Palmitoylation of cysteines close to TMDs can affect the conformation of the TMD and can influence their trafficking from the ER (Blaskovic, Blanc, & van der Goot, 2013; Chum et al., 2016). Also for some transmembrane proteins, mutations leading to defects in palmitoylation can result in ubiquitination of neighboring lysines, causing ER retention and increased rates of degradation (Blaskovic et al., 2013). We observed a large reduction in both the surface and total expression of P2X7 after incubating HEK293 cells with the palmitic acid analogue 2-BP for 16 h (Robinson & Murrell-Lagnado, 2013). This is consistent with the idea that palmitoylation of P2X7 is an important determinant of its trafficking properties and stability, but could also show that a protein that regulates P2X7 receptor trafficking requires palmitoylation for normal function. In conclusion, while it seems likely that palmitoylation of P2X7 occurs, its role in targeting the receptor to rafts is still unclear.

6. THE EFFECTS OF CHOLESTEROL ON P2X7 RECEPTOR FUNCTION AND SIGNALING

There are several ways in which the recruitment of the P2X7 receptor to lipid rafts and the cholesterol content of the rafts could affect receptor signaling: first, by affecting the plasma membrane expression of the receptor, second, by affecting receptor function such as the gating of the channel pore, and third, by concentrating signaling molecules that act either as downstream effectors of the receptor or as regulators of the receptor.

Little is known about what controls the trafficking of P2X7 receptors to and from the plasma membrane. We and others have seen variability in surface expression dependent upon the isoform of the receptor and upon the cell type. When heterologously expressed in mammalian cell lines such as HEK293 cells or HeLa cells, we see a predominant distribution within the endoplasmic reticulum, particularly for the human P2X7 receptor. Rodent isoforms show higher expression at the plasma membrane, and when we analyzed the distribution of native P2X7 receptors in rodent immune cells we saw a predominantly cell surface expression (Boumechache, Masin, Edwardson, Gorecki, & Murrell-Lagnado, 2009). Others have also reported an intracellular distribution of P2X7 in human lymphocytes and monocytes, whereas cell surface expression increases when monocytes differentiate into macrophages (Gu et al., 2000; Gudipaty, Humphreys, Buell, & Dubyak, 2001). The sterol content of the membranes along the secretory pathway differs, being about 5 mol% for the ER membrane and 40 mol% at the plasma membrane (Dufourc, 2008). Interactions between P2X7 and cholesterol is therefore expected to affect the rate at which the receptor traffics along this pathway. Receptor stability and rates of endocytosis are also likely to be affected by the targeting of P2X7 to lipid rafts.

6.1 Sensitivity of P2X7 Receptor Function to Changes in Membrane Cholesterol

One of the most striking characteristics of currents mediated by the P2X7 receptor is the sensitization of the current that occurs during prolonged or repetitive application of a high concentration of agonist. Sensitization in response to a sustained application of agonist is often seen as a distinctive biphasic inward current comprised of an instantaneous component, that typically peaks within hundreds of milliseconds, followed by a slower component that takes tens of seconds to peak (Roger, Pelegrin, & Surprenant, 2008; Yan, Li, Liang, Tomić, & Stojilkovic, 2008). This larger current also develops with repeated short applications of agonist. There are species differences in the rate at which facilitation of the current develops, with the human receptor responding more slowly than the rat receptor (Roger, Gillet, Baroja-Mazo, Surprenant, & Pelegrin, 2010). We tested the effects of cholesterol depletion and cholesterol loading on the properties of different P2X7 receptor isoforms expressed in HEK293 cells, using meβCD either alone or loaded with cholesterol. Incubation times were kept short (10 min) and the concentration used was relatively low (5 mM) to preferentially remove raft rather than extra-raft cholesterol (Robinson & Murrell-Lagnado, 2013). The effect of depleting cholesterol on the human P2X7 receptor was a dramatic acceleration in the rate of receptor sensitization (Fig. 2). Whereas currents evoked by repetitive 5-s applications of BzATP normally increased in amplitude slowly during the first 10–15 applications, following meβCD treatment, the response to the first application of BzATP was much larger and subsequent responses were of similar magnitude. For the mouse P2X7 receptor, we generally do not observe the slow buildup in the amplitude of the currents with repetitive application of BzATP, suggesting that the sensitization process occurs more rapidly than for the human receptor. Following cholesterol depletion, the amplitude of the

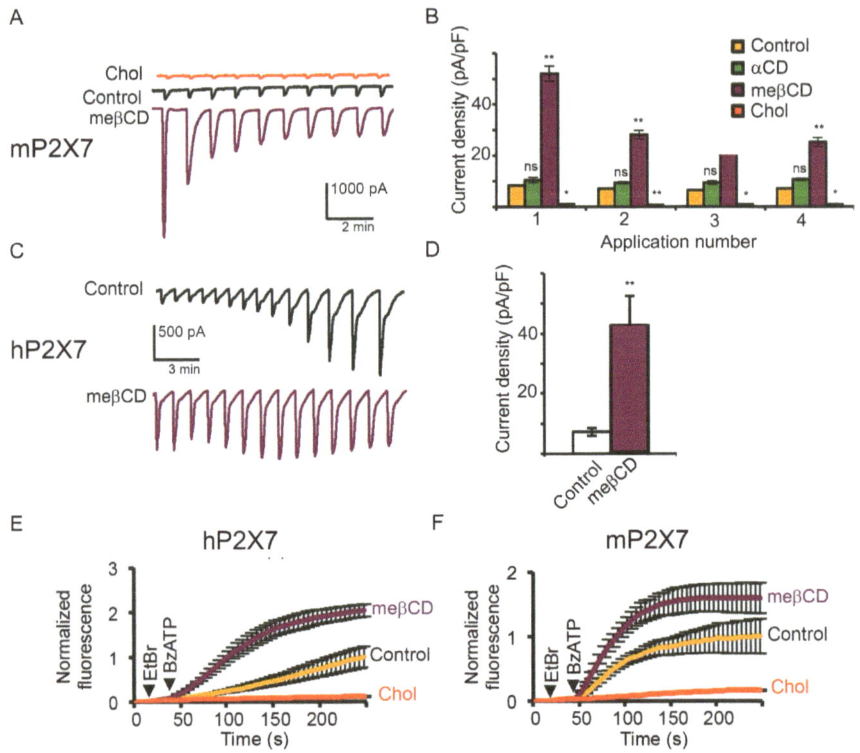

Fig. 2 Acute changes in cholesterol affect receptor whole-cell currents and large pore formation. (A) Whole-cell patch clamp recordings from HEK293 cells expressing the mouse P2X7 receptor stimulated with repeated 5-s applications of 300 μM BzATP followed by a 1-min wash. A 10-min preincubation with 5 mM meβCD potentiated inward currents recorded at −60 mV in a NaCl-based extracellular solution. In contrast, preincubation with 100 μg/mL water-soluble cholesterol inhibited receptor currents. (B) Mean current densities after one or more applications of BzATP in untreated cells or cells treated as indicated ($n=3$–18 cells, $^{*}P<0.01$, $^{**}P<0.001$). (C) Example traces of a whole-cell patch clamp recording from cells expressing the human P2X7 receptor stimulated in the same way as the mouse receptor. (D) Mean current density for first application of agonist ($n=18$ cells). (E, F) P2X7 receptor-mediated ethidium uptake was potentiated following meβCD treatment and inhibited by loading cells with cholesterol. Uptake of ethidium was measured by an increase in fluorescence using a confocal microscope. 300 μM BzATP was added at the time point indicated. *This research was originally published in Robinson, L. E., Shridar, M., Smith, P., & Murrell-Lagnado, R. D. (2014). Plasma membrane cholesterol as a regulator of human and rodent P2X7 receptor activation and sensitization.* Journal of Biological Chemistry, 289, *31983–31994. © The American Society for Biochemistry and Molecular Biology."*

initial response was dramatically enhanced, but it ran down with subsequent applications of the agonist (Fig. 2A). Conversely loading of cells with cholesterol strongly suppressed both human and mouse P2X7 receptor currents.

Another unusual property of P2X7 receptors is that the sensitized state is associated with a change in the selectivity of the channel. In the non-sensitized state, agonist binding opens a pore that permits small monovalent and divalent cations to permeate, whereas in the sensitized state agonist binding opens a "dilated" pore that also allows the permeation of larger cations, such as NMDG+ and organic cations up to 900 Da (Yan et al., 2008). As a further test of whether cholesterol depletion stabilizes the sensitized/dilated state of the P2X7 receptor, we measured currents carried by NMDG+ and showed that following meβCD treatment, NMDG+ currents triggered by the first application of agonist were much larger (Robinson & Murrell-Lagnado, 2013). There was also a dramatic enhancement in the rate of uptake of ethidium particularly for the human P2X7 receptor and complete inhibition by loading cells with cholesterol (Fig. 2E). For the mouse P2X7 receptor, which activates more rapidly to a sensitized state under normal conditions, the effects of cholesterol depletion were similar although less pronounced (Fig. 2F). All of these measurements of ion fluxes mediated by P2X7 support the idea that cholesterol inhibits pore formation and dilation and that its depletion facilitates the sensitization process.

6.2 The Effects of Cholesterol on Signaling Downstream of the P2X7 Receptor

There are many reports of lipid signaling pathways being activated downstream of P2X7 receptor stimulation, including phospholipase A_2 (PLA_2) and sphingomyelinases (SMases) (Andrei et al., 2004; Bianco et al., 2009). These pathways play a central role in mediating the actions of P2X7. For example, the activation of acid SMase in microglia cells was shown to be necessary for P2X7 receptor-dependent microvesicle shedding and the release of IL-1β (Bianco et al., 2009). In macrophages, the synthesis of leukotrienes through activation of PLA_2 and mobilization of arachidonic acid is required for the P2X7-mediated reduction in the parasitic load of infected cells (Chaves, Marques-da-Silva, Monteiro, Canetti, & Coutinho-Silva, 2014). Although the mechanisms that couple P2X7 receptor activation to these enzymes are unclear, the efficiency of signaling via these pathways is likely to be affected by targeting of P2X7 receptors to lipid rafts, with receptors in lipid rafts producing a greater response compared to the nonraft pool. Evidence in support of this idea has come from studies of P2X7 receptor distribution and function in rat submandibular glands (Garcia-Marcos et al., 2006). In this study, they identified a raft and nonraft pool of receptors and showed that disruption of rafts had no effect on the ATP-evoked Ca^{2+} signaling, but diminished the ATP-evoked increase in PLA_2 activity. They also showed that ATP increased neutral SMase activity, leading to an increase in the ceramide content within lipid rafts but not in nonraft regions.

Not only does targeting to raft or nonraft regions have the potential to modulate the strength of P2X7 receptor signaling, but lipid signaling downstream of P2X7 has the potential to regulate receptor activity via a feedback mechanism. Ceramide produced by the increased activity of SMases, or by de novo synthesis, will displace cholesterol from rafts and is also converted into other bioactive sphingolipids (Megha & London, 2004). Thus, we predict that prolonged activation of P2X7 receptors in lipid rafts would reduce the cholesterol content of the rafts. This perturbation is expected to enhance the activity of raft-embedded P2X7 receptors and could be a positive feedback mechanism involved in activity-dependent enhancement in receptor function. It could also be involved in the activation of phospholipase D (PLD) downstream of P2X7. PLD is localized to rafts, but its activator ARF is in nonraft regions and therefore any change to the cholesterol content of rafts that promotes the migration of PLD to nonraft regions is expected to promote the activation of PLD (Diaz et al., 2002). Activation of PLA_2 downstream of P2X7R will mobilize not only arachidonic acid but also the lysolipids including lysophosphatidic acid and lysophosphatidylcholine. All of these lipids are reported to activate P2X7 receptors, thus providing another positive regulatory feedback loop that could promote receptor sensitization (Alloisio, Aiello, Ferroni, & Nobile, 2006; Michel & Fonfria, 2007).

7. THE DETERMINANTS OF P2X7 RECEPTOR CHOLESTEROL SENSITIVITY

It is still unclear whether cholesterol exerts its effect on P2X receptor activity directly, by binding to a cholesterol-sensing motif (CSM) within the receptor, or indirectly, via an associated protein and/or by changing the biophysical properties of the membrane. Unlike the cholesterol regulation of P2X1 receptors, the potentiation of P2X7 receptor activity that occurs in response to cholesterol depletion is not affected by the cytoskeletal-stabilizing agent jasplakinolide, which argues against the involvement of the actin cytoskeleton in this effect (Robinson & Murrell-Lagnado, 2013). Other protein regulators of P2X7 could however be involved. For example, CD44 is a raft protein that was recently shown to interact with P2X7 and is a receptor for glycosaminoglycans (GAGs) (Moura et al., 2015). GAGs are positive allosteric modulators of P2X7 receptors, promoting sensitization and pore dilation. Cholesterol depletion triggers the membrane-dependent shedding of CD44 as does the activation of P2X7 receptors and sCD44 is thought to be part of a positive feedback loop contributing to receptor sensitization.

Changes in cholesterol will affect the biophysical properties of the bilayer and this could impact upon protein conformational changes and protein–protein interactions without the involvement of a direct sterol–protein interaction (Subczynski & Wisniewska, 2000). Evidence to support the idea that P2X7 receptor gating is sensitive to the ordered nature of its lipid environment comes from analyzing the effects of the lysolipids on receptor activity. Several of these lipids were shown to enhance P2X7 receptor function and all that were active were amphiphilic and capable of integrating into lipid bilayers (Michel & Fonfria, 2007). Furthermore, for many ion channels and transporters that are regulated by cholesterol here is now good evidence to support a specific sterol–ion channel interaction. Motifs identified as forming cholesterol-binding sites include the cholesterol recognition amino acid consensus (CRAC) motif, $(L/V)X_{1-5}$-Y-$X_{1-5}(K/R)$, the reverse motif known as CARC and the CSM, which involves a positively charged residue (R/K) an aromatic (W/Y) and a hydrophobic residue (I, V, L) (Fantini & Barrantes, 2013). Additionally, regions of hydrophobicity that do not comply with these motifs have also been identified as cholesterol-binding sites, such as those in Kir2 channels (Rosenhouse-Dantsker, Noskov, Durdagi, Logothetis, & Levitan, 2013). These regions are often located within TM helices, at the TMD–cytosolic interface of the channel protein or within regions proximal to TMDs that are presumably intimately associated with the membrane.

We used a mutagenesis approach to identify regions of P2X7R involved in its cholesterol sensitivity (Robinson & Murrell-Lagnado, 2013). We focused on tyrosines that were located within a CRAC/CARC motif of which there are several in the vicinity of the TMDs and in particular a cluster

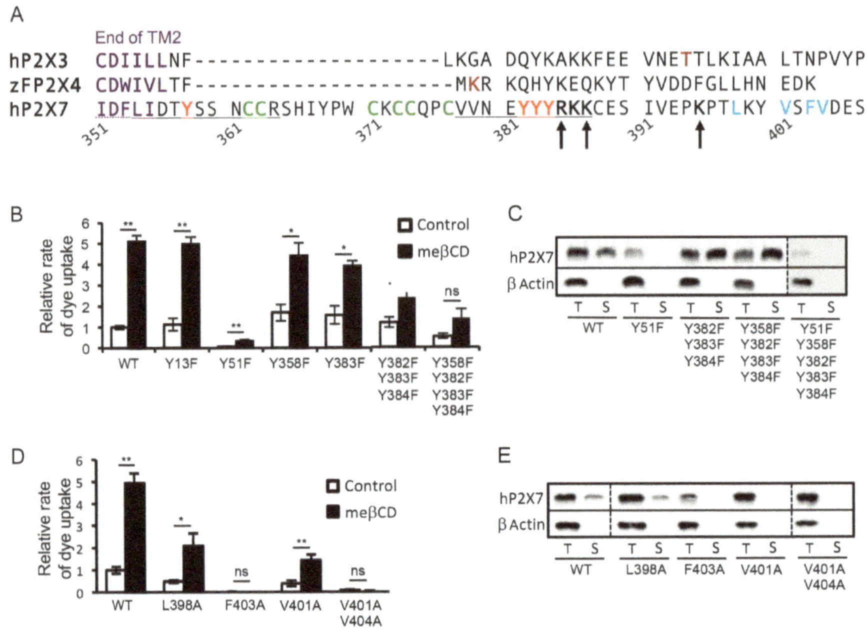

Fig. 3 Cholesterol sensing within the proximal C-terminus of the P2X7 receptor. (A) A comparison of the proximal C-terminal sequence of hP2X7 with those of zFP2X4 and hP2X3 shows the additional 18 amino acid cysteine-rich region present in P2X7. Tyrosines that lie within a CRAC motif and were mutated are shown in *red* with the motif underlined in *black*. Mutated hydrophobic residues are shown in *blue* and the PIP_2-binding region of P2X7 is indicated by *arrows*. (B) The relative rates of ethidium uptake, measured between 10 and 50s after BzATP addition, are shown for wild-type and mutated receptors with and without preincubation with meβCD. Mutating all four tyrosines in this region abolished the cholesterol sensitivity of ethidium uptake without affecting plasma membrane expression of the receptor. (C) Western blot analysis shows that the total (T) and surface (S) expression of the quadruple mutated P2X7 receptor was similar to wild type. Surface proteins were biotinylated and subsequently purified prior to their analysis by Western blot. (D, E) Mutating hydrophobic residues to alanines downstream of the PIP_2-binding region disrupted channel function and expression at the plasma membrane, indicating the importance of the hydrophobic nature of this region in receptor structure. *This research was originally published in Robinson, L. E., Shridar, M., Smith, P., & Murrell-Lagnado, R. D. (2014). Plasma membrane cholesterol as a regulator of human and rodent P2X7 receptor activation and sensitization.* Journal of Biological Chemistry, 289, *31983–31994. © The American Society for Biochemistry and Molecular Biology.*

within the proximal C-terminus of the receptor (Y358, Y382, Y383, and Y384) (Fig. 3A). These residues are immediately upstream of positively charged residues that are reported to interact with PIP_2 (R386, K388, and K395) (Zhao et al., 2007). Systematically mutating the tyrosines to phenylalanine to generate single through to quadruple mutants caused a progressive decrease in the sensitivity of receptor-mediated ethidium uptake to meβCD pretreatment, suggesting their involvement in the cholesterol sensitivity of P2X7 (Fig. 3B). This region could be involved in binding cholesterol or in coupling the binding of cholesterol to the gating of the pore. The close proximity between this putative cholesterol-sensing region and the PIP_2-binding region suggests that association with cholesterol could interfere with the binding of PIP_2. PIP_2 is a positive cofactor of all of the functional P2X receptor subtypes (Bernier et al., 2008; Mo et al., 2009), with the exception of P2X5, and therefore its displacement could be responsible for the inhibition of receptor function produced by enhanced levels of cholesterol. Further evidence to support the idea that this proximal C-terminal region associates with and senses membrane lipids comes from mutating hydrophobic residues within this region. There is a cluster of hydrophobic residues located immediately downstream of the putative PIP_2-binding site (L398, V401, F403, and V404) and alanine substitutions here disrupts both channel function and its trafficking to the plasma membrane (Fig. 3C). These results are consistent with the idea that the proximal C-terminal region of P2X7 plays an important role in sensing membrane lipids and coupling this to a change in channel gating.

Can we learn anything more of the structure and function of this region by comparison with the recent structures of hP2X3 receptor? The alignment of the proximal C-terminal regions of hP2X7, hP2X3, and zFP2X4 (Fig. 3A) shows an additional 18-amino acid, cysteine-rich region in P2X7 that is absent in the other receptors, and this will alter the structure of the cytoplasmic cap. Cysteines within this region form one of the clusters required for normal palmitoylation and trafficking of P2X7 (Gonnord et al., 2009). The P2X7 receptor is the most slowly desensitizing of all of the subtypes, and therefore we assume that the cap formed in the open state is stabilized in some way. We and others have tested the effects of deleting this 18-amino acid region and obtained differing results (Jiang et al., 2005; Robinson & Murrell-Lagnado, 2013; Yan et al., 2008). We observed very little ethidium uptake for the hP2X7Δ18 mutant and importantly, the potentiating effect of meβCD treatment was substantially reduced compared to the wild-type receptor (Robinson & Murrell-Lagnado, 2013). The hP2X7Δ18 receptor currents were small, ran down rather than sensitized with repeated application of agonist, and were inhibited and not potentiated by meβCD. These results indicate the importance of this region in the gating of P2X7 and in the stabilization of the open state following cholesterol depletion.

8. FUTURE WORK

In this review, we have focused on the inhibitory regulation of P2X7 receptors by cholesterol but have also briefly touched on other lipids, such as the products of the PLA_2 pathway, which positively regulate the receptor. The underlying molecular mechanisms of this regulation are not yet established, but with the publishing of P2X receptor structures in the close, open, and desensitized states, the field is well placed to elucidate the detailed nature of how lipids directly or indirectly stabilize different conformations of the receptor. There are still several questions to be addressed in order to understand the role of sterols in the regulation of P2X7 in health and disease. First, what is the impact of lipid rafts on P2X7 receptor signaling in different cell types? This requires a better understanding of the mechanisms targeting P2X7 to lipid rafts. For example, if caveolin-1 plays a key role, then we expect there to be a higher proportion of receptors within rafts in fibroblasts or epithelial cells compared to immune cells. Second, to what extent does the activation of P2X7 receptors lead to remodeling of lipid rafts, and what are the implications of this remodeling on receptor function? We have proposed that increased levels of ceramide triggered by P2X7 receptor stimulation could feedback to alter the structure of rafts and the properties of the receptors within rafts. Does this contribute to the activity-dependent sensitization of the P2X7 receptor? Finally, how does the remodeling of lipid rafts associated with age and inflammation affect P2X7 receptor signaling? If sterol regulation of P2X7 exerts a greater influence on receptor function with age or with inflammation, then therapeutic or dietary interventions that alter the structure and composition of rafts, such as the ω-3 polyunsaturated fatty acids, might have beneficial effects upon P2X7-mediated disease pathogenesis.

ACKNOWLEDGMENTS

I thank Dr. Andrew Thompson for assistance with Fig. 1 and for discussions. This work have been supported by the BM1406 COST Action.

REFERENCES

Adinolfi, E., Cirillo, M., Woltersdorf, R., Falzoni, S., Chiozzi, P., Pellegatti, P., et al. (2011). Trophic activity of a naturally occurring truncated isoform of the P2X7 receptor. *The FASEB Journal*, *24*(9), 3393–3404. http://dx.doi.org/10.1096/fj.09-153601.

Alloisio, S., Aiello, R., Ferroni, S., & Nobile, M. (2006). Potentiation of native and recombinant P2X7-mediated calcium signaling by arachidonic acid in cultured cortical astrocytes and human embryonic kidney 293 cells. *Molecular Pharmacology*, *69*(6), 1975–1983. http://dx.doi.org/10.1124/mol.105.020164.

Allsopp, R. C., Lalo, U., & Evans, R. J. (2010). Lipid raft association and cholesterol sensitivity of P2X1-4 receptors for ATP: Chimeras and point mutants identify intracellular amino-terminal residues involved in lipid regulation of P2X1 receptors. *The Journal of Biological Chemistry*, *285*(43), 32770–32777. http://dx.doi.org/10.1074/jbc.M110.148940.

Andrei, C., Margiocco, P., Poggi, A., Lotti, L. V., Torrisi, M. R., & Rubartelli, A. (2004). Phospholipases C and A2 control lysosome-mediated IL-1 beta secretion: Implications for inflammatory processes. *Proceedings of the National Academy of Sciences of the United States of America*, *101*(26), 9745–9750.

Barth, K., Weinhold, K., Guenther, A., Linge, A., Gereke, M., & Kasper, M. (2008). Characterization of the molecular interaction between caveolin-1 and the P2X receptors 4 and 7 in E10 mouse lung alveolar epithelial cells. *The International Journal of Biochemistry & Cell Biology*, *40*(10), 2230–2239. http://dx.doi.org/10.1016/j.biocel.2008.03.001.

Barth, K., Weinhold, K., Guenther, A., Young, M. T., Schnittler, H., & Kasper, M. (2007). Caveolin-1 influences P2X7 receptor expression and localization in mouse lung alveolar epithelial cells. *The FEBS Journal*, *274*(12), 3021–3033. http://dx.doi.org/10.1111/j.1742-4658.

Bernier, L. P., Ase, A. R., Chevallier, S., Blais, D., Zhao, Q., Boue-Grabot, E., et al. (2008). Phosphoinositides regulate P2X4 ATP-gated channels through direct interactions. *The Journal of Neuroscience*, *28*(48), 12938–12945. http://dx.doi.org/10.1523/JNEUROSCI.3038-08.2008.

Bianco, F., Perrotta, C., Novellino, L., Francolini, M., Riganti, L., Menna, E., et al. (2009). Acid sphingomyelinase activity triggers microparticle release from glial cells. *The EMBO Journal*, *28*(8), 1043–1054. doi:emboj200945.

Blaskovic, S., Blanc, M., & van der Goot, F. G. (2013). What does S-palmitoylation do to membrane proteins? *The FEBS Journal*, *280*(12), 2766–2774. http://dx.doi.org/10.1111/febs.12263.

Boumechache, M., Masin, M., Edwardson, J. M., Gorecki, D. C., & Murrell-Lagnado, R. (2009). Analysis of assembly and trafficking of native P2X4 and P2X7 receptor complexes in rodent immune cells. *The Journal of Biological Chemistry*, *284*(20), 13446–13454. http://dx.doi.org/10.1074/jbc.M901255200.

Burnstock, G., & Verkhratsky, V. (2012). *Mechansims of ATP release and inactivation. In Purinergic Signalling and the Nervous System* (pp. 79–118). Berlin, Heidelberg: Springer. http://dx.doi.org/10.1007/978-3-642-28863-0_4.

Chaves, M. M., Marques-da-Silva, C., Monteiro, A. P., Canetti, C., & Coutinho-Silva, R. (2014). Leukotriene B4 modulates P2X7 receptor-mediated Leishmania amazonensis elimination in murine macrophages. *Journal of Immunology*, *192*(10), 4765–4773. http://dx.doi.org/10.4049/jimmunol.1301058.

Chen, X. Q., Wang, B., Wu, C., Pan, J., Yuan, B., Su, Y. Y., et al. (2012). Endosome-mediated retrograde axonal transport of P2X3 receptor signals in primary sensory neurons. *Cell Research*, *22*(4), 677–696. http://dx.doi.org/10.1038/cr.2011.197.

Chen, X. Q., Zhu, J. X., Wang, Y., Zhang, X., & Bao, L. (2014). CaMKIIalpha and caveolin-1 cooperate to drive ATP-induced membrane delivery of the P2X3 receptor. *Journal of Molecular Cell Biology*, *6*(2), 140–153. http://dx.doi.org/10.1093/jmcb/mju011.

Chum, T., Glatzova, D., Kvicalova, Z., Malinsky, J., Brdicka, T., & Cebecauer, M. (2016). The role of palmitoylation and transmembrane domain in sorting of transmembrane adaptor proteins. *Journal of Cell Science*, *129*(1), 95–107. http://dx.doi.org/10.1242/jcs.175190.

Diaz, O., Berquand, A., Dubois, M., Di Agostino, S., Sette, C., Bourgoin, S., et al. (2002). The mechanism of docosahexaenoic acid-induced phospholipase D activation in human lymphocytes involves exclusion of the enzyme from lipid rafts. *The Journal of Biological Chemistry*, *277*(42), 39368–39378. http://dx.doi.org/10.1074/jbc.M202376200.

Dufourc, E. J. (2008). Sterols and membrane dynamics. *Journal of Chemical Biology*, *1*(1–4), 63–77. http://dx.doi.org/10.1007/s12154-008-0010-6.

Fairbairn, I. P., Stober, C. B., Kumararatne, D. S., & Lammas, D. A. (2001). ATP-mediated killing of intracellular mycobacteria by macrophages is a P2X(7)-dependent process inducing bacterial death by phagosome-lysosome fusion. *Journal of Immunology*, *167*(6), 3300–3307.

Fantini, J., & Barrantes, F. J. (2013). How cholesterol interacts with membrane proteins: An exploration of cholesterol-binding sites including CRAC, CARC, and tilted domains. *Frontiers in Physiology*, *4*, 31. http://dx.doi.org/10.3389/fphys.2013.00031.

Fessler, M. B., & Parks, J. S. (2011). Intracellular lipid flux and membrane microdomains as organizing principles in inflammatory cell signaling. *Journal of Immunology (Baltimore, Md.: 1950)*, *187*(4), 1529–1535. http://dx.doi.org/10.4049/jimmunol.1100253.

Gangadharan, V., Nohe, A., Caplan, J., Czymmek, K., & Duncan, R. L. (2015). Caveolin-1 regulates P2X7 receptor signaling in osteoblasts. *American Journal of Physiology Cell Physiology*, *308*(1), C41–C50. http://dx.doi.org/10.1152/ajpcell.00037.2014.

Garcia-Marcos, M., Perez-Andres, E., Tandel, S., Fontanils, U., Kumps, A., Kabre, E., et al. (2006). Coupling of two pools of P2X7 receptors to distinct intracellular signaling pathways in rat submandibular gland. *Journal of Lipid Research*, *47*(4), 705–714. http://dx.doi.org/10.1194/jlr.M500408-JLR200.

Gentile, D., Natale, M., Lazzerini, P. E., Capecchi, P. L., & Laghi-Pasini, F. (2015). The role of P2X7 receptors in tissue fibrosis: A brief review. *Purinergic Signal*, *11*(4), 435–440. http://dx.doi.org/10.1007/s11302-015-9466-3.

Gnanasekaran, A., Sundukova, M., Hullugundi, S., Birsa, N., Bianchini, G., Hsueh, Y. P., et al. (2013). Calcium/calmodulin-dependent serine protein kinase (CASK) is a new intracellular modulator of P2X3 receptors. *Journal of Neurochemistry*, *126*(1), 102–112. http://dx.doi.org/10.1111/jnc.12272.

Gnanasekaran, A., Sundukova, M., van den Maagdenberg, A. M., Fabbretti, E., & Nistri, A. (2011). Lipid rafts control P2X3 receptor distribution and function in trigeminal sensory neurons of a transgenic migraine mouse model. *Molecular Pain*, *7*, 77. http://dx.doi.org/10.1186/1744-8069-7-77.

Gonnord, P., Delarasse, C., Auger, R., Benihoud, K., Prigent, M., Cuif, M. H., et al. (2009). Palmitoylation of the P2X7 receptor, an ATP-gated channel, controls its expression and association with lipid rafts. *The FASEB Journal*, *23*(3), 795–805. http://dx.doi.org/10.1096/fj.08-114637.

Gu, B. J., Zhang, W. Y., Bendall, L. J., Chessell, I. P., Buell, G. N., & Wiley, J. S. (2000). Expression of P2X(7) purinoceptors on human lymphocytes and monocytes: Evidence for nonfunctional P2X(7) receptors. *American Journal of Physiology. Cell Physiology*, *279*(4), C1189–C1197.

Gudipaty, L., Humphreys, B. D., Buell, G., & Dubyak, G. R. (2001). Regulation of P2X(7) nucleotide receptor function in human monocytes by extracellular ions and receptor density. *American Journal of Physiology. Cell Physiology*, *280*(4), C943–C953.

Hattori, M., & Gouaux, E. (2012). Molecular mechanism of ATP binding and ion channel activation in P2X receptors. *Nature*, *485*(7397), 207–212. http://dx.doi.org/10.1038/nature11010.

Head, B. P., Patel, H. H., & Insel, P. A. (2014). Interaction of membrane/lipid rafts with the cytoskeleton: Impact on signaling and function: Membrane/lipid rafts, mediators of cytoskeletal arrangement and cell signaling. *Biochimica et Biophysica Acta*, *1838*(2), 532–545. http://dx.doi.org/10.1016/j.bbamem.2013.07.018.

Jelassi, B., Chantome, A., Alcaraz-Perez, F., Baroja-Mazo, A., Cayuela, M. L., Pelegrin, P., et al. (2011). P2X(7) receptor activation enhances SK3 channels- and cystein cathepsin-dependent cancer cells invasiveness. *Oncogene*, *30*(18), 2108–2122. http://dx.doi.org/10.1038/onc.2010.593.

Jiang, L. H., Rassendren, F., Mackenzie, A., Zhang, Y. H., Surprenant, A., & North, R. A. (2005). N-methyl-D-glucamine and propidium dyes utilize different permeation

pathways at rat P2X(7) receptors. *American Journal of Physiology. Cell Physiology, 289*(5), C1295–C1302.
Kawate, T., Michel, J. C., Birdsong, W. T., & Gouaux, E. (2009). Crystal structure of the ATP-gated P2X(4) ion channel in the closed state. *Nature, 460*(7255), 592–598. http://dx.doi.org/10.1038/nature08198.
Khakh, B. S., & North, R. A. (2006). P2X receptors as cell-surface ATP sensors in health and disease. *Nature, 442*(7102), 527–532.
Lalo, U., Roberts, J. A., & Evans, R. J. (2011). Identification of human P2X1 receptor-interacting proteins reveals a role of the cytoskeleton in receptor regulation. *The Journal of Biological Chemistry, 286*(35), 30591–30599. http://dx.doi.org/10.1074/jbc.M111.253153.
Li, J., & Fountain, S. J. (2012). Fluvastatin suppresses native and recombinant human P2X4 receptor function. *Purinergic Signal, 8*(2), 311–316. http://dx.doi.org/10.1007/s11302-011-9289-9.
Lingwood, D., & Simons, K. (2010). Lipid rafts as a membrane-organizing principle. *Science (New York, N.Y.), 327*(5961), 46–50. http://dx.doi.org/10.1126/science.1174621.
Mansoor, S. E., Wei, L., Oosterheert, W., Shekhar, M., Tajkhorshid, E., & Gouaux, E. (2016). X-ray structures define human P2X3 receptor gating cycle and antagonist action. *Nature, 538*, 66–71.
Megha, & London, E. (2004). Ceramide selectively displaces cholesterol from ordered lipid domains (rafts): Implications for lipid raft structure and function. *The Journal of Biological Chemistry, 279*(11), 9997–10004. http://dx.doi.org/10.1074/jbc.M309992200.
Michel, A. D., & Fonfria, E. (2007). Agonist potency at P2X7 receptors is modulated by structurally diverse lipids. *British Journal of Pharmacology, 152*(4), 523–537. http://dx.doi.org/10.1038/sj.bjp.0707417.
Mo, G., Bernier, L. P., Zhao, Q., Chabot-Dore, A. J., Ase, A. R., Logothetis, D., et al. (2009). Subtype-specific regulation of P2X3 and P2X2/3 receptors by phosphoinositides in peripheral nociceptors. *Molecular Pain, 5*, 47. http://dx.doi.org/10.1186/1744-8069-5-47.
Moura, G. E., Lucena, S. V., Lima, M. A., Nascimento, F. D., Gesteira, T. F., Nader, H. B., et al. (2015). P2X7 receptor activity regulation: The role of CD44 proteoglycan GAG chains. *Cell Death & Disease6*, e1997. http://dx.doi.org/10.1038/cddis.2015.340.
North, R. A. (2002). Molecular physiology of P2X receptors. *Physiological Reviews, 82*(4), 1013–1067.
Pike, L. J. (2004). Lipid rafts: Heterogeneity on the high seas. *The Biochemical Journal, 378*(2), 281–292. http://dx.doi.org/10.1042/BJ20031672.
Robinson, L. E., & Murrell-Lagnado, R. D. (2013). The trafficking and targeting of P2X receptors. *Frontiers in Cellular Neuroscience*, 7, 233. http://dx.doi.org/10.3389/fncel.2013.00233.
Roger, S., Gillet, L., Baroja-Mazo, A., Surprenant, A., & Pelegrin, P. (2010). C-terminal calmodulin-binding motif differentially controls human and rat P2X7 receptor current facilitation. *The Journal of Biological Chemistry, 285*(23), 17514–17524. http://dx.doi.org/10.1074/jbc.M109.053082.
Roger, S., Pelegrin, P., & Surprenant, A. (2008). Facilitation of P2X7 receptor currents and membrane blebbing via constitutive and dynamic calmodulin binding. *The Journal of Neuroscience, 28*(25), 6393–6401. http://dx.doi.org/10.1523/JNEUROSCI.0696-08.2008.
Rosenhouse-Dantsker, A., Noskov, S., Durdagi, S., Logothetis, D. E., & Levitan, I. (2013). Identification of novel cholesterol-binding regions in Kir2 channels. *The Journal of Biological Chemistry, 288*(43), 31154–31164. http://dx.doi.org/10.1074/jbc.M113.496117.
Shemon, A. N., Sluyter, R., Fernando, S. L., Clarke, A. L., Dao-Ung, L. P., Skarratt, K. K., et al. (2006). A Thr357 to Ser polymorphism in homozygous and compound

heterozygous subjects causes absent or reduced P2X7 function and impairs ATP-induced mycobacterial killing by macrophages. *The Journal of Biological Chemistry*, *281*(4), 2079–2086.

Subczynski, W. K., & Wisniewska, A. (2000). Physical properties of lipid bilayer membranes: Relevance to membrane biological functions. *Acta Biochimica Polonica*, *47*(3), 613–625.

Surprenant, A., & North, R. A. (2009). Signaling at purinergic P2X receptors. *Annual Review of Physiology*, *71*, 333–359. http://dx.doi.org/10.1146/annurev.physiol.70.113006.100630.

Vacca, F., Amadio, S., Sancesario, G., Bernardi, G., & Volonte, C. (2004). P2X3 receptor localizes into lipid rafts in neuronal cells. *Journal of Neuroscience Research*, *76*(5), 653–661. http://dx.doi.org/10.1002/jnr.20069.

Vial, C., & Evans, R. J. (2000). P2X receptor expression in mouse urinary bladder and the requirement of P2X(1) receptors for functional P2X receptor responses in the mouse urinary bladder smooth muscle. *British Journal of Pharmacology*, *131*(7), 1489–1495.

Vial, C., & Evans, R. J. (2005). Disruption of lipid rafts inhibits P2X1 receptor-mediated currents and arterial vasoconstriction. *The Journal of Biological Chemistry*, *280*(35), 30705–30711.

Vial, C., Fung, C. Y., Goodall, A. H., Mahaut-Smith, M. P., & Evans, R. J. (2006). Differential sensitivity of human platelet P2X1 and P2Y1 receptors to disruption of lipid rafts. *Biochemical and Biophysical Research Communications*, *343*(2), 415–419.

Wiley, J. S., Sluyter, R., Gu, B. J., Stokes, L., & Fuller, S. J. (2011). The human P2X7 receptor and its role in innate immunity. *Tissue Antigens*, *78*(5), 321–332. http://dx.doi.org/10.1111/j.1399-0039.2011.01780.x.

Yan, Z., Li, S., Liang, Z., Tomić, M., & Stojilkovic, S. S. (2008). The P2X7 receptor channel pore dilates under physiological ion conditions. *The Journal of General Physiology*, *132*(5), 563–573. http://dx.doi.org/10.1085/jgp.200810059.

Zhao, Q., Yang, M., Ting, A. T., & Logothetis, D. E. (2007). PIP(2) regulates the ionic current of P2X receptors and P2X(7) receptor-mediated cell death. *Channels (Austin, TX)*, *1*(1), 46–55.

CHAPTER NINE

Depletion of Membrane Cholesterol Suppresses *Drosophila* Transient Receptor Potential-Like (TRPL) Channel Activity

Maximilian Peters, Ben Katz, Shaya Lev, Rachel Zaguri, Rita Gutorov, Baruch Minke[1]

Institute for Medical Research Israel-Canada (IMRIC), Edmond and Lily Safra Center for Brain Sciences (ELSC), Faculty of Medicine, The Hebrew University, Jerusalem, Israel
[1]Corresponding author: e-mail address: baruchm@ekmd.huji.ac.il

Contents

Abstract

Cholesterol is an essential compound of higher eukaryotic cell membranes and a known modulator of ion channel activity. Changes in phospholipids and cholesterol composition of cell membranes are known to alter the activity of ion channels. However, there is little knowledge on the effects of cholesterol on transient receptor potential (TRP)

Current Topics in Membranes, Volume 80
ISSN 1063-5823
http://dx.doi.org/10.1016/bs.ctm.2017.05.005

channels. In this study, we explore the effects of cholesterol depletion on the *Drosophila* photoreceptor channel TRP-like (TRPL), when expressed in tissue culture cells. Depletion of membrane cholesterol with methyl-β-cyclodextrin (MβCD) induced fast (<100 s) suppression of spontaneous TRPL channel activity, a typical state of expressed TRPL channels in *Drosophila* S2 cells. An equally fast suppression of receptor-induced TRPL channel activity in HEK293 cells, downstream of phospholipase C (PLC) activation, was also induced by MβCD. Biochemical experiments showed binding of TRPL to immobilized cholesterol, suggesting direct binding of cholesterol to TRPL. Exploring the effects of several mutations in a putative cholesterol-binding site of TRPL was inconclusive as some did not render the channel insensitive to cholesterol depletion while others rendered the channel inactive. We conclude that (i) cholesterol is essential for TRPL channel activity, (ii) TRPL channels interact with cholesterol, and (iii) the binding site of cholesterol in TRPL differs from the putative binding site of TRPV1. Thus, the fast and strong effects of cholesterol depletion on the TRPL channel activity suggest that cholesterol is an important component of fly photoreceptor signaling membrane.

1. INTRODUCTION

Cholesterol, an essential component of animal cell membranes, is too hydrophobic to form a sheet structure on its own, but it is intercalated among the membrane phospholipids. The bulky, rigid lipophilic ring system of cholesterol interacts with the lipophilic part of phospholipids, while the hydroxyl group interacts with their polar head groups, resulting in a higher local membrane lipid packing. The net effect of cholesterol on membrane fluidity varies, depending on the lipid composition and the temperatures. At low temperature, cholesterol restricts the random movement of the polar heads of the fatty acyl chains, which are closest to the outer surfaces of the leaflets, but it separates and disperses their tails, causing the inner regions of the bilayer to become slightly more fluid, while at high temperature the bilayer becomes less fluid. In addition to having direct effects on membrane phospholipids (Ohvo-Rekilä, Ramstedt, Leppimäki, & Slotte, 2002), cholesterol was also shown to directly and physically interact with membrane proteins, e.g., the β-adrenergic receptor requires cholesterol for its functional structure (Hanson et al., 2008). Cholesterol functions have been attributed to several possible effects: binding to sterol recognition sites in proteins, ability to alter membrane physical properties and a role in lateral compartmentation, and dynamic anchoring/targeting of signaling molecules (Svobodova & Groschner, 2016).

Cholesterol–protein interactions have been extensively studied and specific sterol recognition elements such as the cholesterol recognition amino acid consensus have been suggested (Levitan, Singh, & Rosenhouse-Dantsker, 2014). Thus, a consensus motif, named the CRAC domain, has been defined. This *C*holesterol *R*ecognition/interaction *A*mino acid *C*onsensus sequence is a short linear amino acid motif, consisting of (L/V)-X_{1-5}-(Y)-X_{1-5}-(K/R) (whereby X is any amino acid except Pro (Li & Papadopoulos, 1998)). A motif with the opposite orientation is called CARC domain, which has exactly the same amino acid sequence as CRAC, but in the reverse orientation. The CARC domain was also implicated in cholesterol binding (Baier, Fantini, & Barrantes, 2011). These motifs have a fairly broad definition resulting in a high rate of false-positive predictions, e.g., bacteria that are not known to metabolize cholesterol still have CRAC motifs (Palmer, 2004). Nevertheless, in the absence of better predictions, these motifs help in guiding the search for putative cholesterol-binding sites (CBSs). Interestingly, transmembrane domain 5 (TM5) of the *Drosophila* TRPC channel, TRPL, has a CARC motif, while mammalian TRPV1 has a CRAC motif. The mammalian TRPV1 channel showed sensitivity to increased levels of cholesterol in the plasma membrane. Increasing cholesterol concentrations in excised patches containing TRPV1 resulted in a decrease of TRPV1 currents (Picazo-Juárez et al., 2011).

For mammalian TRPC channels, it was demonstrated that TRPC1 interacted with the cholesterol-binding scaffold protein caveolin-1 (Cav-1) (Lockwich et al., 2000), which controls channel activity via an N-terminal Cav-1-binding motive and a C-terminal caveolin-scaffolding consensus binding domain. In line with TRPC1 activity being governed by association with Cav-1, TRPC1-mediated currents were abolished in cells lacking Cav-1 (Murata et al., 2007), which targets TRPC1 into the cholesterol-rich membrane environment of caveolae (Lockwich et al., 2000). Interestingly, the function of TRPC6 channels located at the podocytes of Bowman's capsule in the kidneys was crucially dependent on targeting the channel to cholesterol-rich domains via the cholesterol-binding adaptor podocin (Anderson, Kim, Hagmann, Benzing, & Dryer, 2013). The suggested reversible targeting of TRPCs into lipid rafts, and their exposure to environments of different lipid compositions, specifically to phosphoinositides and cholesterol, may cause modulation of channel gating processes by: protein lipidation, specific lipid-binding events, or altering membrane architecture and mechanics (Svobodova & Groschner, 2016). Evidence for such a mechanism came from the observation that podocin targeting of TRPC channels is essential for mechanosensitivity of the protein complex (Huber et al., 2006).

Studies of *Drosophila* photoreceptors showed that the scaffold protein with multi-PDZ domains, INAD, and its target proteins undergo light-induced recruitment to detergent-resistant membrane (DRM) rafts. Reduction of ergosterol, a key component of lipid rafts in *Drosophila* (see Section 6), resulted in a loss of INAD-signaling complexes associated with DRM fractions (Sanxaridis et al., 2007). Genetic analysis demonstrated that translocation of INAD-signaling complexes to DRM rafts requires activation of the phototransduction cascade, while constitutive activation of the TRP channels resulted in recruitment of complexes to DRM rafts in the dark. Mutations affecting INAD and TRP showed that PDZ4 and PDZ5 domains of INAD, as well as the INAD–TRP interaction, are required for translocation of components to DRM rafts. Finally, selective recruitment of phosphorylated, and therefore activatable, eye-PKC to DRM rafts suggests that DRM domains are likely to function in signaling, rather than trafficking (Sanxaridis et al., 2007).

It was previously hypothesized that light-induced activation of phospholipase C (PLC), converting PIP_2 with the large polar head group into diacylglycerol having a very small polar group, acts as modifier of membrane lipid–channel interactions in a mechanotransduction gating mechanism of *Drosophila* TRP (Parnas et al., 2009). Indeed, it was demonstrated by combined patch-clamp recordings and atomic force microscopy that by cleaving PIP_2, PLC generates rapid physical changes in the lipid bilayer. These physical changes led to contraction of the *Drosophila* photoreceptors' microvilli, showing that the resultant mechanical forces have a kinetics of the light response and it is eliminated by the PLC mutant *norpA*, in a putative mechanotransduction process. This process may contribute together with pH changes to the gating of the TRP and TRPL channels (Hardie & Franze, 2012). One prediction of this model is that any change in surface membrane lipid packing should alter TRP and TRPL channel activity.

Depletion of membrane cholesterol can alter lipid packing in the surface membrane, while cholesterol can be rapidly depleted from biological membranes by using methyl-β-cyclodextrin (MβCD). The enzymatic degradation of starch results in cyclic sugars with variable ring sizes where the polar groups are pointing outward, while the hydrophobic part forms the inner part. Depending on the ring size, hydrophobic molecules can be selectively complexed and solubilized. Thus, MβCD is specific for cholesterol and allows rapid removal of cholesterol from living cells. Based on the light-induced recruitment of the *Drosophila* INAD-signaling complex to DRM rafts, the putative mechanosensitive model of *Drosophila* TRP gating, and the effects of cholesterol on the TRPV1 channel, we investigated in this study the effects of cholesterol modulation on *Drosophila* TRPL channel activity. We found that depletion of membrane cholesterol virtually abolished both spontaneous and receptor-induced TRPL-dependent current when expressed in tissue culture cells. Our biochemical experiments showed binding of TRPL to immobilized cholesterol, suggesting direct binding of cholesterol to TRPL. Since a CBS was suggested for TRPV1, we also mutated the corresponding amino acids in TRPL in order to test whether cholesterol binding to the homologue protein area has a functional effect. Exploring the effects of several mutations in a putative CBS of TRPL was inconclusive, as some did not render the channel insensitive to cholesterol depletion, while others rendered the channel inactive, suggesting variability in the CBS of different TRP channels.

2. METHODS

2.1 Molecular Biology

2.1.1 Generation of Plasmids

In order to generate pmt.-b-TRPL-V5-HIS and pcdna-TRPL-V5-HIS plasmids, PCR was performed in order to remove initial TRPL stop codon. This stop codon was replaced by two GG residues in order to keep the reading frame followed by a short linker, a V5-tag and 6xHIS tag. The PCR product was then purified and digested with *Kpn*I and *Eco*RI, and ligated with pmt.-b plasmid for expression in *Drosophila* Schneider S2 cell line. For mammalian expression the PCR product was digested with *Eco*RI and *Not*I, and ligated with pcDNA4 plasmid. In order to generate pcDNA4-TRPL-TEV-eGFP, PCRs were performed in order to remove the initial TRPL stop codon. This stop codon was replaced with a short hydrophilic linker (GNSGS), followed by a TEV-protease site (ENLYFQ), another flexible hydrophilic linker (SGSGSGSGS) and an eGFP. The final PCR product was then digested with *Eco*RI and *Not*I restriction enzymes and ligated with the pcDNA4 plasmid.

2.1.2 Site-Directed Mutagenesis

All of the TRPL mutations were generated by site-directed substitution mutagenesis using different primers containing the desired mutations in the pmt.-b-TRPL-V5-HIS plasmid (Table 1). The resulting TRPL mutant fragments were digested with *Kpn*I and *Eco*RI restriction enzymes and ligated with the pcDNA4-TRPL-eGFP plasmid.

Table 1 Oligonucleotides Used

Primer Name	Sequence
TRPL_F558G/V_f	/5phos/AA ATT TTT C**GK** TAT CTA CAC CCT GGT GCT GTT CGC C
TRPL_Y560F_f	/5Phos/AA ATT TTT CTT TAT CT**T** CAC CCT GGT GCT GTT CGC C
TRPL_T561I_f	/5phos/AA ATT TTT CTT TAT CTA CA**T** CCT GGT GCT GTT
TRPL_L562M/I_f	/5phos/AA ATT TTT CTT TAT CTA CAC CAT **K**GT GCT GTT CGC C
TRPL_TM5_r	/5phos/TA CTA TAT CGA TGA CCA TGC GAC CTA GGG AGA T

Mutations are in bold, *K*—keto (G or T).

2.2 Tissue Culture

S2 cells were cultured at 25°C and all experiments were performed on cells stably overexpressing the TRPL construct in the presence of a hygromycin helper plasmid (in a 19:1 ratio). Transfections were performed using Escort IV (Sigma-Aldrich) and stable transformants were selected by growing the cells in the presence of hygromycin for 3–4 weeks. The growth medium was Schneider medium, supplemented with 10% fetal calf serum (FCS), penicillin–streptomycin (100 U/mL, 100 μg/mL), and hygromycin (300 μg/mL). HEK293 TREX cells (HEK293 stably transfected with a tetracycline-responsive repressor) (Invitrogen) were grown in Dulbecco's Modified Eagle Medium, supplemented with 10% FCS, penicillin–streptomycin (100 U/mL, 100 μg/mL), L-glutamine (2 mM), and blasticidin (5–10 μg/mL). Cells were transfected with TransIT (Mirus Bio LLC) in a DNA transfection reagent ratio of 1:3 according to the manufacturer protocol. For cotransfections, 20% of the TRPL plasmid DNA was replaced with pcDNA3-hM1R (Missouri S&T cDNA Resource Center, GenBank Accession No. AF498915).

2.3 Electrophysiology

All electrophysiological experiments were carried out 1 day after induction of TRPL expression by $CuSO_4$ (1 mM) for S2 cells or a saturating concentration of tetracycline (2.5 μg/mL) for HEK293 cells. HEK293 cells were seeded 1 day before transfection and S2 cells were seeded 1 day before $CuSO_4$ induction on poly-L-lysine coated 16-mm cover slips. Whole-cell recordings were performed at 21°C using patch pipettes of 3–5 MΩ resistance (for HEK293 cells) and 7–12 MΩ resistance (for S2 cells), pulled from fiber-filled borosilicate glass capillaries. Junction potential was nulled prior to seal formation. Signals were amplified using an AxoPatch 200B (Molecular Devices, Sunnyvale, CA) patch-clamp amplifier. Voltage-clamp pulses were generated, and data were captured using a Digidata 1440A interfaced to a computer running the pClamp 10 software (Molecular Devices). Membrane potential was held at 0 mV and current was measured in response to voltage ramps from −150 to +150 mV every 5 s. Responses were analyzed off-line using Clampfit 10.2 software (Molecular Devices). The following stock solutions were used (stock concentration, final concentration, solvent): linoleic acid (30 mM, 60 μM, EtOH), carbachol (Calbiochem) (100 mM, 100 μM, H_2O), and $GdCl_3$ (1 M, 10 mM, H_2O). For the used solutions, see Table 2.

2.4 Chemicals

All chemicals for tissue culture were purchased from Biological Industries, Kibbutz Beit Haemek, Israel. All other chemicals were purchased from Sigma-Aldrich unless otherwise stated. Cholesterol-loaded MβCD was prepared by dissolving MβCD in H_2O, heating to 70°C, adding 20 μL portions

Table 2 Solutions

Intracellular Solutions in mM

Cell Line	Potassium D-Gluconate	TES	$MgCl_2$	MgATP	NaGTP	pH
S2	130	10	2	4	0.4	7.15

Cell Line	CsCl	$CaCl_2$	$MgCl_2$	MgATP	NaGTP	HEPES	EGTA	pH
HEK293	110	1	2	2	0.3	10	10	7.4

Extracellular Solutions in mM

Cell Line	NaCl	KCl	$MgCl_2$	TES	L-Proline	L-Alanine	EGTA	pH
S2	150	5	4	10	25	5	0.5	7.15

Cell Line	NaCl	KCl	$MgCl_2$	HEPES	D-Glucose	pH
HEK293	145	5	1	10	10	7.4

of a saturated solution of cholesterol in acetone, vortexing, and reheating until saturation was achieved and all acetone was evaporated. MβCD was mixed directly with the extracellular solution to achieve a final concentration of 10 mM.

2.5 Bioinformatics

Pymol (The PyMOL Molecular Graphics System, Version 1.3 Schrödinger, LLC) was used to structurally align the transmembrane region of rTRPV1 (R579-I599) with a homology model (Fernández-Ballester & Ferrer-Montiel, 2008 used by Picazo-Juárez et al., 2011).

2.6 Two-Photon Imaging

For measurements of the translocation of the eGFP-tagged Pleckstrin homology (PH) domain in HEK293 cells, images were acquired in a Zeiss LSM 7 MP Multiphoton Imaging System using a Zeiss Objective W "Plan-Apochromat" 20 × /1.0 DIC M27 water immersion objective. The cells were excited using a hands-free operation ultrafast Ti:Sapphire laser system with tuning range of 680–1080 nm, at 800 nm. Optical sections were recorded from the middle of the cells at a rate of 8–15 s, depending on the experiment. *Z*-axis positioning was verified using transmitted optical images. Cell integrity was verified morphologically at the beginning and end of each experiment. The cells were treated as described for the electrophysiological experiments.

2.7 Protein Lipid Overlay Assay

Nitrocellulose membranes containing different concentrations of lipids (in Fig. 3C, the only concentration of 10 mM lipids is shown) including cholesterol, phosphatidylserine, and triolein were incubated, 1 h at room temperature with a blocking solution containing 0.8 g BSA in 20 mL PBS-Tween (PBS-T). S2-TRPL stable line cells were centrifuged for 5 min × 300 × *g* @ 4°C and the supernatant was removed. Cells were solubilized on ice using 200 μL lysis buffer containing 1% Triton X-100, 25 mM Tris–HCl (pH 7.5), 150 mM NaCl, and 5 mM EDTA, with protease inhibitor cocktail (1:100, P8340, Sigma) overhead 1 h @ 4°C. Lysate was centrifuged for 20 min × 12,000 × *g* @ 4°C and supernatant containing the TRPL protein was laid on the membrane nitrocellulose 1 h at room temperature. The membranes were incubated with an anti-HIS (1:1000) antibody overnight. After washing three times in PBS-T, membranes were incubated with an antimouse IgG as secondary antibody conjugated to peroxidase (1:1000, Sigma). Spots were visualized by the enhanced chemiluminescence.

3. CHOLESTEROL DEPLETION SUPPRESSED CONSTITUTIVE TRPL-DEPENDENT CURRENT IN TISSUE CULTURE S2 CELLS

In an attempt to characterize the *Drosophila* light-activated TRPC channels, TRPL channels were heterologously expressed in tissue-cultured S2 (Parnas et al., 2009; Fig. 1) and HEK293 cells (Lev, Katz, Tzarfaty, & Minke, 2012; see later). When expressed in the *Drosophila* S2 cells, TRPL shows a basal constitutive activity that can be enhanced by the application of the polyunsaturated fatty acid (PUFA), linoleic acid (Lev, Katz, & Minke, 2012; Parnas et al., 2009). When S2 cells expressing TRPL were perfused with MβCD, TRPL-dependent current was rapidly and significantly decreased in less than 100 s (Fig. 1A–C). This relatively fast effect of cholesterol depletion differs from the many minute delays of the reported effects of cholesterol modulations (Singh, Shentu, Enkvetchakul, & Levitan, 2011). The loss of TRPL-dependent current persisted even after prolonged washout of the water-soluble hydrophilic cyclic sugar, MβCD, excluding the possibility that MβCD directly blocks the TRPL channel activity (Fig. 1A and B).

In order to test whether the effects of MβCD arise from cholesterol depletion or from unspecific effects, MβCD was loaded with cholesterol and perfused on TRPL-expressing S2 cells. No significant change was observed on TRPL currents (Fig. 2A, B, and E), suggesting that the suppression of TRPL currents in the presence of MβCD is due to cholesterol depletion. In a minority of cells, a pronounced increase of inward (but not increase of outward) currents was observed during perfusion of MβCD loaded with cholesterol (Fig. 2C and D). A similar effect was previously observed during perfusion of S2 cells with linoleic acid and this observation resulted from removal of divalent open channel block (Parnas et al., 2009). However, during perfusion of MβCD loaded with cholesterol, this effect was rare and thus statistically insignificant (Fig. 2E).

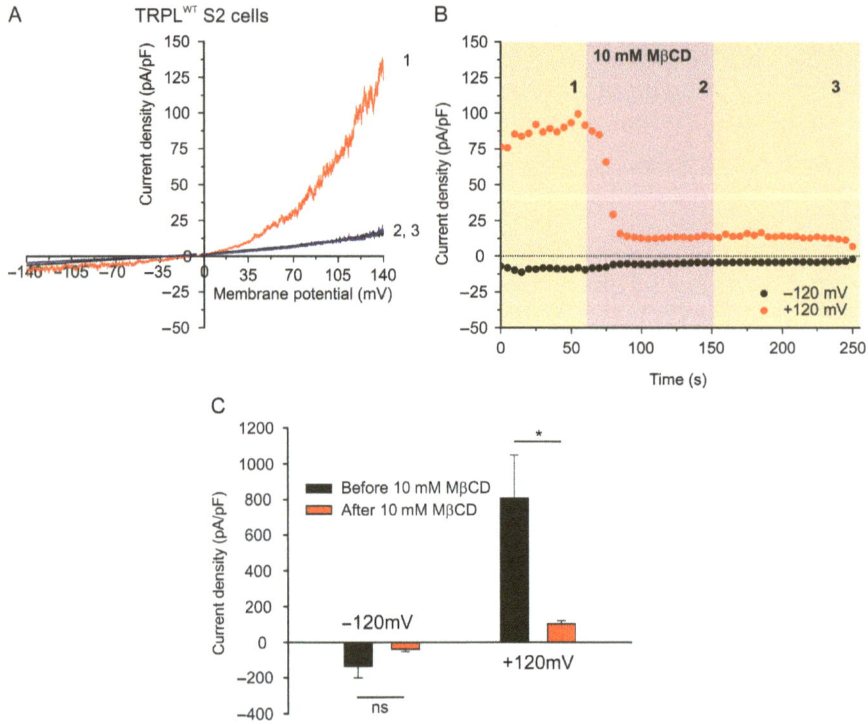

Fig. 1 MβCD blocks constitutive TRPL channels activity. (A) Current–voltage relationships (*I–V* curves) measuring TRPL-dependent currents *I–V* curves obtained in response to voltage ramp (of 1 s duration) from S2 cells expressing TRPL and showing basal channel activity with strong outward rectification, typical for TRPL-dependent current (1). The TRPL channel activity was highly reduced after perfusion with 10 mM MβCD (2) and the effect was irreversible, even after washout of MβCD (3) ($n > 10$). (B) Time course of the MβCD effects on TRPL currents in S2 cells. Current densities are shown as a function of time. Series of *I–V* curves were derived from repeatedly applied voltage ramps every 5 s and currents were measured at ±120 mV holding potentials as a function of time under the various experimental conditions as indicated. The numbers correspond to the numbers on the *I–V* curves in (A). (C) Statistics of the cholesterol depletion experiments in S2 cells. Cholesterol depletion by MβCD had a significant effect on the positive TRPL currents at 120 mV ($n = 5$, values are average ± SEM, paired Student *t*-test, $*P \leq 0.05$).

4. CHOLESTEROL DEPLETION SUPPRESSED RECEPTOR-ACTIVATED AND PLC-MEDIATED TRPL CURRENTS

It has been previously shown that TRPL channels do not show basal constitutive activity when expressed in mammalian HEK293 cells, much

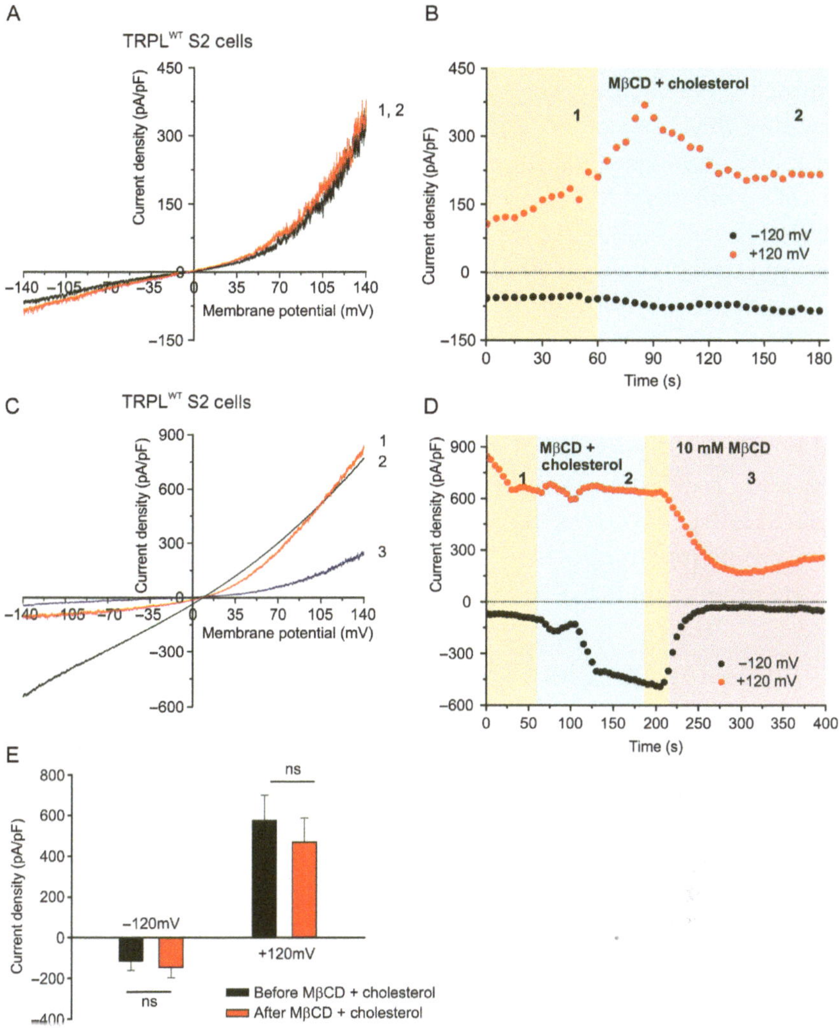

Fig. 2 Control experiments showing that the effect of MβCD is mediated by cholesterol since MβCD loaded with cholesterol has virtually no effect. (A) The basal TPRL current in S2 cells is unaffected by perfusion with cholesterol-loaded MβCD: Current–voltage relationships measuring TRPL-dependent currents in response to voltage ramps recorded from S2 cells expressing TRPL. The *I–V* curve shows basal current with strong outward rectification, typical for TRPL-dependent current (1, *red*). Loading the S2 cell membranes with cholesterol by application of MβCD loaded with cholesterol had minor effects on the TRPL currents (2, *black*), ($n=6$). (B) Time course of the effect of cholesterol loading on TRPL currents in S2 cells. Current densities are shown as a function of time. Series of *I–V* curves were derived from repeatedly applied voltage ramps every 5 s and currents were measured at $\pm 120\,mV$ holding potentials as a function of time under the various

(Continued)

like the native *Drosophila* TRPL and TRP channels in the photoreceptor cells (Lev et al., 2012; see Fig. 3A and B, trace 1). In order to activate the expressed TRPL channels, we coexpressed the mammalian muscarinic receptor 1 (hM1 receptor). Since HEK293 cells have endogenous Gq protein and phospholipase Cβ (PLCβ), activation of hM1receptor by perfusing the cells with the hM1 agonist carbachol (CCH) induced activation of TRPL (Fig. 3A and B, trace 2). This result is similar to that observed in photoreceptor cells in vivo, whereby the TRPL channels are closed in the dark, while light stimulation activates the channels via a PLCβ-mediated GPCR cascade. In this study, application of CCH induced TRPL-mediated currents that were suppressed after application of CCH that was added with MβCD (Fig. 3A and B, trace 3). Finally, the cells were perfused with linoleic acid which failed to restore TRPL currents (Fig. 3A and B, trace 4). The latter result is different from that obtained in S2 cells, whereby linoleic acid operated independently of cholesterol (data not shown). In summary, cholesterol depletion suppressed both the constitutive TRPL activity in S2 cells and the receptor-induced TRPL currents in HEK293 cells.

4.1 Cholesterol Depletion Did Not Affect PLC Activity

In order to examine whether the suppression of TRPL channel activity by cholesterol depletion is a direct effect on the TRPL channels, or it is mediated by inhibition of PLC activity, a fluorescent PIP_2 reporter system was used. The PH domain of PLC-δ was fused to GFP and allowed monitoring membrane PIP_2 levels in intact cells in real time (Balla & Varnai, 2002; Lev et al., 2012; Suh, Inoue, Meyer, & Hille, 2006). Under nonstimulated

Fig. 2—Cont'd experimental conditions as indicated. (C) Some S2 cells showed pronounced enhancements of inward currents when the cells were perfused with MβCD loaded with cholesterol. Current–voltage relationships measuring TRPL-dependent currents in response to voltage ramp from S2 cells expressing TRPL, showing basal currents with strong outward rectification, typical for TRPL-dependent current (1). Application of MβCD loaded with cholesterol induced pronounced enhancements of inward currents, while the TRPL outward currents were virtually unaffected (typical for removal of divalent open channel block (Parnas et al., 2009)). The effect was reversible upon cholesterol depletion with MβCD (3) ($n=3$). (D) Time course of the effect of cholesterol loading on TRPL currents in S2 cells. Current densities are shown as a function of time. Series of *I–V* curves were derived from repeatedly applied voltage ramps every 5 s and currents were measured at ± 120 mV holding potentials as a function of time under the various experimental conditions as indicated. (E) Statistical analysis of all cholesterol-loading experiments in S2 cells. Cholesterol-loaded MβCD did not have a significant effect on TRPL currents ($n=9$, values are average $\pm$ SEM, paired Student *t*-test, $P>0.5$).

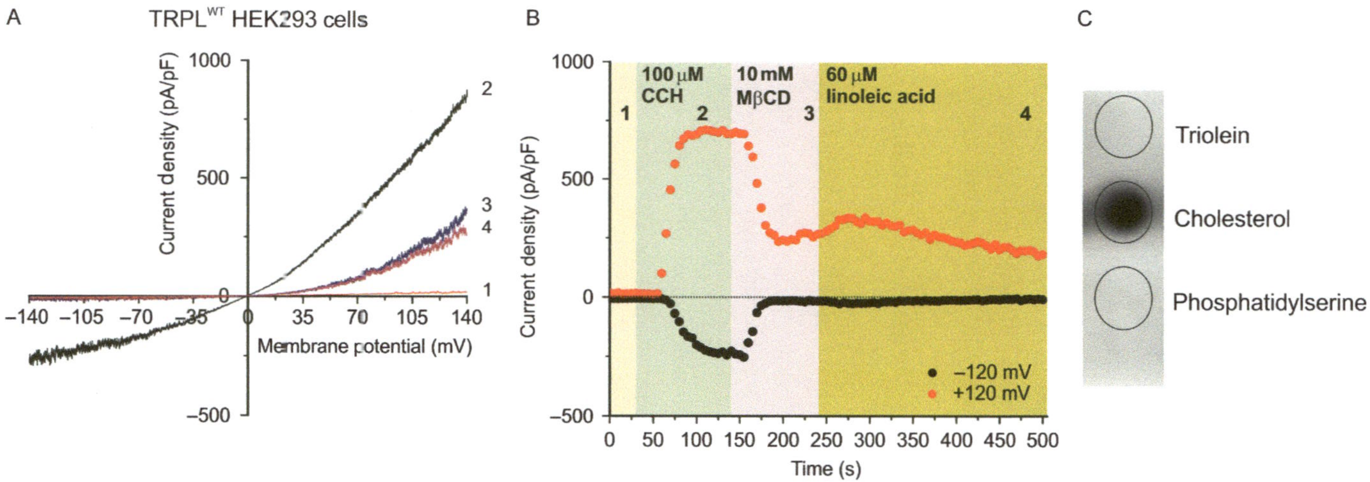

Fig. 3 Cholesterol depletion suppressed receptor-activated TRPL-dependent current. (A) TRPL–GFP did not show any spontaneous activity in HEK293 cells, but it could be readily activated via PLC and blocked by MβCD: Current–voltage relationships measured from HEK293 cells expressing TRPL–GFP, showing no basal channel activity (1). However, coexpression of the hM1 muscarinic receptor and application of carbachol (CCH) activated the expressed TRPL–GFP channels via endogenous PLC-mediated cascade (2) and the TRPL-dependent current in the presence of CCH was suppressed by application of MβCD (3), while subsequent application of linoleic acid, a strong activator of TRPL channels, did not activate the channels after the application of MβCD (4) ($n=4$). (B) Time course of the receptor-activated TRPL-dependent current and the effect of cholesterol depletion on the receptor-activated TRPL currents in HEK293 cells. Current densities are shown as a function of time. Series of *I–V* curves were derived from repeatedly applied voltage ramps every 5 s and currents were measured at ±120 mV holding potentials as a function of time under the various experimental conditions as indicated. (C) Biochemical experiments showing interaction between TRPL and cholesterol. Lysates of S2 cells overexpressing TRPL were incubated with a nitrocellulose membranes on which 10 mM of cholesterol, trioleoylglycerol (triolein), and phosphatidylserine were spotted. After thorough washing, binding was visualized by using antibodies against the HIS-tag of TRPL and a horseradish peroxidase conjugated secondary antibody. The *dark spot* indicates more primary antibody bound to the membrane, revealing stronger binding between TRPL and the lipid. The results indicate a strong, rather specific, interaction between TRPL and cholesterol, but not between phosphatidylserine and trioleoylglycerol.

conditions, the reporter construct PH-GFP is associated with PIP_2 in the plasma membrane (Fig. 4). However, once PIP_2 concentrations are lowered (e.g., by PLC activation), the GFP-tagged protein diffused to the cytosol (Fig. 4). In Fig. 4A and B, the localization of PH-GFP is shown as a function of time before (Fig. 4A and B, the initial ~150 s) and after activation of PLC via the hM1 receptor (Fig. 4A and B after the initial 150 s). These experiments were carried out both in the absence (Fig. 4A) and in the presence of MβCD (Fig. 4B). In both, the PH-GFP construct translocated normally following application of the hM1 activator, CCH, indicating that PLC activity is not affected by cholesterol depletion. These results indicated that the suppression of TRPL activity by cholesterol depletion occurs downstream of PLC, possibly acting directly on the TRPL channel.

4.2 Cholesterol Binding to TRPL as Measured in Biochemical Experiments

We further examined whether the suppression of TRPL activity is due to direct interaction between TRPL and cholesterol, by performing a biochemical binding assay. Accordingly, a lipid overlay assay was developed in S2 cells overexpressing TRPL. Cells were lysed and the lysates were incubated with nitrocellulose membranes on which cholesterol was spotted. Unbound proteins were removed by several wash steps and bound TRPL was detected using antibodies against its HIS-tag. As shown in Fig. 3C, TRPL interacted specifically with cholesterol, but not with the control lipids trioleoylglycerol (triolein) or phosphatidylserine. This result suggests that there is a direct and strong biochemical interaction between TRPL and cholesterol. The binding is not due to a nonspecific hydrophobic interaction because structurally unrelated lipids failed to bind TRPL.

5. MUTATIONS OF THE PUTATIVE CBS AND THE CARC MOTIF

It was demonstrated that TRPV1 is sensitive to increased levels of cholesterol in the plasma membrane. Increasing cholesterol concentrations in excised patches containing TRPV1 resulted in a decrease of TRPV1 currents (Picazo-Juárez et al., 2011). The TM5 domain of rTRPV1 includes its CRAC motif, the putative CBS of TRPV1 (see Section 1). Alignment of the TM5 region of rTRPV1 and the corresponding amino acids of TRPL reveal similarity between the two channels, which includes the CARC motif of TRPL and the suggested CBS of TRPV1, L585, which corresponds

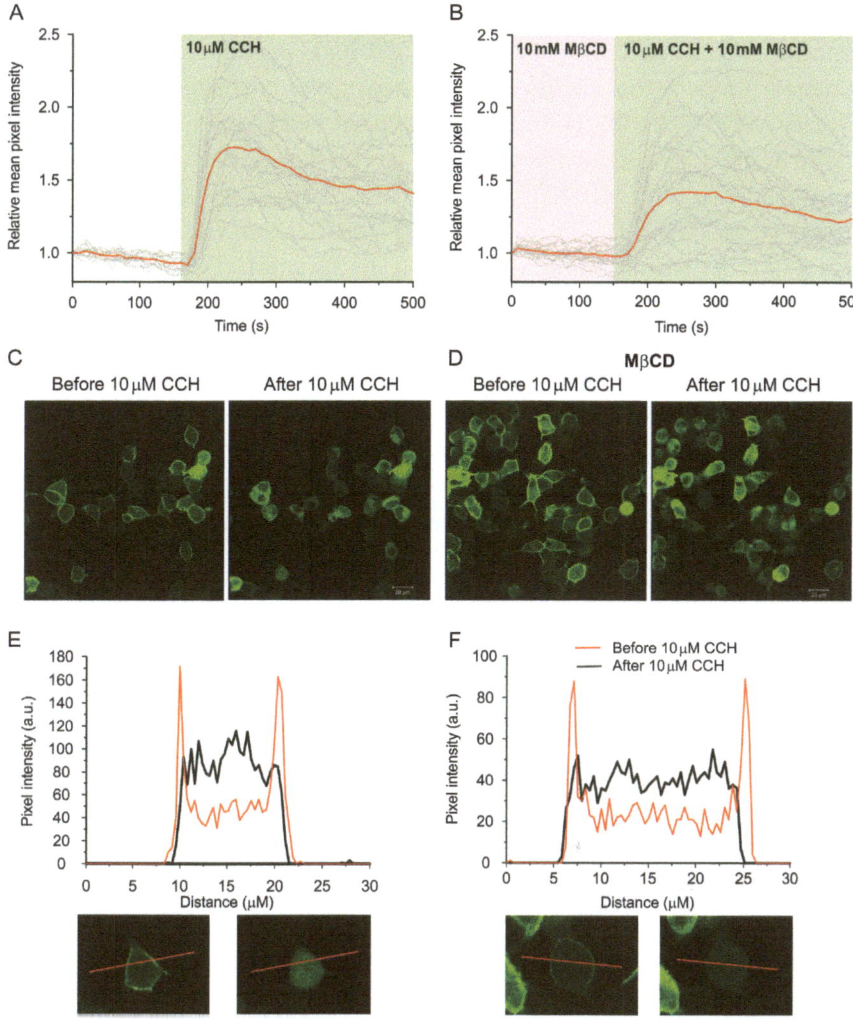

Fig. 4 Cholesterol depletion did not affect receptor-induced PLC activity. No effects of cholesterol depletion on PLC activity as monitored by translocation of the PIP_2 sensor PH–GFP. Representative series of multiphoton images of HEK293 cells coexpressing eGFP-tagged PH domain and hM1 receptor. Application of CCH to the bathing solution, in a concentration that activated the TRPL channels (10 μM CCH, Fig. 3), induced similar translocation of the eGFP-tagged PH domain to the cell body, with and without MβCD, indicating the PLC-mediated hydrolysis of PIP_2 is not affected by MβCD. (A) The time course of fluorescence changes measured in the cytosol before application of MβCD: Graph plotting the relative mean pixels intensity (*red curve*) as a function of time measured from multiphoton images of HEK293 cells expressing PH-GFP and hM1 receptors. Before PLC stimulation by CCH application (*white background*), the GFP–PH is associated with the plasma membrane where most PIP_2 is located and the cell body fluorescence is *(Continued)*

to T561 in TRPL (Table 3). Another putative CBS of rTRPV1, R579, corresponds to K555 of TRPL (Table 3). In TRPV1, introduction of the L585I mutation renders the channel totally insensitive to cholesterol addition (Picazo-Juárez et al., 2011). Introduction of the corresponding T561I mutation in TRPL did not change the I–V curve of TRPL that still revealed the typical constitutive TRPL current, which was suppressed by application of MβCD (Fig. 5A and B). Thus, the T561I mutation did not affect the channel's sensitivity to cholesterol depletion (Fig. 5A and B). The Y560F mutation, which disrupts the CARC motif in TM5 of TRPL (Table 3), also remained sensitive to MβCD (Fig. 5C and D). Mutating the two amino acids of the putative CBS, namely TRPL F558V ($n=6$) and L562M ($n=5$), resulted in inactive channels in S2 cells (data not shown) and could therefore not be tested for sensitivity to cholesterol depletion.

Table 3 Alignment of Transmembrane Domain 5 of rTRPV1 and TRPL

rTRPV1	**R**FMFV**Y**L**V**FLFGF
TRPL	**K**FFFI**Y**T**L**VLFAF
TRPL Position	555 ··· 561

The putative cholesterol-binding site of TRPV1 and the corresponding amino acids of TRPL are *underlined*, and the CARC motif domain of TRPL is marked in *bold*.

Fig. 4—Cont'd low (for quantification, see E, F). Once PLC is activated and PIP_2 is hydrolyzed (*green background*), the PH-GFP translocates to the cytosol (for quantification, see E, F) and there is a marked increase in fluorescence intensity at the cytosol. The individual single-cell measurements are shown by noisy *dim gray traces*. (B) The time course of fluorescence changes measured in the cytosol after application of MβCD: Similar graph as in (A), but measured following application of MβCD. (C) Multiphoton images of HEK293 cells expressing PH-GFP and hM1 receptor without application of MβCD: *Left*: GFP fluorescence of cells before application of CCH, little PH-GFP translocation was observed (for quantification, see E, F). *Right*: GFP fluorescence of cells after perfusion with the M1 agonist CCH. Translocation of PH-GFP is observed (for quantification, see E, F). MβCD was not applied ($n>50$). (D) Multiphoton images of HEK293 cells expressing PH-GFP and hM1 receptor after application of MβCD: Similar images of HEK293 cells expressing TRPL PH-GFP and hM1 receptor before (*left*) and after application of CCH (*right*). MβCD was applied, but it did not affect PH-GFP translocation ($n>50$). (E, F) Graphs plotting the PH-GFP fluorescence intensity as a function of cell position: Fluorescence intensity of images showing cross sections of two representative cells (multiphoton images are shown later) along the *red line*, before application of CCH (*red curve*), and after application of CCH (*black curve*) in the absence of MβCD (E) and after application of MβCD (F).

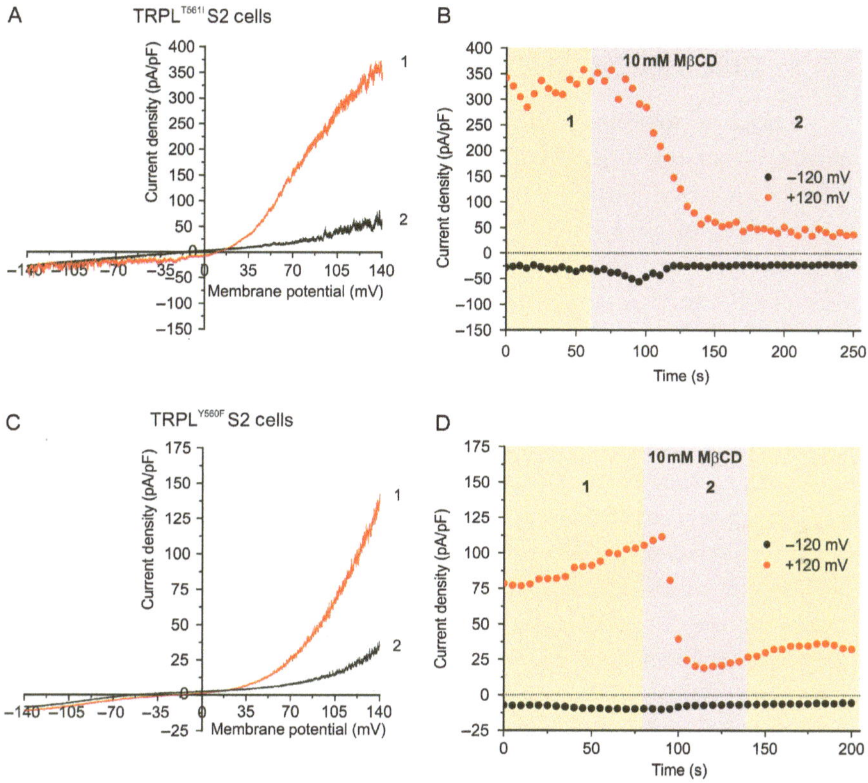

Fig. 5 TRPL mutants T561I and Y560F expressed in S2 cells remained sensitive to cholesterol depletion by MβCD. (A) Current–voltage relationships (*I–V* curves) measuring T561I-TRPL-dependent currents: *I–V* curves measuring constitutive TRPL-dependent current in the T561I mutant of TRPL in response to voltage ramp, as measured from S2 cells (1). The TRPL mutant current was suppressed by application of MβCD (2) ($n=4$). (B) Time course of the T561I-TRPL mutant current and its suppression by MβCD in S2 cells: Current densities are shown as a function of time. Series of *I–V* curves were derived from repeatedly applied voltage ramps every 5 s and currents were measured at ± 120 mV holding potentials as a function of time under the various experimental conditions as indicated ($n=3$). (C) Current–voltage relationships (*I–V* curves) measuring Y560F-TRPL-dependent currents: *I–V* curves measuring constitutive TRPL-dependent current in theY560F mutant of TRPL in response to voltage ramp, as measured from S2 cells (1). The TRPL mutant current was suppressed by application of MβCD (2) ($n=4$). (D) Time course of the Y560F-TRPL mutant current and its suppression by MβCD in S2 cells: Current densities are shown as a function of time. Series of *I–V* curves were derived from repeatedly applied voltage ramps every 5 s and currents were measured at ± 120 mV holding potentials as a function of time under the various experimental conditions as indicated ($n=4$).

6. DISCUSSION AND CONCLUSIONS

Evidence now exists substantiating membrane cholesterol as a crucial determinant of ion channel functionality. Cholesterol effects have been attributed to its binding to sterol recognition sites in membrane proteins, for the ability of cholesterol to change membrane physical properties, or for its role in compartmentalization and dynamic anchoring of signaling molecules (Levitan et al., 2014; Svobodova & Groschner, 2016). Direct association of cholesterol with ion channels was demonstrated for a few channel proteins (e.g., Kir and Orai1 channels) as evidenced by the existence of a saturable sterol-binding site demonstrated by ligand-binding assays (Levitan et al., 2014) and cholesterol–channel interactions shown by FRET microscopy, respectively (Derler et al., 2016). The mechanisms by which cholesterol governs channel functions appear to involve either ligand gating mechanisms, changes in bilayer physical properties or in coordination of cellular targeting of the channel complex.

One of the interesting biochemical differences between insects and mammals is their change in lipid metabolism. Insects produce little to no arachidonic acid, an important PUFA in mammals, and have evolutionarily lost receptors associated with this fatty acid as well (McPartland, Di Marzo, De Petrocellis, Mercer, & Glass, 2001). Insects, unlike mammals, are unable to synthesize sterols (Clark & Block, 1959) and *Drosophila melanogaster* is totally dependent on external supply of sterols, such as ergosterol from yeast (Bos, Burnet, Farrow, & Woods, 1976). Nevertheless, several enzymes of *D. melanogaster* do not differentiate between endogenous ergosterol and exogenous cholesterol (Phillips, Woodruff, Liang, Patten, & Broadie, 2008). In this sense, the observed effects of cholesterol depletion by MβCD on TRPL currents can be easily dismissed as an epiphenomenon because *Drosophila* does not use cholesterol. However, all the experiments of this study were performed in *Drosophila* S2 cells or HEK293 cells, which were grown in a medium solely containing cholesterol, but not ergosterol. The observed effects of MβCD are, therefore, due to cholesterol, not ergosterol, depletion. At least for *Drosophila* S2 cells, cholesterol and ergosterol are interchangeable: either cholesterol is enzymatically converted to ergosterol or both sterols function in an identical manner in vivo. Unlike other lipids, membrane cholesterol levels are tightly regulated and, unlike plasma cholesterol levels, do not vary. It is therefore unlikely that modulations of cholesterol levels are a physiological mean to modulate TRPL channel activity. Nevertheless, modulation of cholesterol level in the signaling membrane most likely affects lipid rafts in which the signaling proteins are located (Sanxaridis et al., 2007) and thus are likely to give insight on the still unknown gating mechanism of TRPC channels.

The results presented here show clearly that cholesterol is essential for TRPL channel function and that cholesterol, most likely, directly interacts with the channel. Our still unpublished data on *Drosophila* photoreceptors have indicated that cholesterol depletion strongly affects the light response of the photoreceptor cells. It is unlikely that MβCD exerted its effect via some nonspecific damage to the tissue culture cells (Derler et al., 2016) because no morphological damage to the cells was observed in the presence of MβCD. When the region of TRPL, which corresponds to the CBS of TRPV1 (Brauchi et al., 2007; Picazo-Juárez et al., 2011), was mutated, mostly nonfunctional channels were obtained. However, the T561I mutation of TRPL (corresponding to L585I mutant of TRPV1), which resulted in cholesterol-insensitive TRPV1 channels, did not alter the cholesterol sensitivity of TRPL. At least two possible interpretations could arise from these results: either this site is essential for cholesterol binding, while without cholesterol binding there is no channel activity, or those mutated amino acids are essential for channel activity regardless of cholesterol binding. When the TRPV1 model, which was used for the computational prediction of cholesterol binding (Brauchi et al., 2007), and the solved TRPV1 structures (Liao, Cao, Julius, & Cheng, 2013) are compared (Fig. 6), it becomes obvious that the side chains that are essential for cholesterol binding (Brauchi et al., 2007; Picazo-Juárez et al., 2011) have a quite different orientation in the two models. Based on the effects of the T561I mutation, which should result in a mutant channel that is not sensitive to cholesterol, it is more likely that this area of TM5 is not the CBS of TRPL. Alternatively, it would be possible that TRPV1 and TRPL have different CBSs and this issue needs further investigations.

While the effect of cholesterol depletion on TRPL channel activity is extremely robust and reproducible, it seems that the consensus on the location of the CBS should be further investigated. Although the exact location of the CBS could not be established in this study, it was nonetheless shown for the first time that TRPL channel activity is affected by modulations of cholesterol levels and that this effect is likely operating at the channel level. Previous studies on *Drosophila* photoreceptors and our unpublished data suggest that modulation of ergosterol levels in the signaling membrane most likely affects lipid rafts in which the signaling proteins are located

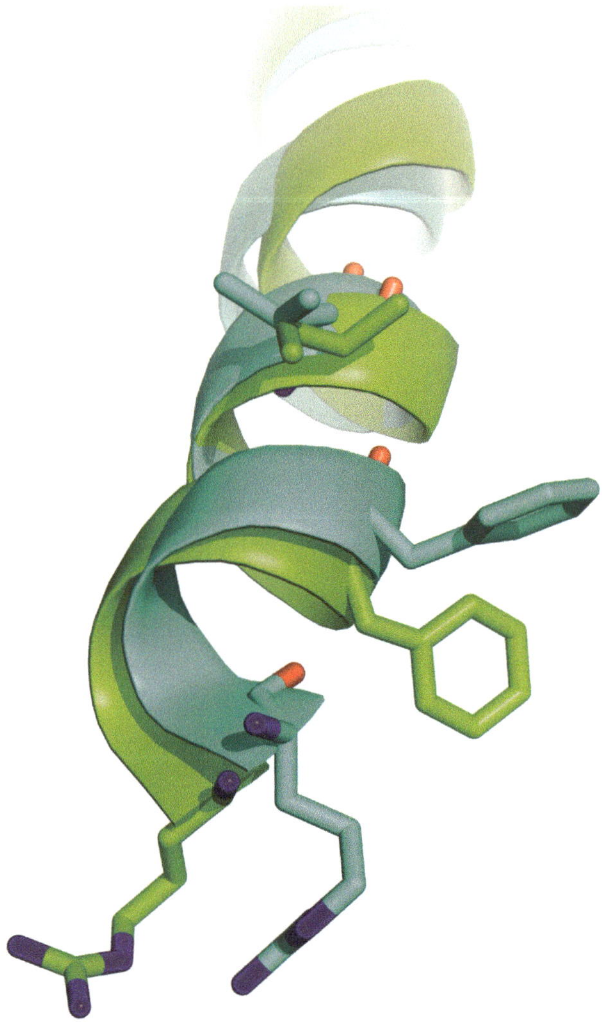

Fig. 6 Alignment of the cholesterol-binding site of a TRPV1 model and the solved TRPV1 structure. A comparison between the TM5 of the TRPV1 model, which was used for the computational prediction of cholesterol binding (*green*, Brauchi et al., 2007), and the solved structures of TRPV1 (*cyan*, Liao, Cao, Julius, & Cheng, 2013) are compared. The comparison shows that the side chains, which are essential for cholesterol binding have a quite different orientation in the two models.

(Sanxaridis et al., 2007). The effects on these signaling proteins are likely the mechanism underlying the observed suppression of expressed TRPL channel activity by depletion of cholesterol. A correlation between modulations of lipid rafts and TRP channel activity in the native tissue constitutes a future powerful tool for investigating the gating mechanism of TRPC channels.

ACKNOWLEDGMENTS

This research was supported by grants from the US-Israel Bi National Science Foundation, the Israel Science Foundation (ISF), and the Deutsch-Israelische Projektkooperation (DIP).

REFERENCES

Anderson, M., Kim, E. Y., Hagmann, H., Benzing, T., & Dryer, S. E. (2013). Opposing effects of podocin on the gating of podocyte TRPC6 channels evoked by membrane stretch or diacylglycerol. *American Journal of Physiology. Cell Physiology*, *305*(3), C276–C289.

Baier, C. J., Fantini, J., & Barrantes, F. J. (2011). Disclosure of cholesterol recognition motifs in transmembrane domains of the human nicotinic acetylcholine receptor. *Scientific Reports*, *1*, 69.

Balla, T., & Varnai, P. (2002). Visualizing cellular phosphoinositide pools with GFP-fused protein-modules. *Science's STKE*, *2002*(125), L3.

Bos, M., Burnet, B., Farrow, R., & Woods, R. A. (1976). Development of Drosophila on sterol mutants of the yeast Saccharomyces cerevisiae. *Genetical Research*, *28*(2), 163–176.

Brauchi, S., Orta, G., Mascayano, C., Salazar, M., Raddatz, N., Urbina, H., et al. (2007). Dissection of the components for PIP2 activation and thermosensation in TRP channels. *Proceedings of the National Academy of Sciences of the United States of America*, *104*(24), 10246–10251.

Clark, A. J., & Block, K. (1959). The absence of sterol synthesis in insects. *The Journal of Biological Chemistry*, *234*, 2578–2582.

Derler, I., Jardin, I., Stathopulos, P. B., Muik, M., Fahrner, M., Zayats, V., et al. (2016). Cholesterol modulates Orai1 channel function. *Science Signaling*, *9*(412), ra10.

Fernández-Ballester, G., & Ferrer-Montiel, A. (2008). Molecular modeling of the full-length human TRPV1 channel in closed and desensitized states. *The Journal of Membrane Biology*, *223*(3), 161–172.

Hanson, M. A., Cherezov, V., Griffith, M. T., Roth, C. B., Jaakola, V. P., Chien, E. Y., et al. (2008). A specific cholesterol binding site is established by the 2.8 Å structure of the human beta2-adrenergic receptor. *Structure*, *16*(6), 897–905.

Hardie, R. C., & Franze, K. (2012). Photomechanical responses in Drosophila photoreceptors. *Science*, *338*(6104), 260–263.

Huber, T. B., Schermer, B., Müller, R. U., Höhne, M., Bartram, M., Calixto, A., et al. (2006). Podocin and MEC-2 bind cholesterol to regulate the activity of associated ion channels. *Proceedings of the National Academy of Sciences of the United States of America*, *103*(46), 17079–17086.

Lev, S., Katz, B., & Minke, B. (2012). The activity of the TRP-like channel depends on its expression system. *Channels (Austin)*, *6*(2), 86–93.

Lev, S., Katz, B., Tzarfaty, V., & Minke, B. (2012). Signal-dependent hydrolysis of phosphatidylinositol 4,5-bisphosphate without activation of phospholipase C: Implications on gating of Drosophila TRPL (transient receptor potential-like) channel. *The Journal of Biological Chemistry*, *287*(2), 1436–1447.

Levitan, I., Singh, D. K., & Rosenhouse-Dantsker, A. (2014). Cholesterol binding to ion channels. *Frontiers in Physiology*, *5*, 65.

Li, H., & Papadopoulos, V. (1998). Peripheral-type benzodiazepine receptor function in cholesterol transport. Identification of a putative cholesterol recognition/interaction amino acid sequence and consensus pattern. *Endocrinology*, *139*(12), 4991–4997.

Liao, M., Cao, E., Julius, D., & Cheng, Y. (2013). Structure of the TRPV1 ion channel determined by electron cryo-microscopy. *Nature*, *504*(7478), 107–112.

Lockwich, T. P., Liu, X., Singh, B. B., Jadlowiec, J., Weiland, S., & Ambudkar, I. S. (2000). Assembly of Trp1 in a signaling complex associated with caveolin-scaffolding lipid raft domains. *The Journal of Biological Chemistry*, *275*(16), 11934–11942.

McPartland, J., Di Marzo, V., De Petrocellis, L., Mercer, A., & Glass, M. (2001). Cannabinoid receptors are absent in insects. *The Journal of Comparative Neurology*, *436*(4), 423–429.

Murata, T., Lin, M. I., Stan, R. V., Bauer, P. M., Yu, J., & Sessa, W. C. (2007). Genetic evidence supporting caveolae microdomain regulation of calcium entry in endothelial cells. *The Journal of Biological Chemistry*, *282*(22), 16631–16643.

Ohvo-Rekilä, H., Ramstedt, B., Leppimäki, P., & Slotte, J. P. (2002). Cholesterol interactions with phospholipids in membranes. *Progress in Lipid Research*, *41*(1), 66–97.

Palmer, M. (2004). Cholesterol and the activity of bacterial toxins. *FEMS Microbiology Letters*, *238*(2), 281–289.

Parnas, M., Katz, B., Lev, S., Tzarfaty, V., Dadon, D., Gordon-Shaag, A., et al. (2009). Membrane lipid modulations remove divalent open channel block from TRP-like and NMDA channels. *The Journal of Neuroscience*, *29*(8), 2371–2383.

Phillips, S. E., Woodruff, E. A., Liang, P., Patten, M., & Broadie, K. (2008). Neuronal loss of Drosophila NPC1a causes cholesterol aggregation and age-progressive neurodegeneration. *The Journal of Neuroscience*, *28*(26), 6569–6582.

Picazo-Juárez, G., Romero-Suárez, S., Nieto-Posadas, A., Llorente, I., Jara-Oseguera, A., Briggs, M., et al. (2011). Identification of a binding motif in the S5 helix that confers cholesterol sensitivity to the TRPV1 ion channel. *The Journal of Biological Chemistry*, *286*(28), 24966–24976.

Sanxaridis, P. D., Cronin, M. A., Rawat, S. S., Waro, G., Acharya, U., & Tsunoda, S. (2007). Light-induced recruitment of INAD-signaling complexes to detergent-resistant lipid rafts in Drosophila photoreceptors. *Molecular and Cellular Neurosciences*, *36*(1), 36–46.

Singh, D. K., Shentu, T. P., Enkvetchakul, D., & Levitan, I. (2011). Cholesterol regulates prokaryotic Kir channel by direct binding to channel protein. *Biochimica et Biophysica Acta*, *1808*(10), 2527–2533.

Suh, B. C., Inoue, T., Meyer, T., & Hille, B. (2006). Rapid chemically induced changes of PtdIns(4,5)P2 gate KCNQ ion channels. *Science*, *314*(5804), 1454–1457.

Svobodova, B., & Groschner, K. (2016). Reprint of "Mechanisms of lipid regulation and lipid gating in TRPC channels" *Cell Calcium*, *60*(2), 133–141.

CHAPTER TEN

Sterol Regulation of Voltage-Gated K^+ Channels

Andras Balajthy*, Peter Hajdu†, Gyorgy Panyi*,‡,1, Zoltan Varga*,‡
*Faculty of Medicine, University of Debrecen, Debrecen, Hungary
†Faculty of Dentistry, University of Debrecen, Debrecen, Hungary
‡MTA-DE-NAP B Ion Channel Structure-Function Research Group, RCMM, University of Debrecen, Debrecen, Hungary
[1]Corresponding author: e-mail address: panyi@med.unideb.hu

Contents

Abstract

Cholesterol is an essential lipid building block of the cellular plasma membrane. In addition to its structural role, it regulates the fluidity and raft structure of the membrane and influences the course of numerous membrane-linked signaling pathways and the function of transmembrane proteins, including ion channels. This is supported by a vast body of scientific data, which demonstrates the modulation of ion channels with a great variety of ion selectivity, gating, and tissue distribution by changes in membrane cholesterol. Here, we review what is currently known about the modulation of voltage-gated K^+ (Kv) channels by changes in membrane cholesterol content, considering raft association of the channels, the roles of cholesterol recognition sites, and those of adaptor proteins in cholesterol–Kv channel interactions. We specifically focus on Kv1.3, the

Current Topics in Membranes, Volume 80
ISSN 1063-5823
http://dx.doi.org/10.1016/bs.ctm.2017.05.006

dominant K^+ channel of human T cells. Effects of cholesterol depletion and enrichment and 7-dehydrocholesterol enrichment on Kv1.3 gating are discussed in the context of the immunological synapse and the comparison of the in vitro effects of sterol modifications on Kv1.3 function with ex vivo effects on cells from hypercholesterolemic and Smith–Lemli–Opitz patients.

1. INTRODUCTION

Cholesterol is a prominent component of the plasma membrane and provides a structural stability for the lipid bilayer of mammalian cells. Cholesterol has an influence on the biophysical properties of the plasma membrane as well as on membrane proteins. Cholesterol, along with the sphingolipids, serves as a molecular glue to attach the lipid rafts together, and it classifies proteins as raft-associated and raft-excluded ones. The cholesterol homeostasis of the human body is rigorously controlled and any unbalances can lead to diseases. Dietary hyperlipidemia/hypercholesterolemia is a major health issue, generally leading to atherosclerosis, which nowadays is mainly regarded as an inflammatory immune response (Hansson & Hermansson, 2011). Furthermore, inherited disorders of cholesterol synthesis such as SLOS (Smith–Lemli–Opitz syndrome) that is characterized by several birth defects (microcephaly, cardiac and neuronal deficiencies, behavior problems) (Irons, Elias, Salen, Tint, & Batta, 1993). While the former is associated with an elevated cholesterol level, in the latter the precursors of cholesterol (7-dehydrocholesterol, 8-dehydrocholesterol) are present at high concentration in the plasma and the tissues.

Ion channels of the plasma membrane are transmembrane proteins that form ion-conducting pores to transport ions through the cell membrane. The most ancient and largest group of ion channels are the potassium (K^+) channels, and of these the voltage-gated (depolarization activated) K^+ (Kv) channels are the best-studied subgroup. Kv channels consist of four subunits, they can be either heteromers or homomers, and each subunit contains six alpha-helical segments (S1–S6) connected by intra- or extracellular loops (MacKinnon, 1991). The N- and C-terminal ends of Kv channels reside intracellularly and play a role in association of individual subunits, binding various auxiliary and adaptor proteins (see Section 5.2). Though ion channels and neurons/muscle cells are classically related, many nonexcitable cells such as immune cells express Kv channels (reviewed in Section 3), where they regulate various functions such as migration, cytokine production, Ca^{2+} signaling, and proliferation. During the past two to three decades a large body of information was collected on the interaction of cholesterol/lipids and transmembrane proteins, including ion channels (Andersen et al., 1999; Lundbaek, Birn, Girshman, Hansen, & Andersen, 1996). Alanine-scanning mutational analysis revealed that the S1–S4 segments are in contact with the lipid molecules, and S5–S6 and the pore region are surrounded by the remaining segments and isolated from the lipids (Hong & Miller, 2000; Li-Smerin & Swartz, 2001; Monks, Needleman, & Miller, 1999). Proper functioning of ion channels was found to require the presence of an annular lipid ring firmly attached to the channel protein (Lee, Lee, Chen, & MacKinnon, 2005). Furthermore, many Kv channels were shown to have nonrandom distribution in the membrane, they are localized to noncaveolin lipid rafts or caveolae, and others are excluded from the rafts by mechanisms currently not yet known (Martens et al., 2000; Martens, Sakamoto, Sullivan, Grobaski, & Tamkun, 2001; Panyi et al., 2003; Perez-Verdaguer et al., 2016).

There exist two main theories on how cholesterol modification can modify operation of membrane proteins and ion channels. According to the first, the change in the lipid composition alters the physical–chemical parameters of the membrane (fluidity, stiffness), that generates an increased friction at the interface of the lipids and proteins, and retards the conformational changes (Lundbaek & Andersen, 1994, 1999; Lundbaek et al., 1996). The other model explains the changes in the operation of the channel protein with specific binding sites: the membrane proteins have specific hydrophobic pockets, which can modulate the transition between different states upon cholesterol binding (Fantini & Barrantes, 2013; Rosenhouse-Dantsker, Noskov, Durdagi, Logothetis, & Levitan, 2013). However, these two interactions cannot be separated, usually both influence the biophysical features of the ion channel, and only their relative weight is different from a channel to another.

In this chapter, first we will review the biophysical properties and the physiological role of voltage-gated potassium channels, especially those expressed in nonexcitable cells. Then we summarize the present knowledge obtained with an electrophysiological method (patch-clamp technique) on Kv channels upon in vitro cholesterol level modification. We report in detail about the localization of Kv channels to rafts, and how the ion channel-associated adaptor proteins (with PDZ domain) can interact with cholesterol and lipids. Finally, a review on ion channel-related consequences of sterol metabolism (hypercholesterolemia and SLOS: 7DHC accumulation) will be presented based on our latest results.

2. GENERAL STRUCTURE OF VOLTAGE-GATED K^+ (Kv) CHANNELS

The product of the genes encoding Kv channels is a channel subunit (Kv1.x–Kv12.x), four of which assemble to form a functional channel. These proteins can be divided into two large classes: the conducting and the nonconducting (silent) channels. Conducting tetrameric channels are formed by the Kv1.x, Kv2.x, Kv3.x, Kv4.x, Kv7.x, Kv10.x, Kv11.x, and Kv12.x subunits, while the Kv5.x, Kv6.x, Kv8.x, and Kv9.x proteins are not capable of this (Gutman et al., 2005). These latter subunits form heterotetramers with Kv2.x subunits to create functional channels of various properties. Different types of Kv channels can be expressed by a particular cell, which may be formed by homo- or heterotetramers of subunits belonging to the same family (e.g., Kv1.x), which results in the variability of the biophysical parameters. Variability in Kv channel function is further enhanced by other factors. Auxiliary subunits may attach to the pore-forming α-subunits such as Kvβ, KCHIP (K^+ channel-interacting protein), KCNE, or calmodulin proteins (Pourrier, Schram, & Nattel, 2003). In addition, the aforementioned silent α-subunits, alternative splicing, and post-translational modifications also contribute to Kv channel function diversity (Jan et al., n.d.).

The opening probability of Kv channels depends on the membrane potential. Increasing depolarization opens an increasing fraction of the channels, and the resultant K^+ efflux causes hyperpolarization establishing a negative feedback loop, which leads to channel closing. During prolonged depolarization some Kv channels enter a nonconducting inactivated state, whose extent and rate of development show great variability. Classically Kv channels are divided into two groups based on the kinetic properties of the current they pass: A-type (fast inactivating) and delayed rectifier (slowly or noninactivating) channels. Channels passing typical A-type currents include Kv1.4, Kv3.3, Kv3.4, Kv4.1, Kv4.2, and Kv4.3, which are usually expressed by excitable cells: neurons, muscle, and cardiac muscle cells.

The delayed rectifier channels do not inactivate in the millisecond time range, and as a result, their active period is delayed compared to the voltage-gated Na^+ channels. As such, their main role in excitable cells is the shaping of the repolarization phase of action potentials (Gonzalez et al., 2012; Hodgkin & Huxley, 1952). The typical delayed rectifier channels are Kv1.1, Kv1.2, Kv1.3, Kv1.5, Kv1.6, Kv1.7, Kv1.8, Kv2.x, and Kv7.x.

2.1 Structural Features and Gating of Kv Channels

The voltage-gated K^+ channel of *Drosophila melanogaster*, dubbed *Shaker*, has supplied a vast amount of information about the structure and function of Kv channels. *Shaker* is a close relative of mammalian Kv1.x channels (accordingly named *Shaker* family) and is the most often applied model system, since it is very easy to work with and expresses very well in various expression systems (Tanouye, Ferrus, & Fujita, 1981). Its name originates from the fact that the legs of the *Drosophila* shake uncontrollably during ether anesthesia if the animal carries certain mutant versions of the channel, which modify action potential properties.

The determination of the X-ray crystal structure of the Kv1.2 channel in 2005 by the MacKinnon group (Long, Campbell, & MacKinnon, 2005a) was a great advancement in understanding Kv channel structure, based on which a very detailed model of Kv channels has been developed by today. The tetrameric channel is held together by noncovalent bonds among the subunits (MacKinnon, 1991). Each subunit is composed of six transmembrane α-helical segments and the intra- and extracellular loops connecting these segments. The pore domain (PD) that is responsible for ion conduction is formed by the fifth and sixth transmembrane segments (S5 and S6) and the extracellular loop between them (Doyle et al., 1998). This is the region of the channel where most peptide and small-molecule blockers of Kv channels bind to prevent ion conduction. The pore region contains the selectivity filter that confers Kv channels their high K^+ selectivity. The filter sequence (TVGYG) is highly conserved across K^+ channels with only a few exceptions (e.g., the SVGFG sequence of Kv11.1 and minor differences in two-pore domain potassium channels). The selectivity filter is the narrowest part of the pore where K^+ ions pass in a single file and where the strong selectivity over other ions is achieved. The electronegative carbonyl oxygens of the filter residues line the pore and replace the water molecules lost during the dehydration of K^+ ions leaving the solution. The carbonyl oxygens arranged in a well-defined geometry create binding sites for K^+ ions whose entry into the filter will thus be energetically favorable despite losing their hydrate shell. This explains why Na^+ ions that are smaller than K^+ ions do not significantly permeate through the filter: the carbonyl oxygens "designed" for the size of K^+ ions are too far to coordinate Na^+ ions energetically efficiently. Based on the crystal structures of voltage-gated and the pH-gated bacterial KcsA K^+ channels the pore contains four sequential K^+-binding sites, and the existence of an extracellular rehydration/dehydration site is also suggested (Doyle et al., 1998; Long, Campbell, & MacKinnon, 2005b). During ion permeation, two of the four sites are likely to be occupied and the electrostatic repulsion between the dehydrated ions accomplishes the very high passage rate (10^7–10^8 ions/s) while preserving high selectivity.

At its intracellular end, the selectivity filter widens into a water-filled cavity where K^+ ions are still hydrated. The cavity is closed off by the activation gate at the intracellular side. The activation gate is formed by the crossing of the S6 segments lining the intracellular end of the pore. A highly conserved Pro-Val-Pro motif creates a flexible region, a hinge, where the opening and closing of the activation gate occur as a kink is created in the S6 segments during gating (Bezanilla, 2008).

Kv channels open in response to membrane depolarization, and since many of them are involved in shaping action potentials, their voltage dependence must be steep to enable operation within a narrow membrane potential range and fast gating. Changes in membrane potential are detected by the voltage-sensor domains (VSDs) of the channel, which are made up of the S1–S4 helices in each subunit. Within the VSDs the S4 segments have a special role, as they carry several positively charged amino acid side chains, whose movement is induced by membrane depolarization. The moving charges of S4 generate the measurable "gating current," whose kinetics characterizes the speed of the conformational change associated with gating, and whose integral, the "gating charge," is proportional to the amount of charge moved and thus with the number of gating channels. The conformational change of the VSDs induced by depolarization is transmitted to the PD via a coupling mechanism resulting in the opening of the gate (Chen, Mitcheson, Tristani-Firouzi, Lin, & Sanguinetti, 2001). Based on the measured ionic and gating currents with various voltage protocols, different gating schemes had been suggested over the years to describe the steps of activation, which were comprehensively compared by Zagotta, Hoshi, and Aldrich (1994) on the *Shaker* channel. Comparing their recorded data to simulations based on the gating schemes, they found that the model in which all VSDs must complete two sequential conformational changes to reach the activated state gave the best match. In the model, pore opening was only allowed after all four VSDs had reached the activated state. The

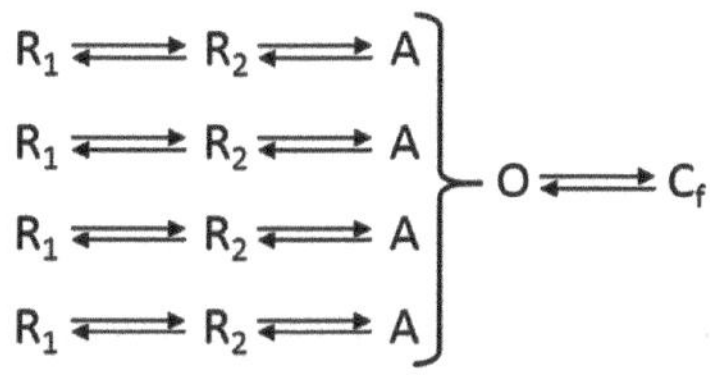

Fig. 1 Scheme of "obligatory" Kv channel gating with tight VSD–PD coupling. The model assumes two sequential voltage-sensor rearrangements in each subunit before the channel pore can open. R_1 and R_2 represent the resting and partially activated states of the VSD, while A represents the fully activated state. Transitions among these states are strongly voltage dependent. The channel pore can only open after the full activation of all four VSDs. A closed state beyond the open state is responsible for fast flickering between the two states in the activated state of the VSDs.

model contains a relatively unstable closed state, which can be reached from the open state. Due to the negligible voltage dependence of this transition, it is likely to cause the fast flipping of the channel between the two states during the activated state of the VSDs (Fig. 1).

The model based on the Kv1.2 crystal structure suggests that the VSD–PD coupling is a direct and rigid link via the continuous polypeptide chain of the S4–S5 linker region (Long et al., 2005b). However, more recent results indicate that this coupling may be much looser and opening of the pore can occur even before the activation of all four VSDs (Osteen et al., 2012). According to this allosteric model, the activation of each VSD increases the probability of pore opening (by a factor of L in the model later), but the activated state of all four is not an absolute requirement (Fig. 2).

Zagotta et al. (1994) examined the suitability of this model to describe *Shaker* gating, but found it inappropriate. This suggests that Kv channels of different families may have different coupling mechanisms: while *Shaker* follows a linear or deterministic scheme, others, such as Kv7.1, are activated by an allosteric mechanism. Rather than the direct rigid VSD–PD link, coupling may require the presence of lipids, or it may not even require the continuity of the polypeptide chain to accomplish gating (Lorinczi et al., 2015; Zaydman et al., 2013), so the great diversity of gating models is not surprising.

Long depolarizations drive many Kv channels into a nonconducting inactivated state, which occurs on the intracellular side by fast N-type inactivation in channels passing A-type currents. This mechanism is similar to the inactivation of the fast-inactivating voltage-gated Na$^+$ channels that is accomplished by the "ball and chain" mechanism. In this, residues at the

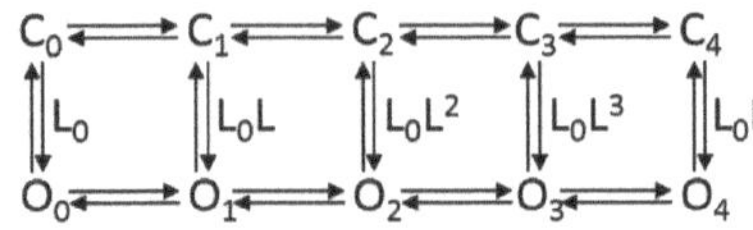

Fig. 2 Scheme of allosteric gating with loose VSD–PD coupling. C_0 represents the closed state with all VSDs in the resting position, and each sequential closed state is reached by the activation of another VSD. In C_4 all VSDs are in the activated state, but the pore is closed. The O_i states represent the same VSD positions but with the pore in the conducting open state. The rate of pore opening with all resting VSDs is L_0, which is increased by a factor of L with each additional VSD activation. Thus, pore opening is possible with any number of activated VSDs, but its probability is increased by each activated VSD.

N-terminal of Kv channels form a ball-like structure tethered to the rest of the channel by a stretch of other residues ("chain"). During depolarization and following the opening of the activation gate the ball binds into the pore via electrostatic interactions, thereby blocking the flow of K^+ ions (Hoshi, Zagotta, & Aldrich, 1990).

Another inactivation mechanism, called the C-type, exists in many Kv channels, which blocks ion flow in the selectivity filter at the extracellular end of the pore (Hoshi, Zagotta, & Aldrich, 1991). The movement of the VSDs induced by depolarization rearranges the hydrogen-bond network of the residues in the selectivity filter, which stabilizes the filter in a state wherein the outermost K^+-binding site is widened. The site widened this way is no longer able to complex the K^+ ion and consequently conduction ceases, and the C-inactivated state is formed. In the absence of K^+ ions the entire filter region widens due to the repulsion of the carbonyl oxygens and the channel is able to conduct Na^+ or other larger cations (Hoshi & Armstrong, 2013). Since the rate of C-type inactivation is determined by the occupancy of the outermost K^+-binding site, any factor influencing this occupancy can modify inactivation kinetics. Such factors include the amino acid residues near the pore entrance, the composition of the solutions (primarily the K^+ concentration) and their pH, and the presence of channel blockers (Lopez-Barneo, Hoshi, Heinemann, & Aldrich, 1993).

3. FUNCTION OF Kv1.3: AN EXAMPLE FOR THE ROLE OF VOLTAGE-GATED ION CHANNELS IN NONEXCITABLE CELLS

The activity of voltage-gated K^+ channels regulates the permeability of the membrane for K^+ ions, thereby influencing the resting potential of cells, the frequency and duration of action potentials in excitable tissue (Dodson & Forsythe, 2004), or membrane potential-driven processes of nonexcitable cells such as proliferation and progression of the cell cycle (Pardo et al., 1999). Generally, the physiological function of voltage-gated K^+ channels is determined by several parameters including the biophysics of channel gating, the single-channel conductance, the specific expression level in a given cell, and the regulation of all these parameters by interacting partners such as auxiliary subunits, and ion channel modulation by transmembrane and intracellular signaling events (phosphorylation/dephosphorylation) and by the physical–chemical environment in the membrane.

This chapter focuses on one specific ion channel, the Shaker-related voltage-gated Kv1.3 channel. This channel is expressed in many tissues, including neurons (Colley, Tucker, & Fadool, 2004), cancer cells (Pardo et al., 1999), and dendritic cells (Zsiros et al., 2009); however, its most significant physiological role was described in the regulation of antigen-dependent activation of T cells (recently reviewed in Feske, Skolnik, & Prakriya, 2012; Feske, Wulff, & Skolnik, 2015).

Kv1.3 is the dominant voltage-gated K^+ channel in human T cells, and its expression in these cells was described by Matteson and Deutsch (1984) and Decoursey, Chandy, Gupta, and Cahalan (1984). This channel is encoded by the KCNA3 gene (Attali et al., 1992) and belongs to the Shaker family of voltage-gated K^+ channels. Accordingly, the structure and gating of this channel conform to the properties of Shaker-related Kv channels described in Section 2. It is generally accepted that the main physiological function of Kv1.3 in T cells is controlling the membrane potential of the T cells during the initial steps of T-cell activation following presentation of the antigens (reviewed in Cahalan, Wulff, & Chandy, 2001; Panyi, Varga, & Gaspar, 2004).

The following properties of this channel and the T-cell membrane are the most important regarding membrane potential control (Fig. 3):

1. *Voltage dependence of steady-state activation and inactivation*: The channel is activated by depolarization of the membrane, the activation threshold of the channels is close to the resting potential of T cells, and the open probability increases very steeply with depolarizations exceeding the activation threshold (Cahalan, Chandy, Decoursey, & Gupta, 1985). At the same time, the voltage dependence of steady-state inactivation of the channels, i.e., the fraction of channels not accessible for opening at a given membrane potential, is not complete at the resting potential of

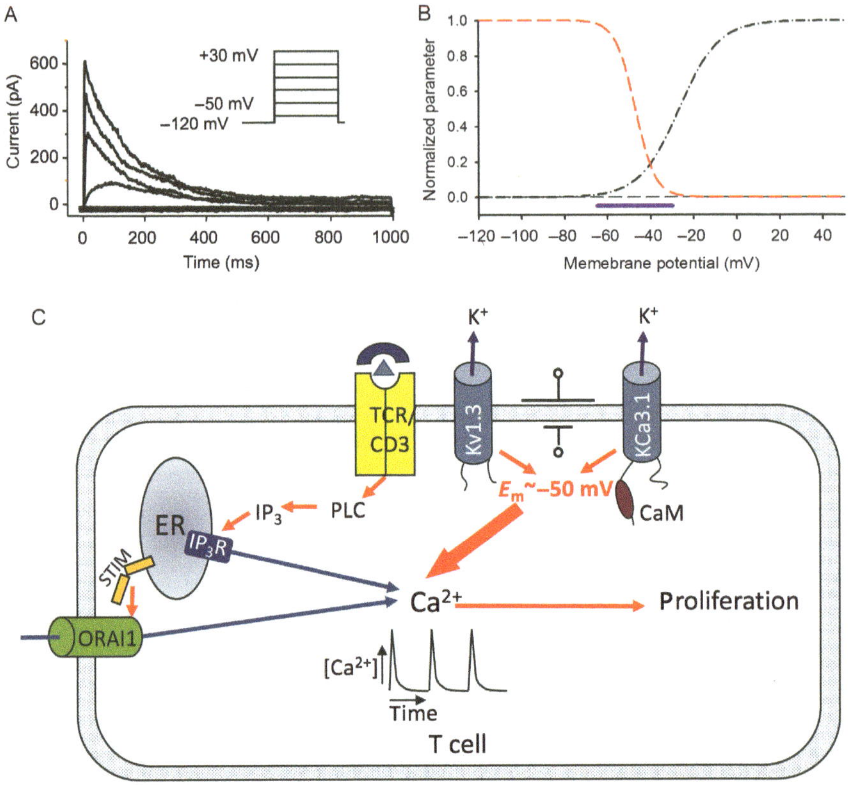

Fig. 3 Properties of Kv1.3 and its role in lymphocyte activation. (A) Whole-cell Kv1.3 currents recorded in a human T cell. The cell was depolarized from a holding potential of −120 mV to different test potentials ranging from −70 to +30 mV in 20 mV steps every 60 s, as indicated in the inset. The activation kinetics of the current progressively becomes faster with increasing depolarizations, and increasingly larger currents are recorded at larger depolarizations. The slow decay following the peak current is the consequence of C-type inactivation. (B) Voltage dependence of steady-state activation (*dash-dot line*) and steady-state inactivation (*dashed line*) of Kv1.3. See text for details. The continuous lines were drawn based on the data by Hajdu, Varga, Pieri, Panyi, and Gaspar (2003). The *solid bar* indicates the membrane potential range where Kv1.3 channels can be open at steady state. (C) Role of cation channels in T-cell activation. Stimulation of the T-cell receptor complex (TCR/CD3) with antigenic peptide (*triangle*) results in the activation of phospholipase Cγ1 (PLC) and the generation of inositol 1,4,5-trisphosphate (IP_3). IP_3 binds to the IP_3 receptor (IP_3R) on the surface of the endoplasmic reticulum (ER) and releases Ca^{2+}. Emptying the ER Ca^{2+} store induces a conformational change in STIM which subsequently activates the ORAI channel located in the plasma membrane which in turn leads to Ca^{2+} influx. Ca^{2+} influx requires a concerted interplay between Ca^{2+} and K^+ channels (Kv1.3 and KCa3.1), the former providing the influx pathway for Ca^{2+} and the latter maintaining a negative membrane potential, which generates the driving force for Ca^{2+} entry. The resulting Ca^{2+} signal is an oscillation of the cytosolic free Ca^{2+} concentration of variable amplitude and frequency (Lewis, 2001) which drives T-cell proliferation. *Red arrows* show regulatory interactions, *blue arrows* show ion fluxes, *CaM*, calmodulin; E_m, membrane potential.

intact T cells (Pahapill & Schlichter, 1992; Verheugen, Vijverberg, Oortgiesen, & Cahalan, 1995). The combination of these results in a window of membrane potentials at which voltage-gated K^+ channels can be open at steady state, and therefore, K^+ fluxes through these channels can contribute to the maintenance of the resting membrane potential.

2. *Inactivation kinetics:* The process of inactivation limits the amount of K^+ current available for membrane potential control. Of the two inactivation processes described in Section 2.1, Kv1.3 inactivates exclusively by the C-type inactivation mechanism (Levy, Panyi, Sheng, Tu, & Deutsch, 1995) as the channel lacks the N-terminal inactivation particle. Therefore, the kinetics of Kv1.3 inactivation is relatively slow (time constant for inactivation, $\tau_i = 200$ ms as opposed to the few milliseconds characteristic for N-type inactivation), allowing significant K^+ efflux through activated channels before entering into the nonconducting state. This also contributes to an efficient control (clamp) of the T-cell membrane potential.
3. *Input resistance:* The lymphocyte membrane has extremely high electrical resistance (10–20 GΩ). Therefore, miniature currents carried by a few open ion channels influence significantly the membrane potential of such cells. Indeed, fluctuation of the resting membrane potential in response to opening of individual ion channels was reported by Maltsev using current clamp (Maltsev, 1990), and by Verheugen et al. who determined resting potential fluctuation from the changes in the reversal potential of the current in cell-attached patch configuration (Verheugen et al., 1995).

In summary, the biophysical properties of Kv1.3 channels and those of the lymphocyte membrane provide the basis for effective clamping of the membrane potential close to the activation threshold of the channels. Accordingly, transfection of the Kv1.3 channel gene alone into CHO cells was sufficient to shift the resting membrane potential to ~−50 mV, as compared to the −5 to −20 mV membrane potential measured in nontransfected cells (Defarias, Stevens, & Leonard, 1995). One main topic of this chapter is how the biophysical parameters of Kv1.3 gating are affected by the modulation of the sterol content of the membrane and how this modulation may result in altered physiological functions of the channel. Therefore, other factors influencing the activity of Kv1.3 (phosphorylation, etc.; see above) are not discussed here, but have been extensively reviewed elsewhere (Panyi, Varga, et al., 2004).

3.1 Kv1.3 and T-Cell Activation

A recent review has discussed 14 ion channels and cation transporters that mediate transmembrane ion fluxes in T cells (Feske et al., 2015). Of these, three ion channels are considered to be the main determinants of the membrane potential regulation of T cells at rest as well as upon antigen stimulation. These include the voltage-gated Kv1.3 and the intermediate conductance Ca^{2+}-activated K^{+} channels (KCa3.1) as well as the calcium release-activated Ca^{2+} channels (CRAC). To understand the physiological role of these channels, we present here a simplified model of T-cell activation during an immune response (reviewed extensively in Cahalan & Chandy, 2009; Cahalan et al., 2001; Feske et al., 2012; Panyi, Varga, et al., 2004).

Professional antigen-presenting cells (APCs) present processed peptides to T cells via MHC molecules in their plasma membrane. The binding of the peptide-loaded MHC molecules to the T-cell receptor (TCR)/CD3 complex and the coreceptors (CD4 or CD8) induces the clustering of signaling molecules: this brings kinases and their substrates into molecular proximity allowing for their efficient operation. The enzyme phospholipase C-γ (PLC-γ1) is recruited in this complex, and following its activation cleaves the membrane-associated phosphatidylinositol 4,5-biphosphate (PIP_2) into 1,2-diacylglycerol and inositol 1,4,5-trisphosphate (IP_3), which diffuses freely in the cytosol. IP_3 then binds to its receptors in the endoplasmic reticulum (ER), which results in the release of Ca^{2+} into the cytosol and a significant rise in $[Ca^{2+}]_i$. This transient rise in $[Ca^{2+}]_i$ is followed by a sustained Ca^{2+} signal which is necessary for initiation of T-cell proliferation. The sustained Ca^{2+} signal is caused by influx of extracellular Ca^{2+} through CRAC channels located in the plasma membrane. Functional CRAC channels are assembled from the ER calcium sensor STIM1/STIM2, which is translocated to the plasma membrane–ER junctions upon emptying of the stores, and the plasma membrane channel ORAI1. Once the channels are opened, the magnitude of Ca^{2+} influx is determined by the electrochemical driving force for Ca^{2+}. It is evident that the depolarizing effect of Ca^{2+}-influx needs to be compensated if the driving force for further Ca^{2+} entry is to be kept high. This counterbalancing K^{+} efflux is mediated by the voltage-gated Kv1.3 and the Ca^{2+}-activated KCa3.1 K^{+} channels. These two channels are activated by Ca^{2+} influx via a feed-forward mechanism: Ca^{2+} entry depolarizes T cells to potentials more positive than the opening threshold of Kv1.3 channels, while the increased $[Ca^{2+}]_i$ directly opens the Ca^{2+}-activated KCa3.1 channels. The concomitant K^+ efflux restores and stabilizes the sufficiently hyperpolarized membrane potential providing optimal conditions for further Ca^{2+} entry and driving the proliferation of the T cells specific for a given antigen.

If this compensating K^+ efflux is downregulated by ion channel blockers or modulation of the biophysical parameters of channel gating by the sterol content of the membrane, the activation of T cells can be inhibited, as we will show it later in this chapter.

4. LIPID RAFTS AND ION CHANNELS

Lipid rafts are specialized membrane domains with a lipid composition different from the bulk of the membrane. While the definition constantly evolves, some features are generally accepted for lipid rafts: enrichment of cholesterol, sphingolipids, and specific proteins, increased resistance to detergents and a liquid-ordered state of the lipids, in which the acyl chains are stretched out much like in the gel phase while still conserving a fluidity similar to the liquid phase (Simons & Ehehalt, 2002). A wide size range has been reported for these microdomains, typically citing 10–200 nm (Pike, 2006); however, their highly dynamic nature precludes precise size determination. Larger platforms can be formed from these nanoscale fluctuating domains, which assemble molecular participants required for specific cellular functions (Lingwood & Simons, 2010). Thus, the main proposed function of lipid rafts is to provide a means for the lateral organization of signaling molecules by controlling their mobility, proximity, and trafficking. A special type of lipid rafts is the caveolae, invaginations of the plasma membrane shaped by caveolin proteins, which are also involved in signal transduction and endocytosis.

Ion channels, as membrane proteins and participants in signal transduction pathways, can also interact with lipid rafts, which may consequently influence their function. There have been several reports about Kv channels being associated with or excluded from rafts. Cyclodextrins, which are cyclic oligosaccharides having a hydrophilic external surface and a hydrophobic internal cavity, are routinely used to manipulate the cholesterol content of the plasma membrane. Most often methyl-β-cyclodextrin (MβCD) is used, which can extract cholesterol selectively over other lipids from the plasma membrane (Gimpl, Klein, Reilander, & Fahrenholz, 1995; Irie, Fukunaga, & Pitha, 1992; Klein, Gimpl, & Fahrenholz, 1995). However, when employed as preloaded with cholesterol or its analogues, it can also specifically enrich the sterol content of the membrane. This provides a tool to disrupt or rearrange the raft structure of membranes and observe its effect on ion channel function. However, it must be noted that cholesterol extraction by cyclodextrins is not selective for lipid rafts (Mahammad & Parmryd, 2008), so altered channel function cannot be strictly regarded as a consequence of a change in raft structure.

The earliest reports on raft association of Kv channels demonstrated that Kv2.1 is targeted to noncaveolar lipid rafts, while Kv1.5 targeted to caveolar lipid rafts. In addition, cholesterol depletion altered raft composition and modified the gating of Kv2.1 (Martens et al., 2000, 2001). Later, a role for caveolin in regulating Kv1.5 trafficking to cholesterol-rich rafts was demonstrated (McEwen, Li, Jackson, Jenkins, & Martens, 2008). Other studies found that Kv1.5 behaved differently when overexpressed in heterologous systems compared to that in native cells. In the former case, Kv1.5 tended to localize to rafts, while in the latter it appeared not to target to these microdomains (Felipe, Soler, & Comes, 2010). In leukocytes, the tendency of Kv1.3/1.5 heteromers to localize to rafts depended on the stoichiometry of the subunits with higher Kv1.3 ratios favoring raft targeting, while higher Kv1.5 ratios favor exclusion of the channels from rafts. The variable behavior of Kv1.5 is likely to be affected by the association of auxiliary subunits.

Although reports on Kv1.3 found contradicting effects of cholesterol depletion, all studies agreed that Kv1.3 localizes to rafts and that disruption of rafts alters Kv1.3 gating (Bock, Szabo, Gamper, Adams, & Gulbins, 2003; Hajdu, Varga, Pieri, Panyi, & Gaspar, 2003; Pottosin, Valencia-Cruz, Bonales-Alatorre, Shabala, & Dobrovinskaya, 2007). Interestingly, recently a domain interacting with caveolin-1 was identified on Kv1.3, suggesting its targeting to caveolar rafts, which appeared very sensitive to various factors and thus regulate the surface localization of the channel (Perez-Verdaguer et al., 2016).

Originally, Kv4.2 was found to be excluded from rafts (Martens et al., 2000), but another study showed a fraction of Kv4.2 channels to be raft associated, similar to Kv1.4 (Wong & Schlichter, 2004). The scaffolding protein PSD95 aided the raft association of Kv1.4, but not that of Kv4.2.

Confirming this finding, more recently a partial distribution of Kv4.2 and its auxiliary subunit KChiP2 to rafts was described. MBCD treatment caused a redistribution of channels out of rafts and reduced the current carried by the channels (Rudakova, Wagner, Frank, & Volk, 2015). Examination of raft association of Kv channels in pancreatic cells revealed that in α-cells Kv4.1/4.3 channels, while in β-cells Kv2.1, but not Kv1.4, channels target to rafts, and in both cases depletion of cholesterol by MBCD decreased raft association of the channels (Xia et al., 2004, 2007).

Association of the members of the Eag family to lipid rafts was also demonstrated. Kv10.1 was found to be distributed between raft and non-raft membrane regions at various ratios, which was influenced by cholesterol and the cytoskeleton. Cholesterol depletion increased current density of Kv10.1 channels without affecting other aspects of gating (Jimenez-Garduno et al., 2014). Both Kv11.1 and its regulatory subunit KCNE2 were also detected in rafts and channel gating was modified by both cholesterol depletion and loading of the membrane (Balijepalli et al., 2007).

Although certain channels appear more or less prone to target to rafts, these highly variable and sometimes conflicting data indicate that the raft association of a particular Kv channel is likely to be cell specific. Various factors, such as the precise lipid composition, including cholesterol content, the presence of scaffolding proteins, and auxiliary subunits, influence the affinity of the channel for rafts. For example, the distribution and clustering of Kv2.1 and Kv1.4 channels were significantly different even between atrial and ventricular myocytes (O'Connell, Whitesell, & Tamkun, 2008). In addition to these factors, depletion and loading of cholesterol may have a direct effect on Kv channels via cholesterol-binding sites, not only through the disruption or assembly of lipid rafts.

4.1 Ion Channels in the Immunological Synapse

A special lipid microenvironment for Kv1.3 and other channels involved in the regulation of T-cell activation can be the immunological synapse (IS). This signaling platform is dynamically assembled upon presentation of the antigen to the T cell by professional APCs. The IS signaling platform contains TCR/CD3 complex along with the auxiliary, signal transducer and scaffolding proteins (extensively reviewed in Bromley et al., 2001; Davis,

2002; Dustin, 2014). Lipid rafts are found to accumulate at the IS, as was reported by staining the cells with cholera toxin B subunit (CTB), which binds to ganglioside M1 (GM1), a marker of raft membranes (Janes, Ley, & Magee, 1999; Janes, Ley, Magee, & Kabouridis, 2000; Viola, Schroeder, Sakakibara, & Lanzavecchia, 1999). High plasma membrane lipid order in the IS was also found using a phase-sensitive membrane dye together with a variety of spectrally resolved microscopy techniques (Owen et al., 2010). We have shown that Kv1.3 is colocalized with the TCR/CD3 complex (Panyi et al., 2003) and recruited into the IS formed between cytotoxic T cells and target cells (Panyi, Vamosi, et al., 2004). Later the KCa3.1 and CRAC channels (STIM/ORAI) were also shown to be localized in the IS (Lioudyno et al., 2008; Nicolaou, Neumeier, Peng, Devor, & Conforti, 2007). Thus, it seems that the most important ion channels regulating T-cell activation are dynamically associated with the IS. Furthermore the colocalization of Kv1.3 was shown with CTB-labeled lipid rafts using confocal microscopy and fluorescence cross-correlation (Panyi, Vamosi, et al., 2004). Considering the very unique microenvironment in the IS (both protein and lipid), one might imagine that the function of the channels would be dramatically altered upon recruitment into the IS. Unfortunately, very limited information is available about this in the literature due to the difficulty inherent with the experimental system using conventional electrophysiology: one cannot separately study the function of the channels in the IS since the APC impedes access of the pipette to the IS and use of the cell-attached patch recording. One indication that the function of Kv1.3 might be altered upon recruitment to the IS, however, came from our laboratory (Toth, Szilagyi, Krasznai, Panyi, & Hajdu, 2009). We performed whole-cell patch-clamp on mouse cytotoxic T cells either alone or following the formation of the IS (conjugated cell). We found that the activation kinetics of the Kv1.3 current was slower, whereas the inactivation kinetics was faster in cells conjugated in IS as compared to the standalone cells. In addition, the voltage dependence of steady-state activation was shifted toward the depolarizing potentials in T cells engaged in IS. As the IS is also a signaling platform, we separated the postsynthetic modification of the channel by phosphorylation using kinase inhibitors. We concluded that modulation of the inactivation kinetics is rather related to regulation of the channel by phosphorylation, whereas the change in the activation kinetics upon IS formation may be attributed to redistribution of the channels into distinct membrane domains.

5. ELECTROPHYSIOLOGY OF Kv1.X CHANNELS UPON MEMBRANE CHOLESTEROL MANIPULATION

We described in Section 3 that Kv1.3 channels are critical for the activation and proliferation of T lymphocytes. The change in the function of Kv1.3 can lead to inadequate Ca^{2+} signaling and prevention of the myriad subsequent steps required for an adaptive immune response. On the other hand, the relatively easy access to the human peripheral T cells makes them an attractive model for studying the effect of various factors on the function of voltage-gated ion channels. Also via monitoring the downstream steps, the physiological consequences of the Kv1.3 modulation can be traced.

In the beginning of the investigation of the role of cholesterol in signaling, the use of various cyclodextrins that make it possible to manipulate the membrane cholesterol level was applied in vitro. First Hajdu et al. applied MβCD and its cholesterol complexed form (MβCD/CHOL) to modify the cholesterol content of the T lymphocyte membrane (Hajdu et al., 2003). One-hour incubation of activated T cells with MβCD/CHOL at concentrations 0.72 and 1 mM could modify the biophysical parameters of the membrane as revealed by fluorescent polarization anisotropy: the increase in the membrane viscosity was observed upon cholesterol loading with MβCD/CHOL, while MβCD treatment had a negligible effect on membrane fluidity (application of higher MβCD/CHOL or MβCD prevented the stable gigaseal formation during patch-clamp recording). Upon loading the cells with cholesterol (0.72 and 1 mM MβCD/CHOL) a remarkable change in the kinetic parameters was detected: the activation of the Kv1.3 current slowed approximately 1.5-fold or more, and the inactivation rate significantly decreased (ca. 50% slower). Furthermore, cholesterol loading shifted the steady-state activation of the Kv1.3 channels toward positive membrane potentials (ca. 7 mV), but the equilibrium inactivation was not changed. It was also reported that cholesterol loading led to a biexponential activation of the Kv1.3 current, which could be explained with presence of an activated-closed state and not with coexistence of two channel populations in different lipid environments. This hypothesis is supported by several facts: (1) the ratio of slow and fast current components is not dependent on the depolarizing potential (shift in steady-state activation is not due to two populations in different lipid milieu) and (2) inactivation kinetics did not exhibit biphasic decay. The cholesterol removal had no influence on the gating properties of Kv1.3 except a slight speedup of the inactivation kinetics. The MβCD treatment did not influence equilibrium parameters of activation and inactivation. The activity of Kv1.3 channels that was characterized by current density (peak current at +50 mV divided by the whole-cell capacitance) increased upon cholesterol depletion (only for 1 mM MβCD), whereas cholesterol loading suppressed channel activity hallmarked by the reduced current density, similar to the reports on the volume-regulated anion channel and the Kir2.1 channel (Levitan, Christian, Tulenko, & Rothblat, 2000; Romanenko et al., 2004). Pottosin et al. studied the effect of MβCD treatment on the Kv1.3 current in Jurkat T lymphoma cell line (Pottosin et al., 2007). They incubated the cells with 1 or 2 mM MβCD (for 1 h as in Hajdu et al.) and observed a vast reduction in cellular cholesterol level (63% and 75% at 1 and 2 mM, respectively). Cholesterol removal also led to the decrease in the proliferative capacity of Jurkat cells, which partially can be attributed to the collapse of signaling pathways due to the collapse of the lipid raft system. However, they did not report the increase or decrease in the current density of Kv1.3 channels, which could reflect the change in the channel activity. Unlike in T lymphocytes, Kv1.3 channels of Jurkat T cells were affected by the drastic decrease in the cholesterol level: the activation time constant of the whole-cell current increased ca. 2.5-fold after 2 mM but not 1 mM MβCD incubation. Also a right shift in steady-state activation (11 mV) and inactivation (6 mV) curves was reported upon the higher concentration MβCD treatment. Interestingly, they found that after cholesterol depletion with MβCD the kinetic and equilibrium parameters of activation returned to the value equal to that measured prior to treatment. Bock et al. reported that formation of large membrane platforms induced by ceramide addition or stimulating CD95 receptors of Jurkat T cells inhibited Kv1.3 activity (Bock et al., 2003). They also disrupted rafts with two cyclodextrins, β-cyclodextrin and β-methyl-cyclodextrin at 2 mM (60 min incubation) and 5 mM (10 min incubation), respectively, and could observe the Kv1.3 channel activity drop to 60%–75%, as determined from the whole-cell peak current at +70 mV. However, no changes on the biophysical parameters were analyzed or studied.

These three studies clearly demonstrate that the raft-associated Kv1.3 channels are in intimate relationship with the lipid environment. Only one paper reported about the effect of excess cholesterol in the membrane on Kv1.3 channels, and in parallel with previous findings on different ion channels (VRAC, BKCa, Kir) the activity of Kv1.3 was suppressed upon cholesterol elevation, which may be assigned to the altered protein–lipid interaction or the restructuring/reorganization of raft system upon cholesterol treatment (Hajdu et al., 2003). Moreover, the change in the raft structure itself—fusion of small rafts into large platform—can impair the channel activity, which emphasizes the importance of membrane integrity in Kv1.3 channel operation. The removal of cholesterol in Jurkat cells caused more prominent impact on Kv1.3 current than in primary/native T cells (practically no effect); however, we should note that cholesterol removal with MβCD concentration higher than 1 mM resulted in modification of the Kv1.3 current, which vetoed electrophysiological recording in T lymphocytes. The cholesterol depletion in Jurkat E6-1 cells manifested in the change of kinetic and equilibrium parameters of activation and inactivation but not in the current density/amplitude, while the other Jurkat cell line responded with reduced peak current upon cholesterol extraction. This latter discrepancy emphasizes that there could be very distinct outcomes of the same treatments probably due to differences in passage number, cell cycle, adaptor protein expression, etc. On the other hand, differences of Kv1.3 operation obtained upon cholesterol manipulation in primary and cultured cells promote the use of primary cells over cell lines, so that no ambiguous conclusions about the physiological effect are drawn.

Besides Kv1.3, many more Kv channels were shown to target to lipid rafts and their function is modified by cholesterol removal/loading. Martens et al. showed for the first time that Kv2.1 channels expressed in the mouse Ltk$^-$ cell line are localized to noncaveolin lipid rafts (Martens et al., 2000). When cholesterol was depleted with 2-hydroxypropyl-β-cyclodextrin, which destroyed the raft system, they obtained that Kv2.1 steady-state inactivation curve was largely shifted toward positive membrane potentials (ca. 35 mV increase in the half-maximal voltage). Surprisingly, the test potential dependence of equilibrium activation was not affected by cholesterol removal. They also reported that Kv1.5 channels are colocalized with caveolin on the cell surface, in domains distinct from Kv2.1 (Martens et al., 2001). Treatment of the cell membrane with 2-hydroxypropyl-β-cyclodextrin or fumonisin (inhibitor of sphingomyelin biosynthesis) shifted half-maximal voltage of equilibrium activation and inactivation by 10 mV to negative potentials. However, in both studies the biophysical parameters of the Kv4.2 channel, which is not associated with lipid rafts, were not influenced by cholesterol depletion. In contrast, in rat left ventricular myocytes the Kv4.2 channels with their auxiliary subunit KChIP2 are partially localized to lipid rafts. MβCD treatment induced decrease in whole-cell current and redistribution of Kv4.2 channels to nonraft domains. These two latter cases clearly prove that membrane embedded ion channels can respond to cholesterol modification differently/oppositely in native or heterologous expression systems.

The neuronal Kv10.1 or Eag1 was also shown to partition into two subpopulations in mouse neurons and HEK cells: one localized to and the other was excluded from detergent insoluble membranes (Jimenez-Garduno et al., 2014). Cholesterol extraction had multiple effects on Kv10.1: (1) reduced the fraction of raft-targeted Eag1 channels and (2) increased the whole-cell peak current but did not change any other biophysical parameters.

The cardiac Kv11.1 or hERG channels together with Kv7.1 (KvLQT1 or KCNQ1) turned out to be localized to cholesterol- and sphingolipid-enriched membrane regions in both canine ventricular myocytes and HEK293 cells (Balijepalli et al., 2007). Interestingly, a long QT-related mutant of hERG (G601S) is not present in rafts. Furthermore, use of MβCD resulted in a positive shift in the test potential dependence of steady-state activation, and deactivation of the current was also accelerated in HEKs. However, in mouse myocytes only the faster deactivation kinetics was detected upon cholesterol removal. Chun et al. described that increase in the membrane cholesterol of HEK cells decreased the activation rate of hERG channels but had no effect on the deactivation and inactivation kinetics (Chun, Oh, Park, Cho, & Chung, 2013). However, cholesterol did not directly modify hERG channel function: activation of PLCβ1 reduces the level of PIP_2 (phosphatidylinositol-4,5-bisphosphate), which uncovers a cross talk between two lipids, cholesterol and PIP_2, in the regulation of hERG.

5.1 Cholesterol Recognition Sites May Mediate the Sterol–Protein Interactions of Kv1.3

Cholesterol can regulate the operation of ion channels indirectly, either via modifying the physical properties of the cell membrane (fluidity, thickness, viscosity, etc.) or via modulating the composition and assembly of lipid rafts (Levitan, Fang, Rosenhouse-Dantsker, & Romanenko, 2010). In recent years, much evidence has been gathered supporting an alternative regulatory mechanism where specific and direct cholesterol–protein interactions are responsible for the sterol sensitivity of the ion channels (Levitan, Singh, & Rosenhouse-Dantsker, 2014). In line with this, several ion channels have been reported to contain one or more of the most widely studied cholesterol recognition sites, the CRAC and CARC motifs (cholesterol recognition amino acid consensus, CRAC: **L/V**-$X_{1–5}$-**Y**-$X_{1–5}$-**R/K** and CARC:

K/R-X_{1-5}-**Y/F**-X_{1-5}-**L/V**, where X is any amino acid) (Picazo-Juarez et al., 2011; Robinson, Shridar, Smith, & Murrell-Lagnado, 2014; Singh et al., 2012). These sequences follow a very simple linear algorithm where well-defined amino acid residues linked with theoretically unconcerned regions responsible for the cholesterol binding (Fantini & Barrantes, 2013). Care must be taken when considering the significance of these motifs due to the flexible requirements of the algorithms (Epand, 2006). However, it is notable that a simple point mutation or deletion of these selected motifs could totally abolish CHOL sensitivity of ion channels (Fantini & Barrantes, 2013; Robinson et al., 2014). Surprisingly, not just the transmembrane segments but the intracellular C-terminal end of the channels also carries functional cholesterol recognition motifs (Fantini & Barrantes, 2013; Robinson et al., 2014). Screening the amino acid sequence of Kv1.3, we found multiple CRAC and CARC sequences (Fig. 2). Certain motifs overlapped with the conservative parts of the channels like the activation gate or the voltage sensor. Others were mapped on the N- or C-terminal as Singh and his colleagues predicted in a former study (Singh et al., 2012). Mutational analysis of the voltage sensor or activation gate would be hardly interpretable since these regions are very sensitive to any modifications including single-point mutations (Hackos, Chang, & Swartz, 2002; Lee, Banerjee, & MacKinnon, 2009). In contrast, Kv1.3 tolerates the mutations or removal of the entire C-terminus well. We published that the removal of the last 84 amino acid residues (including CARC 4 (K474-Y477-V478) and 5 sequences (K517-K518-F572-V575)) of the C-terminus has no significant effect on the time- and voltage-dependent properties of Kv1.3 gating (Szilagyi, Boratko, Panyi, & Hajdu, 2013). Therefore, we aimed to investigate the biological relevance of the two CARC sequences located on the distal C-terminus using the previously characterized ΔC-Kv1.3 construct. Transiently transfected CHO cells were loaded with 120 μM CHOL. Our results showed that the ΔC-Kv1.3 channels were resistant to the CHOL loading of the cell membrane. In contrary to the wild-type Kv1.3 neither the $V_{1/2}$ nor the activation kinetics of ΔC-Kv1.3 changed after the sterol loading of the membrane. We proposed that the lack of CARC4 and CARC5 motifs is responsible for the abolished sterol sensitivity (Fig. 4). To prove this concept, we designed specific point mutations along the CARC4 and CARC5 sequences where the key amino acid residues were changed to alanine (mCARC4: K526A-Y529A-V531A, mCARC5: K517A-K518A-F572A-V575A) (Fig. 5). As expected, the cholesterol sensitivity of the mutated channels was different from the wild-type Kv1.3. CHOL loading was

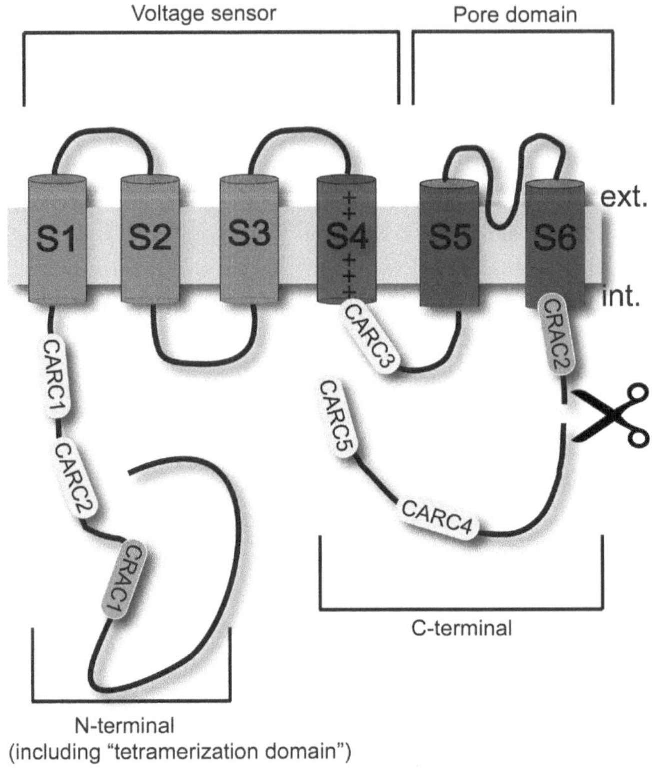

Fig. 4 Schematic representation of the cholesterol recognition motifs identified along the Kv1.3 sequence. Each subunit contains at least seven putative cholesterol recognition sites. Two CARC sequences are located on the distal end of the C-terminus. As we previously demonstrated removal of the last 84 amino acid residues (including the CARC4 and CARC5 sequences) abolishes cholesterol sensitivity of Kv1.3.

ineffective on the activation kinetics of the $mKv1.3_{CARC4}$ and just slightly modified the $V_{1/2}$. In case of the $mKv1.3_{CARC5}$ CHOL loading had modest effect on the $V_{1/2}$ and had negligible influence on the activation kinetics. We propose that CARC motifs of the C-terminal interact with membrane cholesterol of the inner leaflet via their hydrophobic residues and mediate sterol sensitivity of Kv1.3. A similar conclusion was reached by Robinson et al. who envisioned that the C-terminal of the P2X7 ion channel dips back into the plasma membrane and contacts the membrane via palmitoyl groups and hydrophobic residues. Furthermore, in the absence of a well-defined structure of the C-terminus, we cannot exclude the possibility that mutations of the putative cholesterol-binding pockets indirectly affect alternative regulation of the channel activity by cholesterol.

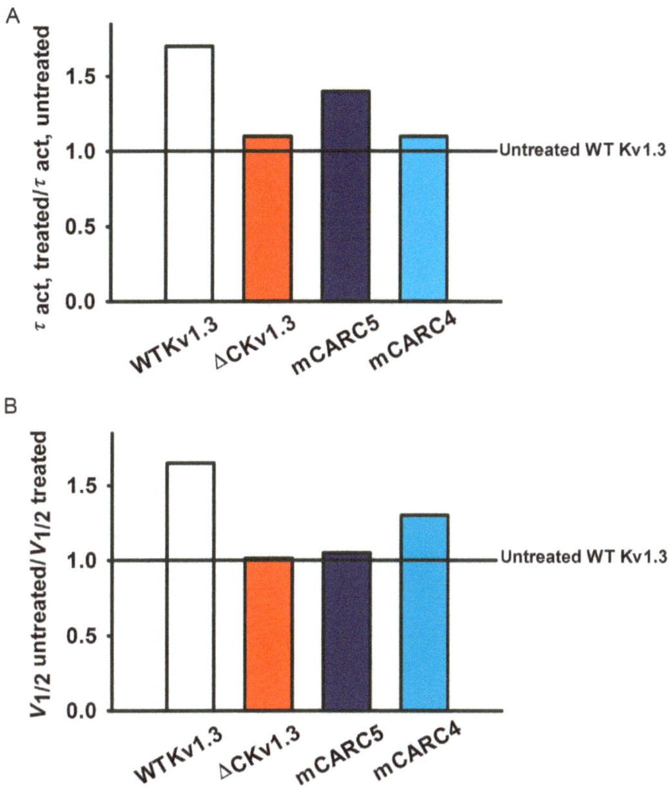

Fig. 5 The effect of 420 μM CHOL loading on the gating properties of various Kv1.3 constructs. CHO cells were transiently transfected with the following constructs: wild-type Kv1.3, ΔCKv1.3 (which lacks the last 84 amino acid residues of the C-terminus), a CARC4 mutant (mCARC4, K526A-Y529A-V531A), and a CARC5 mutant (mCARC5: K517A-K518A-F572A-V575A) and loaded with 420 μM CHOL. To determine the activation time constant and the midpoint of the steady-state activation ($V_{1/2}$) we applied the same protocols as we discussed in Fig. 7. This figure demonstrates the relative change in the activation time constant and $V_{1/2}$ due to the cholesterol loading in case of a particular construct.

5.2 Adaptor Proteins and Cholesterol, a Possible Communication Through Cholesterol Binding

Voltage-gated potassium channels often associate with various adaptor proteins, which can influence their function and localization. These intracellular "anchoring" proteins, first described in neurons, play roles in trafficking to the plasma membrane, facilitating/hindering lateral membrane diffusion, promoting clusterization, or coupling to various signaling proteins (Burke et al., 1999; Gong et al., 1999; Hajdu et al., 2015; Hanada, Lin, Chandy,

Oh, & Chishti, 1997; Jugloff, Khanna, Schlichter, & Jones, 2000; Kim, Niethammer, Rothschild, Jan, & Sheng, 1995; Kim & Sheng, 1996; Kuras et al., 2012; Marks & Fadool, 2007; McCormack et al., 1999; Tiffany et al., 2000). It was unveiled that Kv1.3 coupled to synapse-associated protein 97 (SAP97) or hDLG (human homologue of *Drosophila* disc large tumor suppressor protein) and postsynaptic density protein (PSD95 or SAP90)—two members of MAGUK (membrane-associated guanylate kinase) proteins—in Jurkat T cells (Szilagyi et al., 2013). Both PSD95 and SAP97 possess multiple PDZ domains, which guarantees their interaction with PDZ-binding domain of the C-terminal region. SAP97 is an essential part of Kv1.3-SAP97-Lck signaling complex in the T-cell activation, while PSD95 regulates the localization of Kv1.3 channels to the IS between a T cell and an APC (Hanada et al., 1997; Szilagyi et al., 2013). Cortactin and its orthologue HS1 in T cells, which are actin-binding proteins, can also bind to various ion channels including Kv1.3 where it associates to the SH3-binding domain (PXXP) on the C-terminus. Cortactin hinders the lateral mobility of Kv1.3 upon F-actin polymerization that is a distinctive event occurring in the contact area of a T cell with an APC (Hajdu et al., 2015). HS-1, which also accumulates in the IS on the T-cell side, can facilitate the residency of Kv1.3 in the IS, which also includes their immobilization due to binding to membrane lipids. These proteins can also attract other signaling elements such as PKC, PKA to the channel, which can modify their operation (Gong et al., 1999; Kuras et al., 2012). As PSD95, SAP97, and HS1 redistribute to the IS, they probably have a role in attaching Kv1.3 channels to membrane domains/rafts, probably via specific lipid–protein interactions.

In recent years, several studies were published that uncovered or predicted the interaction of adaptor proteins and membrane lipids (Chen et al., 2012; Epand, 2006; Erlendsson & Madsen, 2015; Listowski, Leluk, Kraszewski, & Sikorski, 2015; Pan et al., 2007; Sheng et al., 2012; Wu et al., 2007; Zimmermann et al., 2002). Cholesterol-binding motifs (cholesterol recognition amino acid consensus, CRAC: **L/V**-X_{1-5}-**Y**-X_{1-5}-**R/K** and CARC: **K/R**-X_{1-5}-**Y/F**-X_{1-5}-**L/V**, where X is any amino acid) can be found in these adaptor proteins, especially within their PDZ domain (Listowski et al., 2015). An intensive computational research project revealed that approximately 2000 PDZ domains have membrane-binding properties, and many of these have an affinity in or below the micromolar range (Chen et al., 2012). The PDZ domain of PSD95, SAP97, PICK1 (protein interacting with C kinase 1), ZO-1 (zonula occludens 1 or tight junction protein-1 (TJP1)), NHERF1 (sodium-hydrogen antiporter 3 regulator 1) and many others was shown to bind model plasma membranes (Sheng et al., 2012). This interaction was selective for cholesterol; i.e., it was weakened by the replacement of cholesterol with other derivatives or upon its depletion from the membrane. Interestingly, it was described that the protein-binding pocket, which is in charge of the interaction with the PDZ-binding domain, and the lipid-interacting sequence are separated. A loss-of-function mutation in the lipid-binding motif of the PICK1 PDZ domain, which associates with the GluR2 subunit of the AMPA receptor, prevented its trafficking to the synapse (Pan et al., 2007). Moreover, the lack of PICK1 interaction with the cell membrane led to the mistargeting of AMPA receptors in neurons.

All the information collected about the interaction between ion channels and adaptor proteins and the cholesterol-binding ability of PDZ domains of adaptor proteins may open a new approach to the understanding of the protein–cholesterol/lipid interaction. To our best knowledge, there is no study, which describes the consequence of adaptor protein–ion channel interactions on cholesterol sensitivity. We suppose these adaptor proteins may participate in anchoring the ion channels to rafts or other cell surface structures, which influence their biophysical parameters.

6. BIOLOGICAL RELEVANCE OF THE CHOLESTEROL SENSITIVITY OF Kv1.3 IN T CELLS

As we discussed earlier, in vitro manipulation of membrane cholesterol altered the operation of various ion channels including the Kv1.3 channels of T cells (Hajdu et al., 2003). These results raised the question whether the in vitro studies have any physiological relevance? Theoretically, human diseases (type 2 diabetes, inborn errors of cholesterol biosynthesis) or pathological conditions (diet-induced plasma hypercholesterolemia), which alter the cholesterol content of the cells, may result in altered ion channel function and a consequent dysfunction of the ion channel-regulated physiological processes. One of the widely accepted methods to study cholesterol-related abnormalities is to induce plasma hypercholesterolemia by feeding rodents or primates on high-cholesterol diet and then isolate the cells of interest, whose sterol profile changed due to the diet. The advantage of this approach is that it mimics the high-cholesterol food intake, which is a frequent cause of the human dyslipidemias. Several animal models were investigated by electrophysiological methods using diet-induced hypercholesterolemia (reviewed by Bukiya & Rosenhouse-Dantsker, 2015). However, useful tools for animal models are to study human-related medical conditions/problems, and they cannot substitute the examination of primary cells/tissues. In contrast to animal models, only limited information is available about the relationship between K^+ channels and cholesterol-related diseases. To gain insight into the relevance of in vitro experiments and animal models, we aimed to determine the function of Kv1.3 channels isolated from patients with hypercholesterolemia or SLOS. It has been published that in hypercholesterolemia the total cholesterol content of the lymphocytes increased (Seres et al., 1996). Our hypothesis was that the increased cholesterol content of the HC cells alters the Kv1.3 function similar to in vitro cholesterol loading (Somodi et al., 2013). Hypercholesterolemic lymphocytes were isolated from two patient groups: the total cholesterol level of group I patients was between 5.2 and 10 mmoL/L ($n=11$), and patients with total cholesterol more than 10 mmoL/L were categorized into group II ($n=9$). Using a patch-clamp technique, we studied the expression level, the kinetic, and the equilibrium parameters of the gating of Kv1.3. Our experiments with primary HC cells had a different outcome compared to in vitro loading. The activation kinetics of the current at 0 mV was significantly faster for groups I and II, while at +50 mV only for group II than in control cells. We suppose that the slightly accelerated activation kinetics have no significant effect on the ion channel-regulated physiological processes. We found that neither the whole-cell peak current of Kv1.3 channels nor the current density changed in hypercholesterolemia. The midpoint and slope of the voltage dependence of steady-state activation were the same for all groups (1.5 mg/mL in vitro CHOL loading significantly decreased the peak current and current density and shifted the midpoint of the voltage dependence of steady-state activation to positive potentials (Hajdu et al., 2003)). We hypothesized that the discrepancy between the effect of HC and cholesterol loading on Kv1.3 operation could be the consequence of the different membrane cholesterol levels of the studied samples. Therefore, we compared the cholesterol content of the control, HC, and cholesterol-loaded cells. We found that the cholesterol content of group II HC cells was approximately 1.5-fold higher than in the normocholesterolemic cells, but it was far below that of the 1.5 mg/mL MbCD/C-treated T cells. It seems that the serum cholesterol level has a moderate effect on the plasma membrane cholesterol level. Therefore, the in vitro cholesterol loading at the applied concentration (1.5 mg/mL) mimics an extreme situation and has no real biological relevance. During the study, we had the opportunity to investigate a patient with extremely high serum cholesterol levels. Analysis of these extreme hypercholesterolemic cells confirmed our conclusion. Similar to the in vitro cholesterol loading, extreme HC cells had decelerated activation kinetics, while the $V_{1/2}$ of steady-state activation shifted to the right (Fig. 6).

SLOS is a rare inherited disease due to the decreased or abolished activity of the 7-dehydrocholesterol reductase enzyme which catalyzes the last step of the endogenous cholesterol biosynthesis (Porter, 2008; Porter & Herman, 2011). The leading clinical symptoms are neurological and anatomical abnormalities, but almost every organ and tissue can be involved (Porter, 2008; Smith, Lemli, & Opitz, 1964). As a result of the enzyme block, the precursor 7-dehydrocholesterol and its derivatives accumulate in the cell membrane (Irons et al., 1993). The increased 7DHC/CHOL ratio may alter the physical properties of the cell membrane and/or modulate the specific sterol–protein interactions. Investigation of the ion channels of SLOS cells can be a useful tool to better understand the effects of sterols on the operation of ion channels. Even so, only limited information is available about the ion channels and transporters in SLOS (Balajthy et al., 2016; Lindegaard et al., 2008; Ren et al., 2011; Tulenko et al., 2006). To demonstrate the significance of optimal sterol composition for proper ion channel operation, we studied the Kv1.3 channels of lymphocytes isolated from eight Hungarian SLOS patients. Our results showed that SLOS significantly decelerates the activation kinetics and shifts the midpoint of voltage dependence of

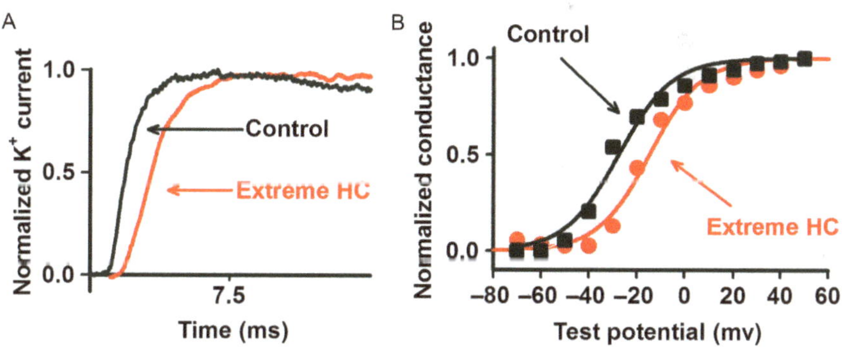

Fig. 6 Effects of the extreme HC on the operation Kv1.3. (A) Representative activation curves of a control and an extreme HC cell after a 15-ms-long depolarizing pulse. As (A) demonstrates, extreme HC decelerates the activation kinetics of Kv1.3. Voltage dependence of equilibrium activation of Kv1.3 channels in an extreme HC (*circle*) and a healthy (*squares*) T cell along with the best-fit Boltzmann functions are shown (B). The midpoint of the steady-state activation shifted toward to the positive membrane potentials in extreme HC.

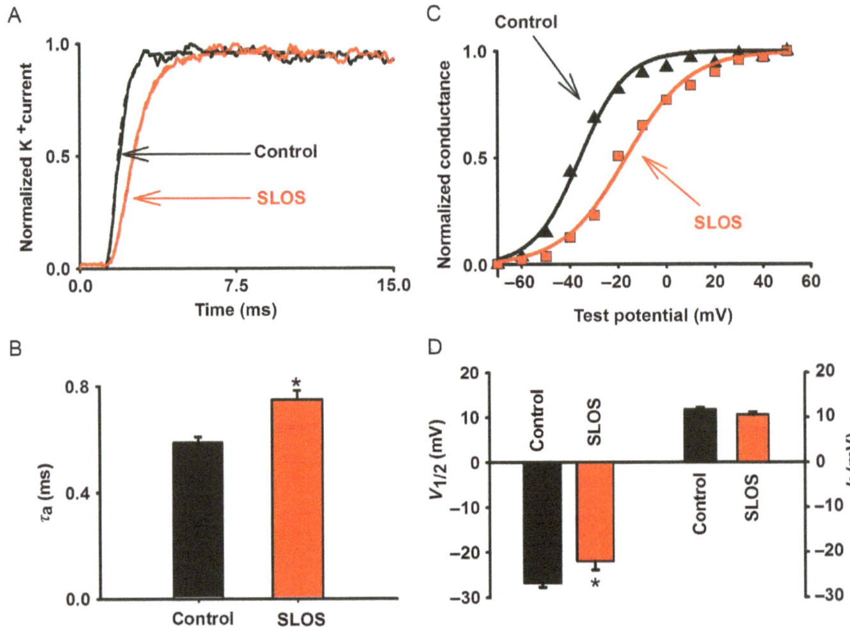

Fig. 7 The effect of SLOS on the gating of Kv1.3 channels. (A) We applied 15-ms-long depolarizing pulses to +50 mV from the holding potential of −120 mV to study the kinetic properties of activation of Kv1.3. Current traces evoked in SLOS and control T cells were fitted using Hodgkin–Huxley n^{4-} model, and the activation time constant (τ_a) was used to describe the activation kinetics. (B) Our results clearly demonstrate that SLOS significantly increases the activation time constant of Kv1.3 channels (*$P = 0.001$). (C) Membrane potential dependence of steady-state activation of the K^+ current was determined as follows: cells were held at −120 mV holding potential and depolarized to the test potentials from −70 to +50 mV in steps of 10 mV. Peak whole-cell conductance ($G(V)$) at each test potential was calculated from the peak current (I_{peak}) at test potential V and the K^+ reversal potential (E_K) using $G(V) = I_{peak}/(V_{test} - E_K)$. The $G(V)$ values were normalized for the maximum conductance (G_{norm}) and plotted as a function of test potential along with the best-fit Boltzmann function. (D) Our results show that the midpoint of the steady-state activation (*$P = 0.03$; $V_{1/2}$, the membrane potential at which 50% of the channels is open at steady state) is significantly more positive for SLOS T cells without a statistically significant change in the slope factor (k).

steady-state activation to the right as compared to the control. Tulenko's lab published similar results using skin fibroblasts from SLOS patients (Fig. 7). They found a +50 mV shift in the voltage activation of BK channels using single-channel, inside-out patch-clamp recordings (Tulenko et al., 2006). We suppose that the increased 7DHC concentration of the cell membrane is the primary cause for the altered channel function. Our model system where MβCD/7DHC complex was used to increase artificially the

7DHC levels clearly proves this theory; namely in vitro 7DHC loading of healthy lymphocytes adequately mimics the effect of SLOS on Kv1.3 operation (Balajthy et al., 2016). Inherited mutations of ion channels often affect the kinetic and equilibrium parameters of ion channel gating causing different channelopathies (Adelman, Bond, Pessia, & Maylie, 1995; Arbour et al., 2008; Syrbe et al., 2015). A simple right shift in the voltage dependence of steady-state activation reduces the number of functionally available ion channels at a given membrane potential and can have pathophysiological consequences similar to the loss-of-function mutations. In line with this, we found reduced activation and proliferation capacity in SLOS T cells, which can be the consequence of the diminished function of Kv1.3. The proliferation of SLOS skin fibroblasts also decreased, which may correlate with decreased activity of BK channels and the Na/K ATPase (Ren et al., 2011; Tulenko et al., 2006). Because the exact mechanism of the altered cellular functions is not clear, we cannot exclude that the nonspecific, toxic effects of 7DHC and its derivatives are responsible for it (Fernandez, Martin, Gomez-Coronado, & Lasuncion, 2005; Moog, Luu, Altmeyer, & Bischoff, 1989; Xu, Sheflin, Porter, & Fliesler, 2012). Taken together, results from our lab and from others indicate that cholesterol-related diseases can have pathological consequences via modifying ion channel operation. Nevertheless, it is important to interpret the results of in vitro studies with caution to avoid overrating their biological relevance.

7. CONCLUSIONS

The regulation of voltage-gated K^+ channels by cholesterol and the cholesterol-dependent membrane microdomains seems to be evident from the review of the literature. The very complex structure of the voltage-gated K^+ channels allows the gating of the channels to be regulated by a variety of mechanisms, among these; the possibility of sterol regulation of the channels was discussed in this chapter. We believe that the regulation of Kv channels by binding of cholesterol or cholesterol derivatives to the protein will be directly demonstrated in the near future, and this may allow specific targeting of the ion channels for efficient modulation of their physiological functions. The altered biophysical functions of Kv1.3 of T cells in the SLO disease might also predict similar effects of 7DHC on the ion channels of the central nervous system. This may contribute to the very severe neurological symptoms of these patients, but further experiments are needed to prove this hypothesis.

ACKNOWLEDGMENTS

This work was supported by GINOP-2.3.2-15-2016-00044 (G.P.), KTIA_NAP_13-2-2015-0009 (Z.V.), Hungarian Scientific Research Fund OTKA K 119417 (G.P.). P.H. is a Lajos Szodoray Fellow and supported by János Bolyai Fellowship. Z.V. is awarded with János Bolyai Fellowship. A.B. was cofinanced by Astellas Pharma Fellowship and National Excellence Program. This research was partly realized in the frames of TÁMOP 4.2.4. A/2-11-1-2012-0001 "National Excellence Program—Elaborating and operating an inland student and researcher personal support system convergence program." The project was subsidized by the European Union and cofinanced by the European Social Fund.

REFERENCES

Adelman, J. P., Bond, C. T., Pessia, M., & Maylie, J. (1995). Episodic ataxia results from voltage-dependent potassium channels with altered functions. *Neuron*, *15*(6), 1449–1454.

Andersen, O. S., Nielsen, C., Maer, A. M., Lundbaek, J. A., Goulian, M., & Koeppe, R. E., 2nd. (1999). Ion channels as tools to monitor lipid bilayer-membrane protein interactions: Gramicidin channels as molecular force transducers. *Methods in Enzymology*, *294*, 208–224.

Bukiya, A., & Rosenhouse-Dantsker, A. (2015). Hypercholesterolemia effect on potassium channels. In S. A. Kumar (Ed.), *Hypercholesterolemia*: InTech, Rijeka, Croatia.

Arbour, L., Rezazadeh, S., Eldstrom, J., Weget-Simms, G., Rupps, R., Dyer, Z., et al. (2008). A KCNQ1 V205M missense mutation causes a high rate of long QT syndrome in a first nations community of northern British Columbia: A community-based approach to understanding the impact. *Genetics in Medicine*, *10*(7), 545–550. http://dx.doi.org/10.1097/GIM.0b013e31817c6b19.

Attali, B., Romey, G., Honore, E., Schmid-Alliana, A., Mattei, M. G., Lesage, F., et al. (1992). Cloning, functional expression, and regulation of two K+ channels in human T lymphocytes. *Journal of Biological Chemistry*, *267*(12), 8650–8657.

Balajthy, A., Somodi, S., Petho, Z., Peter, M., Varga, Z., Szabo, G. P., et al. (2016). 7DHC-induced changes of Kv1.3 operation contributes to modified T cell function in Smith-Lemli-Opitz syndrome. *Pflügers Archiv*, *468*(8), 1403–1418. http://dx.doi.org/10.1007/s00424-016-1851-4.

Balijepalli, R. C., Delisle, B. P., Balijepalli, S. Y., Foell, J. D., Slind, J. K., Kamp, T. J., et al. (2007). Kv11.1 (ERG1) K+ channels localize in cholesterol and sphingolipid enriched membranes and are modulated by membrane cholesterol. *Channels (Austin, TX)*, *1*(4), 263–272.

Bezanilla, F. (2008). How membrane proteins sense voltage. *Nature Reviews Molecular Cell Biology*, *9*(4), 323–332. http://dx.doi.org/10.1038/nrm2376.

Bock, J., Szabo, I., Gamper, N., Adams, C., & Gulbins, E. (2003). Ceramide inhibits the potassium channel Kv1.3 by the formation of membrane platforms. *Biochemical and Biophysical Research Communications*, *305*(4), 890–897.

Bromley, S. K., Burack, W. R., Johnson, K. G., Somersalo, K., Sims, T. N., Sumen, C., et al. (2001). The immunological synapse. *Cancer Immunology Research*, *19*(1), 375–396.

Burke, N. A., Takimoto, K., Li, D., Han, W., Watkins, S. C., & Levitan, E. S. (1999). Distinct structural requirements for clustering and immobilization of K+ channels by PSD-95. *The Journal of General Physiology*, *113*(1), 71–80.

Cahalan, M. D., & Chandy, K. G. (2009). The functional network of ion channels in T lymphocytes. *Immunological Reviews*, *231*(1), 59–87.

Cahalan, M. D., Chandy, K. G., Decoursey, T. E., & Gupta, S. (1985). A voltage-gated potassium channel in human T lymphocytes. *The Journal of Physiology*, *358*, 197–237.
Cahalan, M. D., Wulff, H., & Chandy, K. G. (2001). Molecular properties and physiological roles of ion channels in the immune system. *Journal of Clinical Immunology*, *21*(4), 235–252.
Chen, J., Mitcheson, J. S., Tristani-Firouzi, M., Lin, M., & Sanguinetti, M. C. (2001). The S4-S5 linker couples voltage sensing and activation of pacemaker channels. *Proceedings of the National Academy of Sciences of the United States of America*, *98*(20), 11277–11282. http://dx.doi.org/10.1073/pnas.201250598.
Chen, Y., Sheng, R., Kallberg, M., Silkov, A., Tun, M. P., Bhardwaj, N., et al. (2012). Genome-wide functional annotation of dual-specificity protein- and lipid-binding modules that regulate protein interactions. *Molecular Cell*, *46*(2), 226–237. http://dx.doi.org/10.1016/j.molcel.2012.02.012.
Chun, Y. S., Oh, H. G., Park, M. K., Cho, H., & Chung, S. (2013). Cholesterol regulates HERG K+ channel activation by increasing phospholipase C beta1 expression. *Channels (Austin, TX)*, 7(4), 275–287.
Colley, B., Tucker, K., & Fadool, D. A. (2004). Comparison of modulation of Kv1.3 channel by two receptor tyrosine kinases in olfactory bulb neurons of rodents. *Receptors & Channels*, *10*(1), 25–36.
Davis, D. M. (2002). Assembly of the immunological synapse for T cells and NK cells. *Trends in Immunology*, *23*(7), 356–363.
Decoursey, T. E., Chandy, K. G., Gupta, S., & Cahalan, M. D. (1984). Voltage-gated K+ channels in human T lymphocytes: A role in mitogenesis? *Nature*, *307*, 465–468.
Defarias, F. P., Stevens, S. P., & Leonard, R. J. (1995). Stable expression of human Kv1.3 potassium channels resets the resting membrane potential of cultured mammalian cells. *Receptors & Channels*, *3*(4), 273–281.
Dodson, P. D., & Forsythe, I. D. (2004). Presynaptic K+ channels: Electrifying regulators of synaptic terminal excitability. *Trends in Neurosciences*, *27*(4), 210–217.
Doyle, D. A., Morais Cabral, J., Pfuetzner, R. A., Kuo, A., Gulbis, J. M., Cohen, S. L., et al. (1998). The structure of the potassium channel: Molecular basis of K+ conduction and selectivity. *Science*, *280*(5360), 69–77.
Dustin, M. L. (2014). The immunological synapse. *Cancer Immunology Research*, *2*(11), 1023–1033. http://dx.doi.org/10.1158/2326-6066.CIR-14-0161.
Epand, R. M. (2006). Cholesterol and the interaction of proteins with membrane domains. *Progress in Lipid Research*, *45*(4), 279–294. http://dx.doi.org/10.1016/j.plipres.2006.02.001.
Erlendsson, S., & Madsen, K. L. (2015). Membrane binding and modulation of the PDZ domain of PICK1. *Membranes*, *5*(4), 597–615. http://dx.doi.org/10.3390/membranes5040597.
Fantini, J., & Barrantes, F. J. (2013). How cholesterol interacts with membrane proteins: An exploration of cholesterol-binding sites including CRAC, CARC, and tilted domains. *Frontiers in Physiology*, *4*, 31. http://dx.doi.org/10.3389/fphys.2013.00031.
Felipe, A., Soler, C., & Comes, N. (2010). Kv1.5 in the immune system: The good, the bad, or the ugly? *Frontiers in Physiology*, *1*, 152. http://dx.doi.org/10.3389/fphys.2010.00152.
Fernandez, C., Martin, M., Gomez-Coronado, D., & Lasuncion, M. A. (2005). Effects of distal cholesterol biosynthesis inhibitors on cell proliferation and cell cycle progression. *Journal of Lipid Research*, *46*(5), 920–929. http://dx.doi.org/10.1194/jlr.M400407-JLR200.
Feske, S., Skolnik, E. Y., & Prakriya, M. (2012). Ion channels and transporters in lymphocyte function and immunity. *Nature Reviews Immunology*, *12*(7), 532–547. http://dx.doi.org/10.1038/nri3233.
Feske, S., Wulff, H., & Skolnik, E. Y. (2015). Ion channels in innate and adaptive immunity. *Annual Review of Immunology*, *33*, 291–353. http://dx.doi.org/10.1146/annurev-immunol-032414-112212.

Gimpl, G., Klein, U., Reilander, H., & Fahrenholz, F. (1995). Expression of the human oxytocin receptor in baculovirus-infected insect cells: High-affinity binding is induced by a cholesterol-cyclodextrin complex. *Biochemistry*, *34*(42), 13794–13801.

Gong, J., Xu, J., Bezanilla, M., van Huizen, R., Derin, R., & Li, M. (1999). Differential stimulation of PKC phosphorylation of potassium channels by ZIP1 and ZIP2. *Science*, *285*(5433), 1565–1569.

Gonzalez, C., Baez-Nieto, D., Valencia, I., Oyarzun, I., Rojas, P., Naranjo, D., et al. (2012). K(+) channels: Function-structural overview. *Comprehensive Physiology*, *2*(3), 2087–2149. http://dx.doi.org/10.1002/cphy.c110047.

Gutman, G. A., Chandy, K. G., Grissmer, S., Lazdunski, M., McKinnon, D., Pardo, L. A., et al. (2005). International Union of Pharmacology. LIII. Nomenclature and molecular relationships of voltage-gated potassium channels. *Pharmacological Reviews*, *57*(4), 473–508. http://dx.doi.org/10.1124/pr.57.4.10.

Hackos, D. H., Chang, T. H., & Swartz, K. J. (2002). Scanning the intracellular S6 activation gate in the shaker K+ channel. *The Journal of General Physiology*, *119*(6), 521–531.

Hajdu, P., Martin, G. V., Chimote, A. A., Szilagyi, O., Takimoto, K., & Conforti, L. (2015). The C-terminus SH3-binding domain of Kv1.3 is required for the actin-mediated immobilization of the channel via cortactin. *Molecular Biology of the Cell*, *26*(9), 1640–1651. http://dx.doi.org/10.1091/mbc.E14-07-1195.

Hajdu, P., Varga, Z., Pieri, C., Panyi, G., & Gaspar, R., Jr. (2003). Cholesterol modifies the gating of Kv1.3 in human T lymphocytes. *Pflügers Archiv*, *445*(6), 674–682. http://dx.doi.org/10.1007/s00424-002-0974-y.

Hanada, T., Lin, L., Chandy, K. G., Oh, S. S., & Chishti, A. H. (1997). Human homologue of the Drosophila discs large tumor suppressor binds to p56lck tyrosine kinase and Shaker type Kv1.3 potassium channel in T lymphocytes. *The Journal of Biological Chemistry*, *272*(43), 26899–26904.

Hansson, G. K., & Hermansson, A. (2011). The immune system in atherosclerosis. *Nature Immunology*, *12*(3), 204–212. http://dx.doi.org/10.1038/ni.2001.

Hodgkin, A. L., & Huxley, A. F. (1952). Currents carried by sodium and potassium ions through the membrane of the giant axon of Loligo. *The Journal of Physiology*, *116*(4), 449–472.

Hong, K. H., & Miller, C. (2000). The lipid-protein interface of a Shaker K(+) channel. *The Journal of General Physiology*, *115*(1), 51–58.

Hoshi, T., & Armstrong, C. M. (2013). C-type inactivation of voltage-gated K+ channels: Pore constriction or dilation? *The Journal of General Physiology*, *141*(2), 151–160. http://dx.doi.org/10.1085/jgp.201210888.

Hoshi, T., Zagotta, W. N., & Aldrich, R. W. (1990). Biophysical and molecular mechanisms of Shaker potassium channel inactivation. *Science*, *250*(4980), 533–538.

Hoshi, T., Zagotta, W. N., & Aldrich, R. W. (1991). Two types of inactivation in Shaker K+ channels: Effects of alterations in the carboxy-terminal region. *Neuron*, 7, 547–556.

Irie, T., Fukunaga, K., & Pitha, J. (1992). Hydroxypropylcyclodextrins in parenteral use. I: Lipid dissolution and effects on lipid transfers in vitro. *Journal of Pharmaceutical Sciences*, *81*(6), 521–523.

Irons, M., Elias, E. R., Salen, G., Tint, G. S., & Batta, A. K. (1993). Defective cholesterol biosynthesis in Smith-Lemli-Opitz syndrome. *Lancet*, *341*(8857), 1414.

Jan, L. Y., Attali, B., Nerbonne, J., Sanguinetti, M. C., Trimmer, J. S., Chandy, K. G., et al. (n.d.). Voltage-gated potassium channels. IUPHAR/BPS Guide to PHARMACOLOGY. http://www.guidetopharmacology.org/GRAC/FamilyDisplayForward?familyId=81. Accessed 22.06.2017.

Janes, P. W., Ley, S. C., & Magee, A. I. (1999). Aggregation of lipid rafts accompanies signaling via the T cell antigen receptor. *The Journal of Cell Biology*, *147*(2), 447–461.

Janes, P. W., Ley, S. C., Magee, A. I., & Kabouridis, P. S. (2000). The role of lipid rafts in T cell antigen receptor (TCR) signalling. *Seminars in Immunology*, *12*(1), 23–34.

Jimenez-Garduno, A. M., Mitkovski, M., Alexopoulos, I. K., Sanchez, A., Stuhmer, W., Pardo, L. A., et al. (2014). KV10.1 K(+)-channel plasma membrane discrete domain partitioning and its functional correlation in neurons. *Biochimica et Biophysica Acta*, *1838*(3), 921–931. http://dx.doi.org/10.1016/j.bbamem.2013.11.007.

Jugloff, D. G., Khanna, R., Schlichter, L. C., & Jones, O. T. (2000). Internalization of the Kv1.4 potassium channel is suppressed by clustering interactions with PSD-95. *The Journal of Biological Chemistry*, *275*(2), 1357–1364.

Kim, E., Niethammer, M., Rothschild, A., Jan, Y. N., & Sheng, M. (1995). Clustering of Shaker-type K+ channels by interaction with a family of membrane-associated guanylate kinases. *Nature*, *378*(6552), 85–88.

Kim, E., & Sheng, M. (1996). Differential K+ channel clustering activity of PSD-95 and SAP97, two related membrane-associated putative guanylate kinases. *Neuropharmacology*, *35*(7), 993–1000.

Klein, U., Gimpl, G., & Fahrenholz, F. (1995). Alteration of the myometrial plasma membrane cholesterol content with beta-cyclodextrin modulates the binding affinity of the oxytocin receptor. *Biochemistry*, *34*(42), 13784–13793.

Kuras, Z., Kucher, V., Gordon, S. M., Neumeier, L., Chimote, A. A., Filipovich, A. H., et al. (2012). Modulation of Kv1.3 channels by protein kinase A I in T lymphocytes is mediated by the disc large 1-tyrosine kinase Lck complex. *American Journal of Physiology. Cell Physiology*, *302*(10), C1504–C1512.

Lee, S. Y., Banerjee, A., & MacKinnon, R. (2009). Two separate interfaces between the voltage sensor and pore are required for the function of voltage-dependent K(+) channels. *PLoS Biology*, 7(3), e47. http://dx.doi.org/10.1371/journal.pbio.1000047.

Lee, S. Y., Lee, A., Chen, J., & MacKinnon, R. (2005). Structure of the KvAP voltage-dependent K+ channel and its dependence on the lipid membrane. *Proceedings of the National Academy of Sciences of the United States of America*, *102*(43), 15441–15446. http://dx.doi.org/10.1073/pnas.0507651102.

Levitan, I., Christian, A. E., Tulenko, T. N., & Rothblat, G. H. (2000). Membrane cholesterol content modulates activation of volume-regulated anion current in bovine endothelial cells. *The Journal of General Physiology*, *115*(4), 405–416. http://dx.doi.org/10.1085/jgp.115.4.405.

Levitan, I., Fang, Y., Rosenhouse-Dantsker, A., & Romanenko, V. (2010). Cholesterol and ion channels. *Sub-Cellular Biochemistry*, *51*, 509–549. http://dx.doi.org/10.1007/978-90-481-8622-8_19.

Levitan, I., Singh, D. K., & Rosenhouse-Dantsker, A. (2014). Cholesterol binding to ion channels. *Frontiers in Physiology*, *5*, 65. http://dx.doi.org/10.3389/fphys.2014.00065.

Levy, D. I., Panyi, G., Sheng, Z. F., Tu, L. W., & Deutsch, C. (1995). C-type inactivation determined by multiple subunits. *Biophysical Journal*, *68*, A33.

Lewis, R. S. (2001). Calcium signaling mechanisms in T lymphocytes. *Annual Review of Immunology*, *19*, 497–521.

Lindegaard, M. L., Wassif, C. A., Vaisman, B., Amar, M., Wasmuth, E. V., Shamburek, R., et al. (2008). Characterization of placental cholesterol transport: ABCA1 is a potential target for in utero therapy of Smith-Lemli-Opitz syndrome. *Human Molecular Genetics*, *17*(23), 3806–3813. http://dx.doi.org/10.1093/hmg/ddn278.

Lingwood, D., & Simons, K. (2010). Lipid rafts as a membrane-organizing principle. *Science*, *327*(5961), 46–50. http://dx.doi.org/10.1126/science.1174621.

Lioudyno, M. I., Kozak, J. A., Penna, A., Safrina, O., Zhang, S. L., Sen, D., et al. (2008). Orai1 and STIM1 move to the immunological synapse and are up-regulated during T cell activation. *Proceedings of the National Academy of Sciences of the United States of America*, *105*(6), 2011–2016.

Li-Smerin, Y., & Swartz, K. J. (2001). Helical structure of the COOH terminus of S3 and its contribution to the gating modifier toxin receptor in voltage-gated ion channels. *The Journal of General Physiology*, *117*(3), 205–218.
Listowski, M. A., Leluk, J., Kraszewski, S., & Sikorski, A. F. (2015). Cholesterol interaction with the MAGUK protein family member, MPP1, via CRAC and CRAC-like motifs: An in silico docking analysis. *PLoS One*, *10*(7), e0133141. http://dx.doi.org/10.1371/journal.pone.0133141.
Long, S. B., Campbell, E. B., & MacKinnon, R. (2005a). Crystal structure of a mammalian voltage-dependent Shaker family K+ channel. *Science*, *309*(5736), 897–903.
Long, S. B., Campbell, E. B., & MacKinnon, R. (2005b). Voltage sensor of Kv1.2: structural basis of electromechanical coupling. *Science*, *309*(5736), 903–908.
Lopez-Barneo, J., Hoshi, T., Heinemann, S. H., & Aldrich, R. W. (1993). Effects of external cations and mutations in the pore region on C-type inactivation of Shaker potassium channels. *Receptors & Channels*, *1*(1), 61–71.
Lorinczi, E., Gomez-Posada, J. C., de la Pena, P., Tomczak, A. P., Fernandez-Trillo, J., Leipscher, U., et al. (2015). Voltage-dependent gating of KCNH potassium channels lacking a covalent link between voltage-sensing and pore domains. *Nature Communications*, *6*, 6672. http://dx.doi.org/10.1038/ncomms7672.
Lundbaek, J. A., & Andersen, O. S. (1994). Lysophospholipids modulate channel function by altering the mechanical properties of lipid bilayers. *The Journal of General Physiology*, *104*(4), 645–673.
Lundbaek, J. A., & Andersen, O. S. (1999). Spring constants for channel-induced lipid bilayer deformations. Estimates using gramicidin channels. *Biophysical Journal*, *76*(2), 889–895. http://dx.doi.org/10.1016/S0006-3495(99)77252-8.
Lundbaek, J. A., Birn, P., Girshman, J., Hansen, A. J., & Andersen, O. S. (1996). Membrane stiffness and channel function. *Biochemistry*, *35*(12), 3825–3830.
MacKinnon, R. (1991). Determination of the subunit stoichiometry of a voltage-activated potassium channel. *Nature*, *350*(6315), 232–235. http://dx.doi.org/10.1038/350232a0.
Mahammad, S., & Parmryd, I. (2008). Cholesterol homeostasis in T cells. Methyl-beta-cyclodextrin treatment results in equal loss of cholesterol from Triton X-100 soluble and insoluble fractions. *Biochimica et Biophysica Acta*, *1778*(5), 1251–1258. http://dx.doi.org/10.1016/j.bbamem.2008.02.010.
Maltsev, V. A. (1990). Oscillating and triggering properties of T cell membrane potential. *Immunology Letters*, *26*, 277–282.
Marks, D. R., & Fadool, D. A. (2007). Post-synaptic density perturbs insulin-induced Kv1.3 channel modulation via a clustering mechanism involving the SH3 domain. *Journal of Neurochemistry*, *103*(4), 1608–1627.
Martens, J. R., Navarro-Polanco, R., Coppock, E. A., Nishiyama, A., Parshley, L., Grobaski, T. D., et al. (2000). Differential targeting of Shaker-like potassium channels to lipid rafts. *The Journal of Biological Chemistry*, *275*(11), 7443–7446.
Martens, J. R., Sakamoto, N., Sullivan, S. A., Grobaski, T. D., & Tamkun, M. M. (2001). Isoform-specific localization of voltage-gated K+ channels to distinct lipid raft populations. Targeting of Kv1.5 to caveolae. *The Journal of Biological Chemistry*, *276*(11), 8409–8414. http://dx.doi.org/10.1074/jbc.M009948200.
Matteson, D. R., & Deutsch, C. (1984). K channels in T lymphocytes: A patch clamp study using monoclonal antibody adhesion. *Nature*, *307*, 468–471.
McCormack, T., McCormack, K., Nadal, M. S., Vieira, E., Ozaita, A., & Rudy, B. (1999). The effects of Shaker beta-subunits on the human lymphocyte K+ channel Kv1.3. *The Journal of Biological Chemistry*, *274*(29), 20123–20126.
McEwen, D. P., Li, Q., Jackson, S., Jenkins, P. M., & Martens, J. R. (2008). Caveolin regulates kv1.5 trafficking to cholesterol-rich membrane microdomains. *Molecular Pharmacology*, *73*(3), 678–685. http://dx.doi.org/10.1124/mol.107.042093.

Monks, S. A., Needleman, D. J., & Miller, C. (1999). Helical structure and packing orientation of the S2 segment in the Shaker K+ channel. *The Journal of General Physiology, 113*(3), 415–423.

Moog, C., Luu, B., Altmeyer, A., & Bischoff, P. (1989). Studies on the immunosuppressive properties of 7,25 dihydroxycholesterol—II. Effects on early steps of T-cell activation. *International Journal of Immunopharmacology, 11*(5), 559–565.

Nicolaou, S. A., Neumeier, L., Peng, Y., Devor, D. C., & Conforti, L. (2007). The Ca(2+)-activated K(+) channel KCa3.1 compartmentalizes in the immunological synapse of human T lymphocytes. *American Journal of Physiology. Cell Physiology, 292*(4), C1431–C1439.

O'Connell, K. M., Whitesell, J. D., & Tamkun, M. M. (2008). Localization and mobility of the delayed-rectifer K+ channel Kv2.1 in adult cardiomyocytes. *American Journal of Physiology. Heart and Circulatory Physiology, 294*(1), H229–H237. http://dx.doi.org/10.1152/ajpheart.01038.2007.

Osteen, J. D., Barro-Soria, R., Robey, S., Sampson, K. J., Kass, R. S., & Larsson, H. P. (2012). Allosteric gating mechanism underlies the flexible gating of KCNQ1 potassium channels. *Proceedings of the National Academy of Sciences of the United States of America, 109*(18), 7103–7108. http://dx.doi.org/10.1073/pnas.1201582109.

Owen, D. M., Oddos, S., Kumar, S., Davis, D. M., Neil, M. A., French, P. M., et al. (2010). High plasma membrane lipid order imaged at the immunological synapse periphery in live T cells. *Molecular Membrane Biology, 27*(4–6), 178–189. http://dx.doi.org/10.3109/09687688.2010.495353.

Pahapill, P. A., & Schlichter, L. C. (1992). Modulation of potassium channels in intact human T lymphocytes. *The Journal of Physiology, 445*, 407–430.

Pan, L., Wu, H., Shen, C., Shi, Y., Jin, W., Xia, J., et al. (2007). Clustering and synaptic targeting of PICK1 requires direct interaction between the PDZ domain and lipid membranes. *The EMBO Journal, 26*(21), 4576–4587. http://dx.doi.org/10.1038/sj.emboj.7601860.

Panyi, G., Bagdany, M., Bodnar, A., Vamosi, G., Szentesi, G., Jenei, A., et al. (2003). Colocalization and nonrandom distribution of Kv1.3 potassium channels and CD3 molecules in the plasma membrane of human T lymphocytes. *Proceedings of the National Academy of Sciences of the United States of America, 100*(5), 2592–2597. http://dx.doi.org/10.1073/pnas.0438057100.

Panyi, G., Vamosi, G., Bacso, Z., Bagdany, M., Bodnar, A., Varga, Z., et al. (2004). Kv1.3 potassium channels are localized in the immunological synapse formed between cytotoxic and target cells. *Proceedings of the National Academy of Sciences of the United States of America, 101*(5), 1285–1290.

Panyi, G., Varga, Z., & Gaspar, R. (2004). Ion channels and lymphocyte activation. *Immunology Letters, 92*, 55–66.

Pardo, L. A., del Camino, D., Sanchez, A., Alves, F., Bruggemann, A., Beckh, S., et al. (1999). Oncogenic potential of EAG K(+) channels. *The EMBO Journal, 18*(20), 5540–5547. http://dx.doi.org/10.1093/emboj/18.20.5540.

Perez-Verdaguer, M., Capera, J., Martinez-Marmol, R., Camps, M., Comes, N., Tamkun, M. M., et al. (2016). Caveolin interaction governs Kv1.3 lipid raft targeting. *Scientific Reports, 6*, 22453. http://dx.doi.org/10.1038/srep22453.

Picazo-Juarez, G., Romero-Suarez, S., Nieto-Posadas, A., Llorente, I., Jara-Oseguera, A., Briggs, M., et al. (2011). Identification of a binding motif in the S5 helix that confers cholesterol sensitivity to the TRPV1 ion channel. *The Journal of Biological Chemistry, 286*(28), 24966–24976. http://dx.doi.org/10.1074/jbc.M111.237537.

Pike, L. J. (2006). Rafts defined: A report on the keystone symposium on lipid rafts and cell function. *Journal of Lipid Research, 47*(7), 1597–1598. http://dx.doi.org/10.1194/jlr.E600002-JLR200.

Porter, F. D. (2008). Smith-Lemli-Opitz syndrome: Pathogenesis, diagnosis and management. *European Journal of Human Genetics*, *16*(5), 535–541. http://dx.doi.org/10.1038/ejhg.2008.10.

Porter, F. D., & Herman, G. E. (2011). Malformation syndromes caused by disorders of cholesterol synthesis. *Journal of Lipid Research*, *52*(1), 6–34. http://dx.doi.org/10.1194/jlr.R009548.

Pottosin, I. I., Valencia-Cruz, G., Bonales-Alatorre, E., Shabala, S. N., & Dobrovinskaya, O. R. (2007). Methyl-beta-cyclodextrin reversibly alters the gating of lipid rafts-associated Kv1.3 channels in Jurkat T lymphocytes. *Pflügers Archiv*, *454*(2), 235–244. http://dx.doi.org/10.1007/s00424-007-0208-4.

Pourrier, M., Schram, G., & Nattel, S. (2003). Properties, expression and potential roles of cardiac K+ channel accessory subunits: MinK, MiRPs, KChIP, and KChAP. *The Journal of Membrane Biology*, *194*(3), 141–152. http://dx.doi.org/10.1007/s00232-003-2034-8.

Ren, G., Jacob, R. F., Kaulin, Y., Dimuzio, P., Xie, Y., Mason, R. P., et al. (2011). Alterations in membrane caveolae and BKCa channel activity in skin fibroblasts in Smith-Lemli-Opitz syndrome. *Molecular Genetics and Metabolism*, *104*(3), 346–355. http://dx.doi.org/10.1016/j.ymgme.2011.04.019.

Robinson, L. E., Shridar, M., Smith, P., & Murrell-Lagnado, R. D. (2014). Plasma membrane cholesterol as a regulator of human and rodent P2X7 receptor activation and sensitization. *The Journal of Biological Chemistry*, *289*(46), 31983–31994. http://dx.doi.org/10.1074/jbc.M114.574699.

Romanenko, V. G., Fang, Y., Byfield, F., Travis, A. J., Vandenberg, C. A., Rothblat, G. H., et al. (2004). Cholesterol sensitivity and lipid raft targeting of Kir2.1 channels. *Biophysical Journal*, *87*(6), 3850–3861. http://dx.doi.org/10.1529/biophysj.104.043273.

Rosenhouse-Dantsker, A., Noskov, S., Durdagi, S., Logothetis, D. E., & Levitan, I. (2013). Identification of novel cholesterol-binding regions in Kir2 channels. *The Journal of Biological Chemistry*, *288*(43), 31154–31164. http://dx.doi.org/10.1074/jbc.M113.496117.

Rudakova, E., Wagner, M., Frank, M., & Volk, T. (2015). Localization of Kv4.2 and KChIP2 in lipid rafts and modulation of outward K+ currents by membrane cholesterol content in rat left ventricular myocytes. *Pflügers Archiv*, *467*(2), 299–309. http://dx.doi.org/10.1007/s00424-014-1521-3.

Seres, I., Freyss-Beguin, M., Mohacsi, A., Kozlovsky, B., Simon, J., Devynck, M. A., et al. (1996). Alteration of lymphocyte membrane phospholipids and intracellular free calcium concentrations in hyperlipidemic subjects. *Atherosclerosis*, *121*(2), 175–183.

Sheng, R., Chen, Y., Yung Gee, H., Stec, E., Melowic, H. R., Blatner, N. R., et al. (2012). Cholesterol modulates cell signaling and protein networking by specifically interacting with PDZ domain-containing scaffold proteins. *Nature Communications*, *3*, 1249. http://dx.doi.org/10.1038/ncomms2221.

Simons, K., & Ehehalt, R. (2002). Cholesterol, lipid rafts, and disease. *The Journal of Clinical Investigation*, *110*(5), 597–603. http://dx.doi.org/10.1172/JCI16390.

Singh, A. K., McMillan, J., Bukiya, A. N., Burton, B., Parrill, A. L., & Dopico, A. M. (2012). Multiple cholesterol recognition/interaction amino acid consensus (CRAC) motifs in cytosolic C tail of Slo1 subunit determine cholesterol sensitivity of Ca2+- and voltage-gated K+ (BK) channels. *The Journal of Biological Chemistry*, *287*(24), 20509–20521. http://dx.doi.org/10.1074/jbc.M112.356261.

Smith, D. W., Lemli, L., & Opitz, J. M. (1964). A newly recognized syndrome of multiple congenital anomalies. *The Journal of Pediatrics*, *64*, 210–217.

Somodi, S., Balajthy, A., Szilagyi, O., Petho, Z., Harangi, M., Paragh, G., et al. (2013). Analysis of the K+ current in human CD4+ T lymphocytes in hypercholesterolemic state. *Cellular Immunology*, *281*(1), 20–26. http://dx.doi.org/10.1016/j.cellimm.2013.01.004.

Syrbe, S., Hedrich, U. B., Riesch, E., Djemie, T., Muller, S., Moller, R. S., et al. (2015). De novo loss- or gain-of-function mutations in KCNA2 cause epileptic encephalopathy. *Nature Genetics*, *47*(4), 393–399. http://dx.doi.org/10.1038/ng.3239.

Szilagyi, O., Boratko, A., Panyi, G., & Hajdu, P. (2013). The role of PSD-95 in the rearrangement of Kv1.3 channels to the immunological synapse. *Pflügers Archiv*, *465*(9), 1341–1353. http://dx.doi.org/10.1007/s00424-013-1256-6.

Tanouye, M. A., Ferrus, A., & Fujita, S. C. (1981). Abnormal action potentials associated with the Shaker complex locus of Drosophila. *Proceedings of the National Academy of Sciences of the United States of America*, *78*(10), 6548–6552.

Tiffany, A. M., Manganas, L. N., Kim, E., Hsueh, Y. P., Sheng, M., & Trimmer, J. S. (2000). PSD-95 and SAP97 exhibit distinct mechanisms for regulating K(+) channel surface expression and clustering. *The Journal of Cell Biology*, *148*(1), 147–157.

Toth, A., Szilagyi, O., Krasznai, Z., Panyi, G., & Hajdu, P. (2009). Functional consequences of Kv1.3 ion channel rearrangement into the immunological synapse. *Immunology Letters*, *125*(1), 15–21.

Tulenko, T. N., Boeze-Battaglia, K., Mason, R. P., Tint, G. S., Steiner, R. D., Connor, W. E., et al. (2006). A membrane defect in the pathogenesis of the Smith-Lemli-Opitz syndrome. *Journal of Lipid Research*, *47*(1), 134–143. http://dx.doi.org/10.1194/jlr.M500306-JLR200.

Verheugen, J. A., Vijverberg, H. P., Oortgiesen, M., & Cahalan, M. D. (1995). Voltage-gated and Ca(2+)-activated K+ channels in intact human T lymphocytes. Noninvasive measurements of membrane currents, membrane potential, and intracellular calcium. *The Journal of General Physiology*, *105*(6), 765–794.

Viola, A., Schroeder, S., Sakakibara, Y., & Lanzavecchia, A. (1999). T lymphocyte costimulation mediated by reorganization of membrane microdomains. *Science*, *283*(5402), 680–682.

Wong, W., & Schlichter, L. C. (2004). Differential recruitment of Kv1.4 and Kv4.2 to lipid rafts by PSD-95. *The Journal of Biological Chemistry*, *279*(1), 444–452. http://dx.doi.org/10.1074/jbc.M304675200.

Wu, H., Feng, W., Chen, J., Chan, L. N., Huang, S., & Zhang, M. (2007). PDZ domains of Par-3 as potential phosphoinositide signaling integrators. *Molecular Cell*, *28*(5), 886–898. http://dx.doi.org/10.1016/j.molcel.2007.10.028.

Xia, F., Gao, X., Kwan, E., Lam, P. P., Chan, L., Sy, K., et al. (2004). Disruption of pancreatic beta-cell lipid rafts modifies Kv2.1 channel gating and insulin exocytosis. *The Journal of Biological Chemistry*, *279*(23), 24685–24691. http://dx.doi.org/10.1074/jbc.M314314200.

Xia, F., Leung, Y. M., Gaisano, G., Gao, X., Chen, Y., Fox, J. E., et al. (2007). Targeting of voltage-gated K+ and Ca2+ channels and soluble N-ethylmaleimide-sensitive factor attachment protein receptor proteins to cholesterol-rich lipid rafts in pancreatic alpha-cells: Effects on glucagon stimulus-secretion coupling. *Endocrinology*, *148*(5), 2157–2167. http://dx.doi.org/10.1210/en.2006-1296.

Xu, L., Sheflin, L. G., Porter, N. A., & Fliesler, S. J. (2012). 7-Dehydrocholesterol-derived oxysterols and retinal degeneration in a rat model of Smith-Lemli-Opitz syndrome. *Biochimica et Biophysica Acta*, *1821*(6), 877–883. http://dx.doi.org/10.1016/j.bbalip.2012.03.001.

Zagotta, W. N., Hoshi, T., & Aldrich, R. W. (1994). Shaker potassium channel gating. III: Evaluation of kinetic models for activation. *The Journal of General Physiology*, *103*(2), 321–362. PMID: 8189208.

Zaydman, M. A., Silva, J. R., Delaloye, K., Li, Y., Liang, H., Larsson, H. P., et al. (2013). Kv7.1 ion channels require a lipid to couple voltage sensing to pore opening. *Proceedings of the National Academy of Sciences of the United States of America*, *110*(32), 13180–13185. http://dx.doi.org/10.1073/pnas.1305167110.

Zimmermann, P., Meerschaert, K., Reekmans, G., Leenaerts, I., Small, J. V., Vandekerckhove, J., et al. (2002). PIP(2)-PDZ domain binding controls the association of syntenin with the plasma membrane. *Molecular Cell*, *9*(6), 1215–1225.

Zsiros, E., Kis-Toth, K., Hajdu, P., Gaspar, R., Bielanska, J., Felipe, A., et al. (2009). Developmental switch of the expression of ion channels in human dendritic cells. *Journal of Immunology*, *183*(7), 4483–4492.

Kolibri 360 GmbH
Ferdinand-Jühlke-Straße 7
99095 Erfurt, Deutschland
produktsicherheit@kolibri360.de